1st Colloquim on Noise, Vibration and Comfort (CNVC 2019)

Journal of Physics: Conference Series Volume 1262

Selangor, Malaysia
7 March 2019

Editors:

Mohd Shahrir Mohd Sani
Mohd Azmi Yunus

ISBN: 978-1-5108-9556-0
ISSN: 1742-6588

Printed from e-media with permission by:

Curran Associates, Inc.
57 Morehouse Lane
Red Hook, NY 12571

Some format issues inherent in the e-media version may also appear in this print version.

Institute of Physics
Dirac House, Temple Back
Bristol BS1 6BE UK

Phone: 44 1 17 929 7481
Fax: 44 1 17 920 0979

techtracking@iop.org

Additional copies of this publication are available from:

Curran Associates, Inc.
57 Morehouse Lane
Red Hook, NY 12571 USA
Phone: 845-758-0400
Fax: 845-758-2633
Email: curran@proceedings.com
Web: www.proceedings.com

Table of contents

Volume 1262

1st Colloquium on Noise, Vibration and Comfort

7 March 2019, Selangor, Malaysia

Accepted papers received: 28 May 2019
Published online: 14 August 2019

Preface

Preface The 1st CNVC2019

Committee of 1st Colloquium on Noise, Vibration and Comfort (CNVC 2019)

Peer review statement

Papers

Accuracy and Efficiency of Updated FRF Coupling for the Dynamic Behaviour Investigation of an Assembled Structure

W I I Wan Iskandar Mirza, M N Abdul Rani, M A Yunus and R Omar.....1

Nondeterministic Model Updating of Laser Spot Welded Structure via Response Surface Methodology

M A S Aziz Shah, M A Yunus, M N Abdul Rani, N Salleh, W I I Wan Iskandar Mirza and C Peter.....11

Comparison of crash behaviour prediction of a car crash box using initial and updated finite element model

N.A.Z. Abdullah, M.S.M. Sani and M.S. Salwani.....19

A Comparative Study of CWELD and ACM2 Element Connectors on the Dynamic Behaviour of a Laser Spot Welded Hat-Plate Structure under Initial Stress Influence

C Peter, M N Abdul Rani, H Hashim, M A Yunus, M.N Machmud and M S Mohd Zin.....28

Finite Element Modelling and Updating of a Bolted Structure Using Elements Representing the Stiffness Members, Bolts and Affected Areas of the Joints

R Omar, M N Abdul Rani and M A Yunus.....37

Identification dynamic behaviour of disc brake based on finite element model updating approach

M. S. M. Fouzi, M. S. M. Sani and M. F. R. Ishak.....47

Identifying the Appropriate Frequency Response Function Driving Point of a Car Door Using Finite Element Analysis and Modal Testing

W I I Wan Iskandar Mirza, M N Abdul Rani, M F Musa, M A Yunus, C Peter and M A S Aziz Shah.....57

Noise Source Identification of Automotive Radiator by Using Sound Intensity Mapping Method

N F D Razak, M S M Sani, W H Azmi and B Zhang.....67

Identification of the nonlinearity in mass-spring system via experimental method

A R Bahari, M A Yunus, M N Abdul Rani, M A S Aziz Shah, W I I Wan Iskandar Mirza and N Salleh.....80

Occupational Noise Exposure Among Airport Workers in Malaysia: An Ergonomic Investigation

B. Athirah and M.S. Nurul Shahida.....88

Finite Element Modelling and Model Updating of Riveted Joint Structure

M A S Aziz Shah, M A Yunus, M N Abdul Rani and M. A. Zulkofli.....98

Vibration Analysis on Palm Oil Methyl Ester Biodiesel as a Fuel with The Additional of Butanol

J.M. Zikri, M.S.M. Sani, Erdiwansyah, A.F. Yusop, I. Izzudin and S. Sapee.....106

Finite element modeling and updating of the composite plate structure

R M Yaacob, M A H Hashim and M S M Sani.....116

Experimental investigation of vibrations and noise characterization for spark ignition engine

Erdiwansyah, M.S.M. Sani, R. Mamat, J.M. Zikri, N.F.D Razak and Munawir.....126

Experimental Modal Analysis Procedure for a Laser Spot Welded Hat-Plate Structure

M.S Mohd Zin, M.N Abdul Rani, M.A Yunus, W.I.I Wan Iskandar Mirza and M.S.M Sani.....136

Updating of the Dynamic Behaviour of an Assembled Structure with Bolted Joints in Light of EMA Results

R Omar, M N Abdul Rani, M A Yunus, A A M Isa and Z A Rahman.....145

Potential of Using Natural Fiber for Building Acoustic Absorber: A Review

Esan Martins Taiwo, Khairulzan Yahya and Zaiton Haron.....151

A Framework for Green Concrete Noise Barrier Materials using Palm Oil Clinker

Noor Suhaida Galip, Zaiton Haron, Roslli Noor Mohamad, Khairulzan Yahya, Zanariah Jahya and Muhammad Azril Hezmi.....159

Correlation of numerical and experimental analysis for dynamic behaviour of a 3 blade propeller structure

P Brapakaran, N A Z Abdullah, M S M Fouzi and M S M Sani.....168

Normal mode finite element analysis of aerofoil wing structure with different materials

M N A M Asri, N A Z Abdullah, M S M Fouzi and M S M Sani.....177

Removal of Dust from the Solar Panel Surface using Mechanical Vibrator

Osam H. Attia, Nor Mariah Adam, Azizan As'arry and Khairil Anas Md Rezali.....185

Joining Strategy Analysis of Structural Dynamic in Top Hat Plate of Dissimilar Materials

N.A. Nazri and M.S.M. Sani.....193

Model Updating of Frame Structure with Bolted Joints

M.H.N. Izham and M.S.M. Sani.....201

Tremor Quantification and its Measurements Using Shimmer

Reem Musab, Azizan As'arry, Khairil Anas Md. Rezali, Nawal Aswan Abdul Jalil, Raja Mohd Kamil Raja Ahmad and Mohd Zarhamdy Md Zain.....209

Application of Micro Electro-Mechanical Sensors (MEMS) Devices with Wifi Connectivity and Cloud Data Solution for Industrial Noise and Vibration Measurements

A. S. Sahdom.....216

The Effects of Posture on Suspension Seat Transmissibility during Exposure to Vertical Whole-Body Vibration

S. Aisyah Adam, N. A. Abdul Jalil, K. A.Md Razali and Y. G. Ng.....224

Development of Vibration Test System to Measure Human Biodynamic Responses to Hand-Transmitted Vibration (HTV)

A M Aizuddin, N A Abdul Jalil, K A Md Rezali and Y G Ng.....232

Predicting the Dynamic Stiffness of a Glove Material using Mechanical Impedance Model

K. A. Md Razali, R. Samin, A. As'arry and N. A. A. Jalil.....240

Stability analysis of regenerative vibration in turning operation using I-kaz3D^{TM} signal processing approach

R Samin, M Z Nuawi, S M Haris and J A Ghani.....245

Simulation of Lamb wave interactions with defects in a thin plate

N Ismail, Z M Hafizi, C K E Nizwan and S Ali.....254

Training Feedforward Neural Networks for Structural Health Monitoring of an Aircraft Structure

Tuan A Z Rahman, K A M Rezali, N A Abdul Jalil, A As'arry and Raja Kamil.....260

Biodynamic Modelling of Hand for Glove Transmissibility Prediction using Artificial Neural Networks

Tuan A Z Rahman, K A M Rezali, A As'arry and N A A Jalil.....268

Review on the Vibration Perception Threshold to Early Detect Diabetes Mellitus

M. Taufiq Zaliran, Khairil Anas Md Rezali, Ng Yee Guan, Azizan As'arry and Nawal Aswan Abdul Jalil.....276

Application of the Structural Dynamic Modification Method to Reduce the Vibration of the Vehicle HVAC System

MHA Satar, AZA Mazlan, MH Hamdan, MS Md Isa, S Man, MAR Paiman, MZ Abd Ghapar and MSA Sulaiman.....282

Finite Element Model Updating of Dissimilar Plate with Rivet Joint

H. A. Rahman, N. D. A. Khairi and M. S. M. Sani.....292

IOP Conf. Series: Journal of Physics: Conf. Series **1262** (2019) 011001 doi:10.1088/1742-6596/1262/1/011001

Preface The 1st CNVC2019

The 1st Colloquium on Noise, Vibration and Comfort (CNVC) is the 1st in a series organised by Society of Vibration and Acoustics Malaysia (SVAM) and was successfully held at Kelab Shah Alam, Selangor Malaysia on 7th March 2019. The colloquium covers the particular area of research in noise, vibration and comfort. The colloquium aims to facilitate researchers around the world to publish and share their current research findings.

These proceedings contain 35 scientific papers that have been carefully peer-reviewed and selected from 80 papers submitted by participants from both local and international universities. The papers exhibit concise, clear and lucid expositions that appeal to a wide international readership in science, technology and engineering. The topics of the Colloquium covered, include Industrial Noise and Vibration, Machining Vibration, Noise Vibration and Harshness, Structural Vibration, Acoustics Emission, Building Acoustics and Vibration, Computing and Simulation Techniques in NVC, Emerging Acoustics and Vibration Technology, Energy Harvesting, Environmental Sound and Vibration Issues, Human Vibration, Structural Health Monitoring, Vibration Control and other topics related to NVC.

The papers present original ideas or significant general results supported by clear reasoning and compelling evidence. In addition, methods, theories and practices that are relevant to the fields of research are successfully adopted and demonstrated. In this series, CNVC2019 representatives appointed are a group of people from different experiences and backgrounds in academia and industrial sectors. The plenary and invited speakers presented, discussed, promoted, and disseminated research in all fields of noise, vibration and comfort in constructive ways.

SVAM would like to express its gratitude to the Organizing Committee, Co-Organizer and Sponsors for supporting the colloquium. SVAM also would like to extend its most sincere thanks to all the participants of the colloquium and everyone who contributed to the colloquium and the publication of these proceedings.

Dr. Muhamad Norhisham Abdul Rani, Chairman of CNVC 2019

Assoc. Prof. Dr. Mohd Shahrir Mohd Sani, Chief Editor of CNVC 2019

Ir. Dr. Mohd Azmi Yunus, Editor of CNVC 2019

IOP Conf. Series: Journal of Physics: Conf. Series **1262** (2019) 011002 doi:10.1088/1742-6596/1262/1/011002

Committee of 1st Colloquium on Noise, Vibration and Comfort (CNVC 2019)

Advisor: Assoc. Prof. Ir. Dr. Nawal Aswan Abdul Jalil (Universiti Putra Malaysia)
Chairman: Dr. Muhamad Norhisham Abd. Rani (Universiti Teknologi Mara)
Vice Chairman: Assoc. Prof. Dr. Mohd. Shahrir Mohd. Sani (Universiti Malaysia Pahang)
Treasurer: Ir. Dr. Mohd Azmi bin Yunus (Universiti Teknologi Mara)

Publication Unit
Noor Amzura Abdullah (Universiti Malaysia Pahang)
Nur Farizah Diana Abd. Razak (Universiti Malaysia Pahang)
Mohd. Sahril Mohd. Fouzi (Universiti Malaysia Pahang)

Secretariat, Technical and Logistics Unit
Wan Imaan Izhan Wan Iskandar Mirza (Universiti Teknologi Mara)
Mohd Syazwan Mohd Zin (Universiti Teknologi Mara)
Muhammad Zikri Japri (Universiti Malaysia Pahang)

Sponsorship Unit
Mohamad Azam Shah Bin Aziz Shah (Universiti Teknologi Mara)
Crystal Peter (Universiti Teknologi Mara)

Marketing and Publicity Unit
Dr. Khairil Anas Md Rezali (Universiti Putra Malaysia)
Dr. Azizan As'arry (Universiti Putra Malaysia)

IOP Conf. Series: Journal of Physics: Conf. Series **1262** (2019) 011003 doi:10.1088/1742-6596/1262/1/011003

Peer review statement

All papers published in this volume of *Journal of Physics: Conference Series* have been peer reviewed through processes administered by the proceedings Editors. Reviews were conducted by expert referees to the professional and scientific standards expected of a proceedings journal published by IOP Publishing.

Accuracy and Efficiency of Updated FRF Coupling for the Dynamic Behaviour Investigation of an Assembled Structure

W I I Wan Iskandar Mirza[1], M N Abdul Rani[1*], M A Yunus[1] and R Omar[1,2]
[1] Structural Dynamics Analysis and Validation, Faculty of Mechanical Engineering, Universiti Teknologi MARA (UiTM), 40000 Shah Alam, Selangor, Malaysia
[2] Machinery Technology Centre, SIRIM Industrial Research, 44200 Hulu Selangor, Selangor, Malaysia

*Corresponding author: mnarani@uitm.edu.my

Abstract. The frequency based substructuring (FBS) method allows the combination between the experimental and analytical frequency response function (FRF). The combined FRF is then used for calculation of the dynamic behaviour of an assembled structure which usually consists of a large number structural components or substructures. However, the accuracy of the dynamic behaviour calculated using the FBS method relies heavily on the quality of experimental FRF data of the interfaces on which, in practice, it is very problematic to be attained. Furthermore, in most cases, some parts of rotational FRF data are not included in the FBS method due to the complex measurement process. The exclusion of the rotational FRF data will usually lead to numerous errors in the combined FRF data. Therefore, this paper proposes a new frequency response function (FRF) coupling of the FBS method that may uniquely address the difficulties and improve the quality of predicted results of the FBS method. The proposed coupling was formulated based on the finite element method, model updating method and FRF synthesize method. The finite element model of the physical test substructure was developed and reconciled based on the mode shapes obtained from experimental modal analysis. The FRF synthesize method was performed on the updated finite element model to obtain a complete matrix with a full degree of freedom FRF data containing all the translational and rotational FRF data required. It was found that the proposed coupling has allowed generating full translational and rotational FRF data. The generation has improved significantly the FRF coupling process and the quality of the predicted dynamic behaviour of the FBS method.

1. Introduction

The frequency based substructuring method (FBS) is one of the most powerful dynamic substructuring methods and the FBS method theoretically allows the combination of numerical and experimentally measured frequency response function (FRF) data [1]. The FBS method has been widely used in high-profile engineering analyses to improve the computing efficiency, especially in the structural dynamics and acoustics domain as presented in [2–4]. However, the method has been reported to have often undergone incompatible coupling issues; mainly during the preparation of the experimental model [4–6].

One of the major problems encountered during the preparation of the experimental model for the FBS is the excitation and measurement of rotational FRF data. In the FBS method, the rotational FRF data is essential in order to obtain the full FRF matrix [4] at the coupling the interface. Theoretically, a complete matrix of a single node system is described as follows:

$$\begin{bmatrix} u_t \\ u_\theta \end{bmatrix} = \begin{bmatrix} Y_{tt} & Y_{t\theta} \\ Y_{\theta t} & Y_{\theta\theta} \end{bmatrix} \begin{bmatrix} f_t \\ f_\theta \end{bmatrix} \quad (1)$$

where $\boldsymbol{u}$ and $\boldsymbol{f}$ represent the displacement and force vector respectively. The abbreviation of Y is denoted as a transfer function. The translational and rotational data are denoted by the subscripts t and θ. The above presented matrix clearly indicates that 75% of a single node matrix is built based on the rotational data. Therefore, neglecting 75% of the matrix will give erroneous results on the coupled system because the incomplete matrix is unable to represent the interface flexibility of the physical model.

Recently, several approaches such as equivalent multipoint connection (EMPC) approach [7], coupled and decoupled scheme [8], transmission simulator method [9] and expanding the full FRF matrix from the measured rotational FRF data [10] have been introduced in order to indirectly obtain the rotational FRF data. However, it has been found that the performance of the methods is heavily dependent on the excellent quality of measured data and the high accuracy of FE model on which, in fact, they seem to be very difficult to accomplish [5]. The aim of this research is to introduce a different approach through which the rotational FRF data of the experimental model are obtained by using the FRF synthesis method based on the updated finite element model that may uniquely address the difficulties and improve the quality of predicted results of the FBS method.

This paper puts forward an approach of obtaining the rotational FRF data of the experimental model using the updated FE model based FRF synthesis method. The findings are used to develop an accurate and efficient FRF coupling of the FBS method. The proposed approach is formulated based on the finite element method, model updating technique and FRF synthesis method. The detailed explanation of model updating methods can be found in [11–13]. The proposed approach may offer a unique capability to address the difficulties encountered and is expected to improve the quality of the predicted dynamic behaviour of the FBS method. In addition, the high dependency on the experimental data on which the FBS method relies heavily, may be significantly reduced.

1.1. Description of the Test Assembled Structure

In this work, the applicability and the accuracy of the proposed approach for coupling the finite element model and experimental model were demonstrated by using an assembled structure with bolted joints. The assembled structure consists of two structural components namely an irregular plate and a simple beam as shown in Figure 1. The FRF of the beam was derived from the finite element model, while the FRF of the irregular plate was acquired from the experimental work and proposed method. The test assembled structure was made of steel, 500mm in length and 200mm in width. Both structural components were connected with two bolted joints, namely connection points 1 and 2.

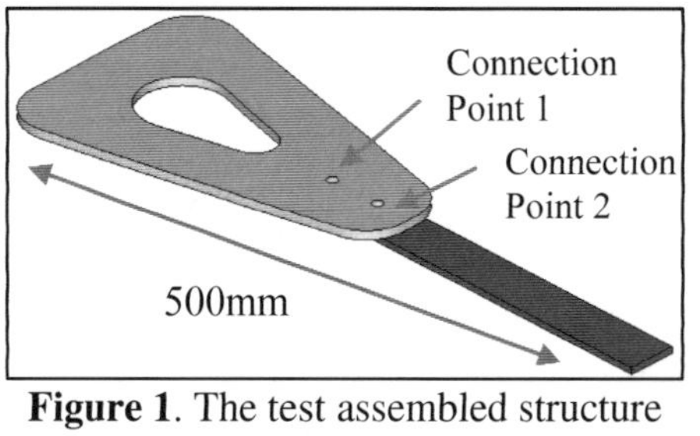

Figure 1. The test assembled structure

The frequency of interest in this work was between 0Hz and 2000Hz. This is because there were several bending and torsional modes occurred at the connection interfaces within the frequency range which is essential for the study of the rotational effects. Prior to performing the FBS analysis using the proposed approach, the FRF of the test assembled structure was measured for validation purposes using the experimental modal analysis (EMA) with free-free boundary conditions as presented in previous studies [4,10,14]. Upon the completion of the FRF measurement, the structure was then disassembled.

On the development of a finite element model of the beam, 3D elements were used and the final FE model is presented in Figure 2. The FE model was used to derive FRFs using the FRF synthesis method [cite a relevant paper covers the method]. For the experimental model (irregular plate), the FRFs were derived from a simplified finite element (SFE) model which is presented in the next subsection.

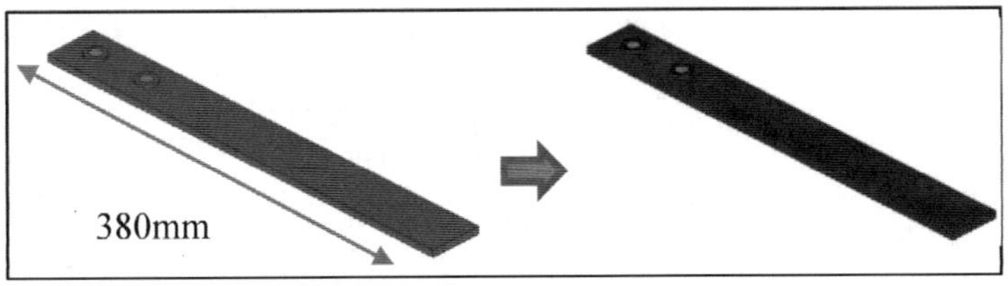

Figure 2. The finite element model for the beam

1.2. Development of the simplified and approximated finite element (SAFE) model for the experimental model

Previous works [4,5,15] reported that the rotational FRF data for the experimental model is normally obtained by performing the modal expansion method or deriving directly from several measured translational FRF data to form a full FRF coupling matrix. In addition, using a full FRF coupling matrix during the coupling process of structural components is proven to lead to a successful FBS analysis as presented in [10]. Therefore, instead of using the modal expansion method to obtain rotational FRF data, this study proposes a new FRF coupling approach developed using the updating method [cite paper here] in the light of experimental data where several mode shapes were used as updating parameters. The updated finite element model of the experimental model was then used to derive and extract the incomplete rotational FRF data required.

Constructing an accurate finite element model of a large complex structure is very difficult and time consuming [16]. Therefore, instead of spending enormous amount of time preparing a finite element model in great detail, this study introduces an approximate modelling approach in which the finite element model of the experimental model was simplified and developed approximately. Figure 3 shows the processes involved in the development of the simplified finite element model of the experimental model (irregular shape). The SFE model of the experimental model was developed using 2D shell elements with a simplified and approximate outline of the experimental model, giving enough details in representing the physical structural component. In this case, it is the irregular shape.

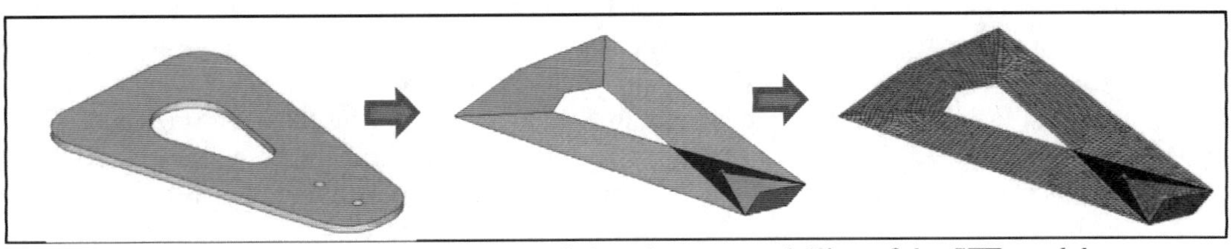

Figure 3. The development and finite element modelling of the SFE model

Theoretically, the developed SFE model allows its parameters to be updated by referring to the experimentally obtained mode shapes. By adopting the SFE method, the mode shapes, natural frequencies and FRF from the updated SFE model can be matched as close as possible to the actual experimental one. The updated SFE model was then used to derive a new FRF matrix which contained both translational and rotational FRFs.

2. Test Setup of the Experimental Model (Irregular Plate)

As mentioned earlier, the SFE model was reconciled based on the experimental natural frequencies and mode shapes of the experimental model of the irregular plate. The plate was tested by using experimental modal analysis to acquire its dynamic behaviour. To minimize any constraints, the experimental work was carried out by suspending the structure using soft springs in order to simulate free-free boundary conditions [17]. Figure 4 shows the test set-up of the irregular plate in which the translational FRF data at the connection interfaces were measured using two 100 mv/g tri-axial accelerometers. The measurement was performed in order to validate the synthesized FRF data of the SFE model.

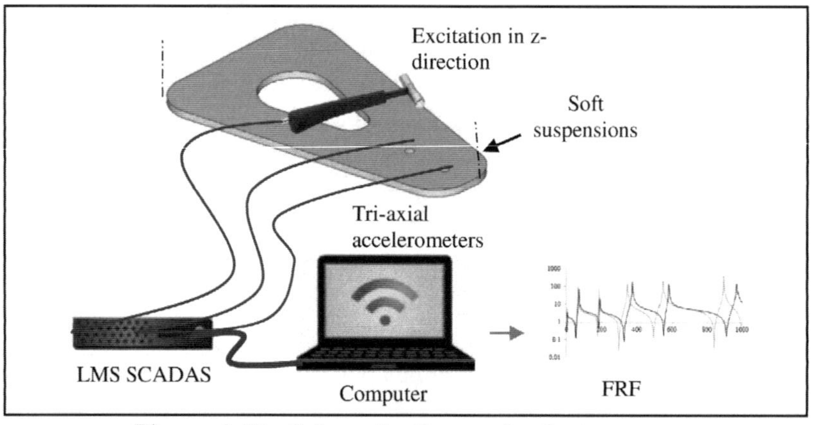

Figure 4. The Schematic diagram for the test setup

3. Model Updating of the SFE Model

The objective of finite element model updating of the SFE model is to reconcile the SFE model with the experimental model by improving the potential design parameters of the SFE model in the light of experimental data to an acceptable level of accuracy. To achieve it, the discrepancies between the measured and predicted responses which are modal data, are minimised [18–20]. The parameterisation can be performed via sensitivity analysis in the form of

$$\mathbf{S} = \Phi_i^T \left[\frac{\delta \mathbf{K}}{\delta \theta_j} - \omega_i \frac{\delta \mathbf{M}}{\delta \theta_j} \right] \Phi_i \tag{2}$$

where matrix $\mathbf{S}$ represents the sensitivity matrix, while Φ, ω and θ represent the eigenvector, eigenvalue and parameter respectively. The i indicates the i-th eigenvalue and j is the j-th parameter. Generally, the sensitivity analysis is used to identify the most sensitive parameters to the predicted natural frequencies. Prior to performing the modal updating method, modal assurance criterion (MAC) analysis was performed to quantify the accuracy of the predicted mode shapes. The information about the MAC analysis can be found in [21]. Then, the objective function containing the natural frequencies and MAC number are defined in order to reduce the discrepancies between the finite element model and experimental data. In this study, 9 mode shapes were used in the updating process. Updating the parameters was performed by using MSC NASTRAN Solution 200 with the same objective function used in [22]. The comparisons between the initial and updated natural frequencies and MAC values of the SFE model are presented in Figure 5 and Table 1.

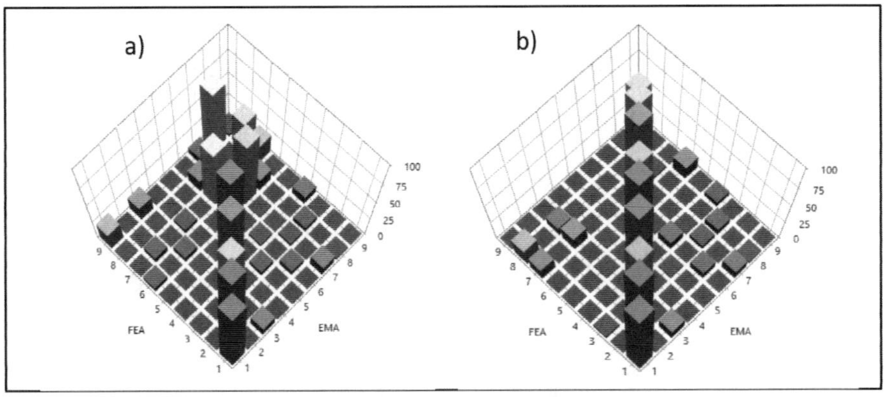

Figure 5. a) MAC results of the initial SFE model b) MAC values of the updated SFE model

From Figure 5 and Table 1, it was found that the MAC values of the 6th to 9th modes of the initial SFE model have recorded uncorrelated, very low and unacceptable values. This indicates that the mode shapes calculated from the initial SFE model are not in good agreement with the measured counterparts especially for the higher modes. The discrepancies occurred because the geometrical dimensions and material properties used for the development of the initial SFE model are based on the simplifications and initial assumptions. As a result, the initial SFE was incapable of representing the actual irregular plate. However, a significant improvement in the natural frequencies and the MAC values of all 9 modes were recorded after the initial SFE was updated. This achievement suggests that the updated SFE model has successfully been used in representing the dynamic behaviour (natural frequencies and mode shapes) of the actual irregular shape.

Table 1. The comparison of the dynamic behaviour between the initial and updated SFE model

Initial SFE Model		**Updated SFE Model**	
1st Mode FEA – 343.02 Hz EMA – 352.19 Hz MAC – 97.4%	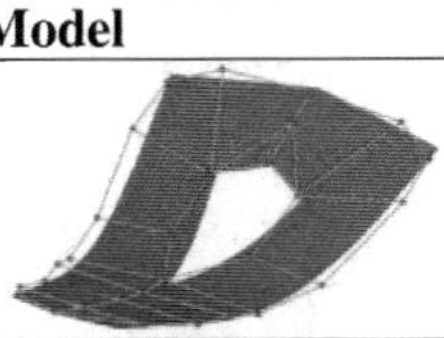	1st Mode FEA – 350.45 Hz EMA – 352.19 Hz MAC – 98.5%	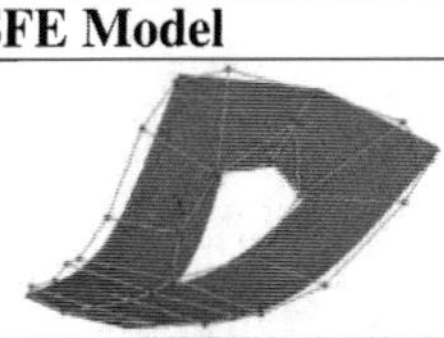
2nd Mode FEA 482.63 Hz EMA – 474.80 Hz MAC - 97.5 %	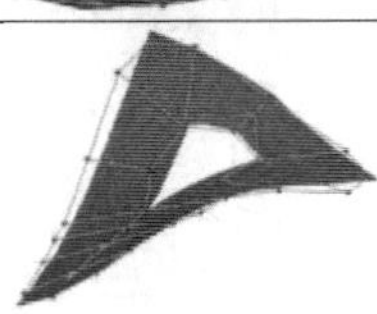	2nd Mode FEA 478.35 Hz EMA – 474.80 Hz MAC - 98.7 %	
3rd Mode FEA – 857.81 Hz EMA – 876.62 Hz MAC - 77.0%		3rd Mode FEA – 887.50 Hz EMA – 876.62 Hz MAC- 81.7%	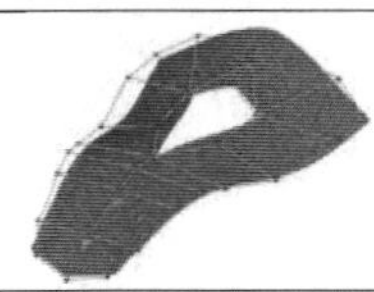
4th Mode FEA – 1033.4 Hz EMA – 1038.0 Hz MAC – 89.3%		4th Mode FEA – 1038.3 Hz EMA – 1038.0 Hz MAC – 94.5%	
5th Mode FEA – 1111.0 Hz EMA – 1128.0 Hz MAC – 93.7%	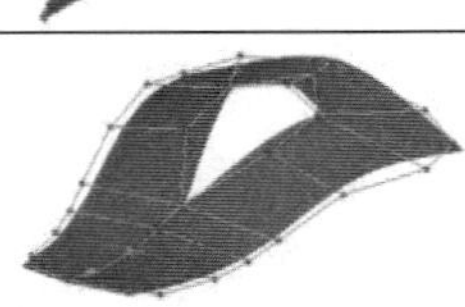	5th Mode FEA – 1130.6 Hz EMA – 1128.0 Hz MAC – 94.7%	

6th Mode FEA – 1922.7 Hz EMA – 1886.7 Hz MAC – 0.3%	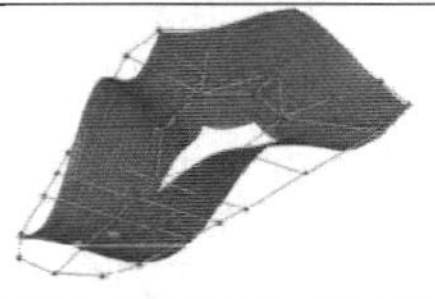	6th Mode FEA – 1889.5 Hz EMA – 1886.7 Hz MAC – 79.6%	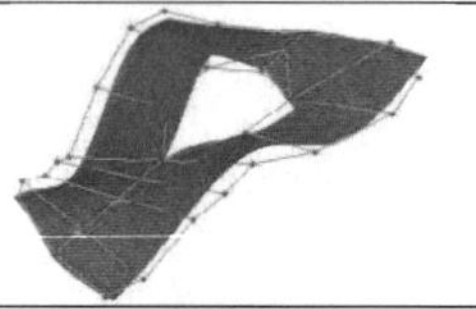
7th Mode FEA – 1974.7 Hz EMA – 1935.2 Hz MAC – 2.0%	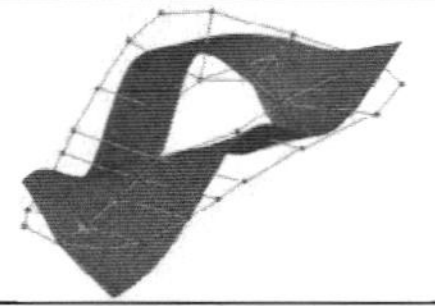	7th Mode FEA – 1914.1 Hz EMA – 1935.2 Hz MAC – 88.1%	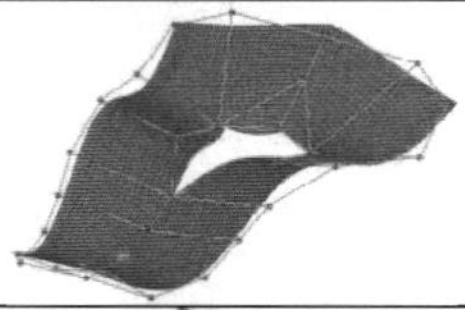
8th Mode FEA – 2083.9 Hz EMA – 2160.5 Hz MAC – 0.7%	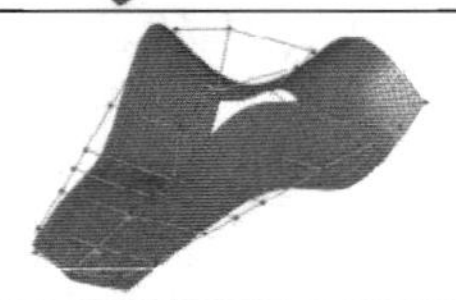	8th Mode FEA – 2146.3 Hz EMA – 2160.5 Hz MAC – 76.5%	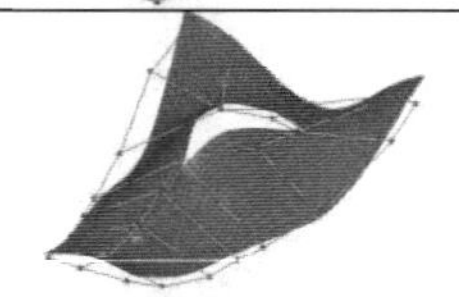
9th Mode FEA – 2088.9 Hz EMA – 2258.4 Hz MAC – 2.2%	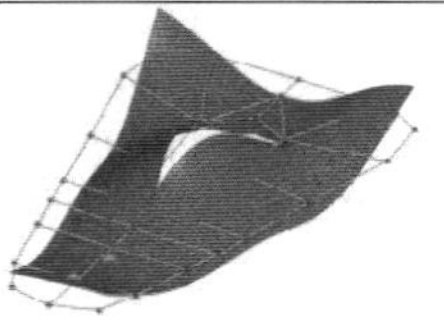	9th Mode FEA – 2231.1 Hz EMA – 2258.4 Hz MAC – 54.9%	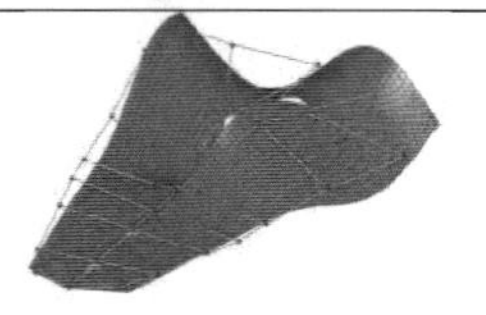

4. Results and Discussion

As stated in the introduction section, the FBS method requires the FRF data instead of modal model data during the coupling process. Before performing the FBS analysis, the translational FRF data of the SFE model were derived and compared with the measured translational FRF counterparts for validation purposes. Later, the rotational FRF data s were derived and then used to form a new FRF coupling matrix with complete degrees of freedom.

4.1. Synthesised FRF Validation

The translational FRF data at the connection points 1 and 2 of the updated SFE model were derived numerically by using the FRF synthesis method based on calculated modes. For this method, the synthesized FRF matrix $H_{syn}(\omega_k)$ and mode shapes are expressed in the form of:

$$H_{syn}(\omega_k) = \sum_{i=1}^{N} \frac{\{\emptyset\}_i\{\emptyset\}_i^T}{(\omega_{n_i}^2 - \omega_k^2) + j2\xi_i\omega_k\omega_{n_i}} \quad (3)$$

where N is the number of calculated modes, $\{\emptyset\}_i$ is the ith mass normalised mode shapes, ω_{n_i} is ith natural frequency and ξ_i is the ith modal damping ratio. The synthesized FRF data were compared with the measured counterparts. Only the FRF data in the z-axis were used in the comparison. This is because all modes of the irregular plate largely are present in the z-direction. Figures 6, 7 and 8 show the comparisons between the FRF data of the z-axis FE and that of EMA at connection points 1, 2 and the cross axis FRF between points 1 and 2.

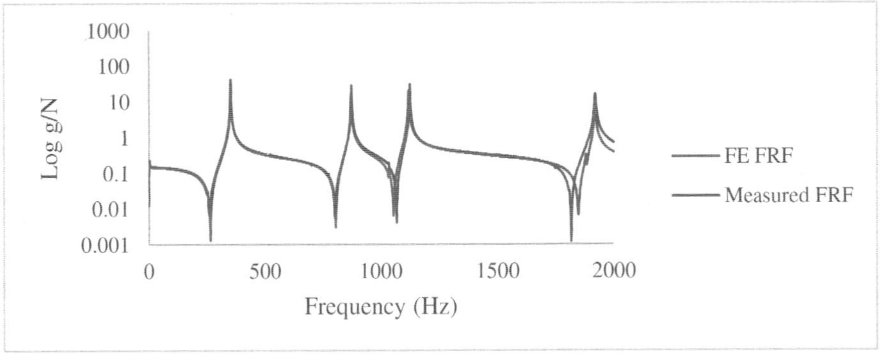

Figure 6. The comparison of z-axis FE and measured FRF at the connection point 1

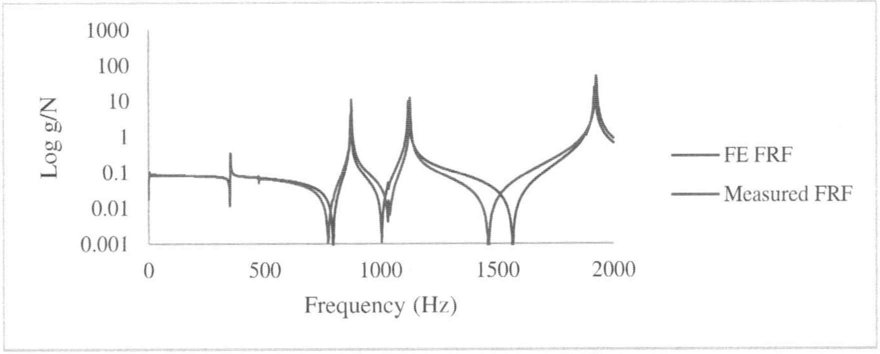

Figure 7. The comparison of z-axis FE and measured FRF at the connection point 2

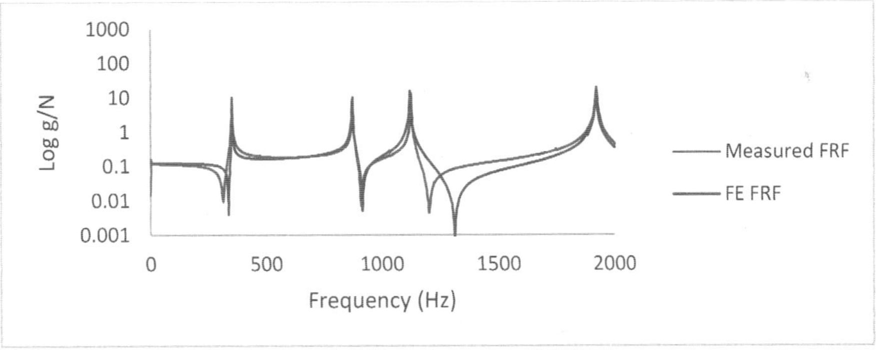

Figure 8. The comparison of cross z-axis FE and measured FRF between connection points 1 and 2

From Figures 6,7 and 8, it can be clearly seen that all the resonance peaks of the synthesised FRF data have shown a perfect match with the measured FRF counterparts, especially at the peaks at connection point 1. Another striking point that can be observed in the synthesised FRF data from Figures 7 and 8 is that the anti-resonance peaks have shown minor discrepancies, particularly, at the peaks within the frequency range of 1000-2000Hz. The discrepancies may be as a result of excluding the effect of structural damping in normal modes analysis of the SFE model. However, the accuracy of the synthesized FRF data obtained from the SFE model is within an acceptable level. This is because the predicted resonance peaks that represent the natural frequencies have shown a strong correlation. The rotational FRF data derived from the SFE model are presented in Figures 9 and 10. Since the rotational FRF data were not measured in this study [23], therefore, the accuracy of the data will be validated with the predicted results of the FBS method.

IOP Conf. Series: Journal of Physics: Conf. Series **1262** (2019) 012001 doi:10.1088/1742-6596/1262/1/012001

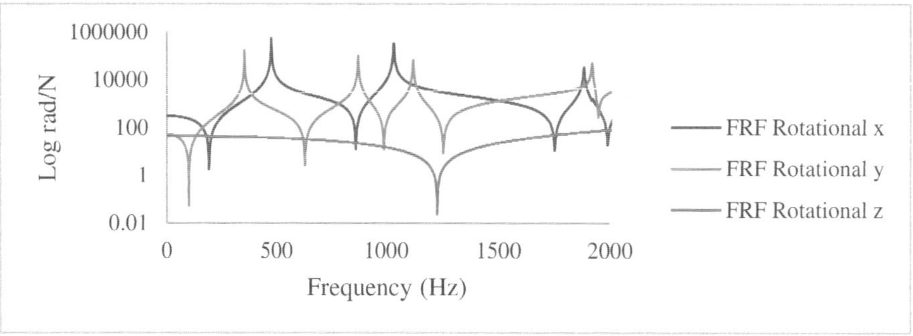

Figure 9. The derived rotational FRF data at connection point 1

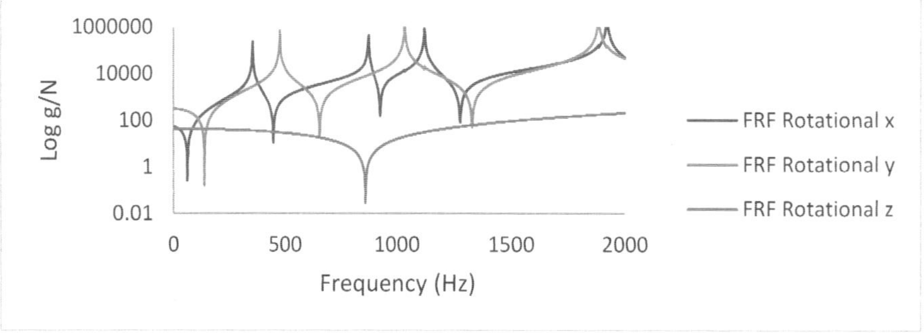

Figure 10. The derived rotational FRF data at connection point 2

4.2. FBS method using the developed SFE model

The FBS method was employed to couple the FRF data derived from the finite element model of the beam and the updated SFE model in order to evaluate the accuracy and efficiency of the proposed FRF coupling matrix. The coupling process of the finite element model and experimental model was carried out using the proposed FRF coupling matrix containing both translational and rotational FRF data required. The validation of the proposed FRF coupling matrix is presented in Figure 11 in which the FRF data of the proposed FRF coupling matrix are compared with that of experimental modal analysis of the assembled structure.

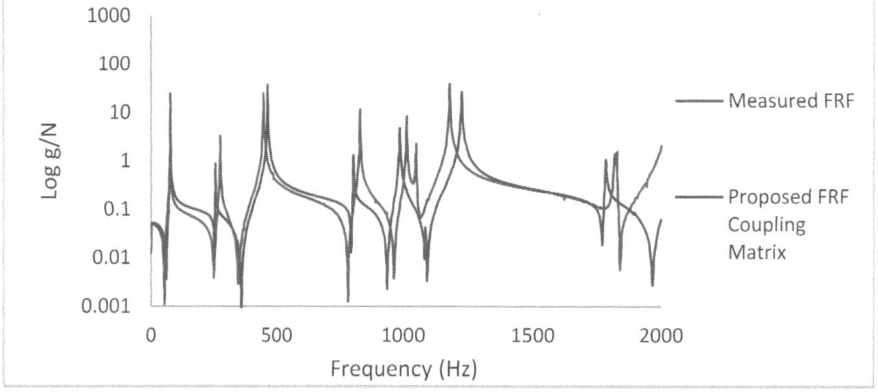

Figure 11. The comparison of the coupled and experimental FRF of the assembled structure

One good point that can be seen from the direct comparison between the proposed coupling matrix and experimental modal analysis as shown in Figure 11 is that there is a strong correlation between both

FRF data within the frequency bandwidth of 0Hz and 2000Hz. The strong correlation in terms of resonance and anti-resonance peaks indicates that the SFE model could be used accurately and efficiently in developing the FRF coupling matrix containing rotational and translational FRF data. Therefore, it is worth noting that the proposed FRF coupling matrix using the SFE model has offered a great capability to accurately predict the dynamic behaviour of the jointed structure.

Another striking point from Figure 11 is that there are only minor discrepancies in the resonance and anti-resonance peaks for the higher modes. The discrepancies are suspected of link with the stiffness determined and the type of the coupling used in modelling the bolted joints, which may be inadequate to accurately represent the actual physical properties of the bolted joints in the assembled structure[24,25]. This is a typical phenomenon encountered in the field of structural dynamics because accurately modelling of joints is very problematic and challenging [17,26].

5. Conclusions

A new FRF coupling of the FBS method for the investigation of the dynamic behaviour of the assembled structure consisting of a beam and an irregular plate has been proposed and evaluated with the experimental results. The comparison of results between the proposed FRF coupling and experimental counterparts reveal that the proposed FRF coupling has a great capability to accurately predict the dynamic behaviour of the assembled structure. This achievement suggests that the proposed FRF coupling has significantly reduced the high dependency of the FBS method on the experimental FRF data which are very difficult to accurately measure. The proposed FRF coupling of the FBS method could be applied quite reliably to other complex assembled structures without a significant declination in accuracy and efficiency.

Acknowledgments

The authors are gratefully indebted to the Malaysian Ministry of Higher Education for providing financial support for this study through the Geran Inisiatif Penyeliaan (GIP) 600-IRMI 5/3/GIP (012/2019). They would also like to extend their sincere gratitude for the support and help given by Associate Professor Dr Siti Akmar Abu Samah and all the SDAV members.

References

[1] Law M and Ihlenfeldt S 2015 A frequency-based substructuring approach to efficiently model position-dependent dynamics in machine tools *Proceedings of the Institution of Mechanical Engineers, Part K: Journal of Multi-body Dynamics* **229** 304–17

[2] Wang J, Zeng X and Gasparini D a. 2008 Dynamic response of high-speed rail foundations using linear hysteretic damping and frequency domain substructuring *Soil Dynamics and Earthquake Engineering* **28** 258–76

[3] Noh H M 2016 Prediction and reduction in vehicle noise by frequency response function-based substructuring *Advances in Mechanical Engineering* **8** 1–13

[4] Mottershead J E, Ghandchi Tehrani M, Stancioiu D, James S and Shahverdi H 2006 Structural modification of a helicopter tailcone *Journal of Sound and Vibration* **298** 366–84

[5] Nicgorski D and Avitabile P 2010 Experimental issues related to frequency response function measurements for frequency-based substructuring *Mechanical Systems and Signal Processing* **24** 1324–37

[6] de Klerk D, Rixen D J, Voormeeren S N and Pasteuning F 2008 Solving the RDoF Problem in Experimental Dynamic Substructuring *Proceedings of the 26th International Modal Analysis Conference - IMAC*

[7] Van der Seijs M, van den Bosch D, Rixen D and de Klerk D 2014 an Improved Methodology for the Virtual Point Transformation of Measured Frequency Response Functions in Dynamic Substructuring *Proceedings of the 4th International Conference on Computational Methods in Structural Dynamics and Earthquake Engineering (COMPDYN 2013)* 4334–47

[8] Law M, Rentzsch H, Ihlenfeldt S and Putz M 2016 Application of Substructure Decoupling

Techniques to Predict Mobile Machine Tool Dynamics: Numerical Investigations *Procedia CIRP* **46** 537–40

[9] Mayes R L, Hunter P S, Simmermacher T W and Allen M S 2008 Combining Experimental and Analytical Substructures with Multiple Connections *26th IMAC: Conference and Exposition on Structural Dynamics 2008: IMAC XXVI* (Orlando, Florida)

[10] Drozg A, Čepon G and Boltežar M 2018 Full-degrees-of-freedom frequency based substructuring *Mechanical Systems and Signal Processing* **98** 570–9

[11] Mohd Zin M S, Rani M N A, Yunus M A, Wan Iskandar Mirza W I I, Mat Isa A A and Mohamed Z 2017 Modal and FRF Based Updating Methods for the Investigation of the Dynamic Behaviour of a Plate *Journal of Mechanical Engineering* **4** 175–89

[12] Mottershead J E and Friswell M I 1993 Model Updating In Structural Dynamics: A Survey *Journal of Sound and Vibration* **167** 347–75

[13] Mottershead J E, Link M and Friswell M I 2011 The sensitivity method in finite element model updating: A tutorial *Mechanical Systems and Signal Processing* **25** 2275–96

[14] Williams A, Chipman C and Avitabile P 2008 Modal and frequency based substructuring using rotational DOF conidetations *26th International Modal Analysis Conference (IMAC XXVI)*

[15] Cuppens K, Sas P and Hermans L 2001 Evaluation of the FRF based substructuring and modal synthesis technique applied to vehicle FE data *Proceedings of ISMA 25* 1143–50

[16] Wan Iskandar Mirza W I I, Rani M N A, Othman M H, Kasolang S and Yunus M A 2016 Reduced order model for model updating of a jointed structure *Journal of Engineering and Applied Sciences* **11** 2383–6

[17] Omar R, Rani M N A, Yunus M A, Mirza W I I W I and Zin M S M 2018 Efficient finite element modelling for the investigation of the dynamic behaviour of a structure with bolted joints *AIP Conference Proceedings* **020082** 020082

[18] Rani M N A, Kasolang S, Othman M H, Yunus M A and Wan Iskandar Mirza W I I 2016 Finite Element Modelling and Modal Based Up- Dating for The Dynamic Behaviour of a Laser Spot Welded Structure *ICSV 2016 - 23rd International Congress on Sound and Vibration: From Ancient to Modern Acoustics* pp 1–8

[19] Abdullah N A Z, Sani M S M, Husain N A, Rahman M M and Zaman I 2017 Dynamics properties of a Go-kart chassis structure and its prediction improvement using model updating approach *International Journal of Automotive and Mechanical Engineering* **14** 3887–97

[20] Fouzi M S M, Jelani K M, Nazri N A and Sani M S M 2018 Finite Element Modelling and Updating of Welded Thin-Walled Beam *International Journal of Automotive and Mechanical Engineering* **15** 5874–89

[21] Pastor M, Binda M and Harčarik T 2012 Modal assurance criterion *Procedia Engineering* **48** 543–8

[22] Rani M N A, Ouyang H, Yunus M A and Aminudin B A 2013 Model Updating for a Thin Steel Sheet Welded Structure *20th International Congress on Sound and Vibration 2013 (ICSV 2013)* **2** 1138–45

[23] Wan Iskandar Mirza W I I, Rani M N A, Yunus M A, Ayub M A, Sani M S M and Mohd Zin M S 2019 Frequency Based Substructuringfor Structure with Double Bolted Joints: A Case Study *International Journal of Automotive and Mechanical Engineering* **16** 6188–99

[24] Omar R, M N A, Wan Iskandar Mirza W I I, Yunus M A and Othman M H 2017 Finite Element Modelling and Updating for Bolted Lap Joints *Journal of Mechanical Engineering* **4** 202–22

[25] Omar R, Rani M N A, Yunus M A, Mat Isa A A, Wan Iskandar Mirza W I ., Mohd Zin M S and Roslan L 2018 Investigation of Mesh Size Effect on Dynamic Behaviour of an Assembled Structure with Bolted Jointsusing Finite Element Method *International Journal of Automotive and Mechanical Engineering* **15** 5695–708

[26] Zahari S N, Sani M S M, Husain N A, Ishak M and Zaman I 2016 Dynamic analysis of friction stir welding joints in dissimilar material plate structure *Jurnal Teknologi* **78** 57–65

IOP Conf. Series: Journal of Physics: Conf. Series **1262** (2019) 012002 doi:10.1088/1742-6596/1262/1/012002

Nondeterministic Model Updating of Laser Spot Welded Structure via Response Surface Methodology

M A S Aziz Shah[1], M A Yunus[1*], M N Abdul Rani[1], N Salleh[2], W I I Wan Iskandar Mirza[1] and C Peter[1]

[1] Structural Dynamics Analysis and Validation, Faculty of Mechanical Engineering, Universiti Teknologi MARA (UiTM), 40000 Shah Alam, Selangor, Malaysia

[2] National Space Agency of Malaysia (ANGKASA), 42700 Banting, Selangor, Malaysia

*Corresponding author: mayunus@salam.uitm.edu.my

Abstract. This paper presents the application of response surface method in finite element model updating, as an approach for improving the accuracy and efficiency of the finite element model of a laser spot welded structure, to reflect the physical responses of the structure. The procedures of implementation for response surface in model updating such as sampling method, selecting the significant updating parameters and constructing a quadratic polynomial response surface are discussed. Initially, the finite element model of the structure was developed using CQUAD4 shell element alongside CWELD element connectors to represent laser spot weld joints. Then, NASTRAN SOL 103 was used to calculate the dynamic behaviour of the model (natural frequencies and mode shapes). The experimental modal analysis was then conducted under free-free boundary conditions via LMS SCADAS to obtain experimental data. The minimisation of the discrepancies of the finite element model was based on objective function that was formed by the residuals between finite element and experimental natural frequencies. Results show that response surface method was efficient to be used in finite element model updating since it is capable to improve the accuracy of the finite element model.

1. Introduction

In recent years, many structural engineering problems have become more laborious to be solved due to the high demands towards complex and unique design of the structure, involving many sophisticated numerical analysis and simulation software to be used for design analysis. One of the most predominant and practical numerical analysis tools is the finite element method. However, the finite element model of a structure is normally constructed based on the engineering assumptions of the structural design whereby it may not truly represent all the aspects of an actual structure [1]–[4]. These may result in a highly disagreement between the predicted result of the finite element model and the actual structure. Discrepancies that are generated from the finite element model basically originated from the uncertainties in simplifying assumptions of the structural geometry, material properties and mechanical joints. It is often required to optimise the finite element model by adjusting the uncertain parameters that may improve the prediction results or responses of an actual structure [5], [6].

Since initial finite element models are always not in good agreement with the actual structure, finite element model updating is such a procedure that can be used to alter or modify the uncertain parameters that appear so that a more realistic model can be achieved. Theoretically, finite element model updating is an inversed problem in which the dynamic characteristic is used to identify and modify the uncertain parameters of finite element model [7]. Previous studies by Ren et al stated that setting up of an objective function, selecting updating parameters and applying robust optimisation algorithm are the crucial steps in model updating [8]. Moreover, Mottershead et al have stated that, finite element model updating is not only emphasizing the satisfactory degree of accuracy of predicted results with actual structure, but the updated parameters must also maintain the physical meaning of the structure so that it can mimic the actual structure accurately and reliably [9].

Basically, there are two methods of model updating which are, direct based model updating and iterative based model updating [9]. Direct based model updating methods are commonly known as one step procedure as it directly updates the finite element model to reduce the discrepancies with experiment result without considering the physical meaning of the updated parameters. Meanwhile, iterative based model updating methods are involved with the identification of sensitivity of responses to the updated parameters so as to preserve the physical meaning of the structure. Many previous works had involved with the use of iterative based model updating by coupling with sensitivity analysis [10]–[14]. Boscato et al stressed that, the sensitivity based model updating required complicated constructions of sensitivity matrices because the finite element models should be tuned and recomputed iteratively during optimisation process [15]. Furthermore, for a large structure, a finite element model contains a very large degree of freedoms and since iterative method is a repeated procedure, it may cause convergence difficulty and also contribute to high computational time [16]. These issues make all the stated methods as not efficient and unpractical to be used for the complex jointed structure such as laser spot welded structure. Therefore, alternative approaches are explored to obtain more efficient method that can be used to improve the correlation of the initial finite element model to the experimental data by using response surface based model updating.

Response surface method is an approach to replace a finite element model by an approximate meta-model. Originally, meta-model is a simplified model of an actual model of finite element model and has been replaced statically to represent the input and output relation. Thus, response surface is much efficient, fast running and has low computational cost since only few yet important parameters are involved in predicting the responses [17], [18]. In the finite element model updating, once the response surface of a structure has been constructed, the process is reduced to the task of finding the smallest value on the response surface and the parameters that correspond to the smallest value are selected to update the model. The aim of this paper is to present a procedure of finite element model updating based on response surface method and to obtain better correlation between the predicted and measured results. In this work, the laser spot welded structure has been used since it contains complex joint connector and it also involves with many uncertainties.

2. Experimental Modal Analysis (EMA)

In the experimental work, the laser spot welded structure which consists of two sub-components: hat-shape plate and flat plate as shown in Figure 1 were tested. The sub-components of the structure were connected by twenty laser spot welds with nominal 5 mm diameter for each and were separated by 60 mm apart. The material used to fabricate the sub-components of the structure are cold rolled mild steel sheets with overall dimension of 560 mm length, 110 mm wide and 15 mm thickness.

The experimental modal analysis was performed to the laser spot welded structure by using impact hammer and roving accelerometers method so as to measure the dynamic characteristic of the structure such as natural frequencies and mode shapes. The frequency bandwidth of interest was set from 1 – 1000 Hz. Prior to measuring the dynamic characteristic of the structure, the structure was set up under free-free boundary conditions whereby four sets of rubber band and strings were used to hang the structure from the specially designed clamps by attaching the strings to the holes on the structure and the other ends of the strings to the clamps (Figure 2). Finally, all the data obtained were processed using LMS SCADAS data acquisition system.

Figure 1. Laser spot welded structure

Figure 2. The structure is suspended using rubber bands and strings

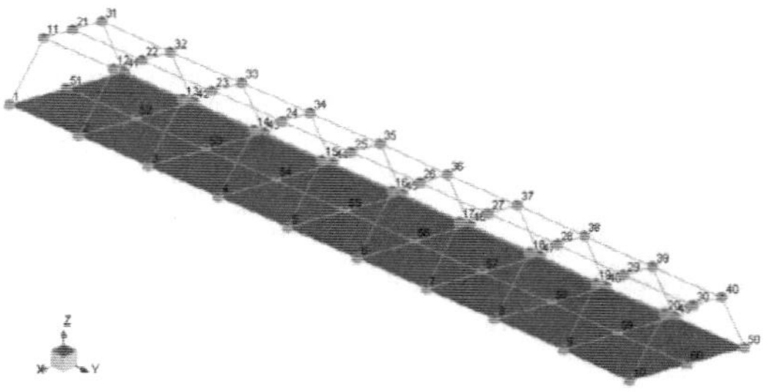

Figure 3. Geometry of a laser spot welded structure

3. Finite Element Modelling of Laser Spot Welded Structure

The finite element model under investigation is a laser spot welded structure that has been constructed so as to replicate component of car body-in-white such as pillar part. In the pre-processing stage of the finite element modelling, the MSC PATRAN was used to develop the finite element model. The structure was modelled using CQUAD4 elements (shell) and the model was meshed into 3 mm meshing size based on the suitability of the meshes size recorded from mesh convergence test (Figure 4). The CWELD element connectors as shown in Figure 5 then were used to represent as laser spot welds and to connect the hat-shape plate and flat shape plate sub-components as recommended from previous studies [19]–[21]. The material properties used in the finite element modelling was based on nominal properties of mild steel as shown in Table 1.

Table 1. Material properties of the laser spot welded structure [21]

Component	Property	Value	Unit
Hat shape plate	Young's Modulus	210	GPa
	Poisson's Ratio	0.30	Unitless
	Mass Density	7700	kg/m^3
Flat shape plate	Young's Modulus	210	GPa
	Poisson's Ratio	0.30	Unitless
	Mass Density	7700	kg/m^3
Laser spot weld (CWELD)	Young's Modulus	210	GPa
	Poisson's Ratio	0.30	Unitless
	Mass Density	7700	kg/m^3

Figure 4. Finite element model of laser spot welded structure

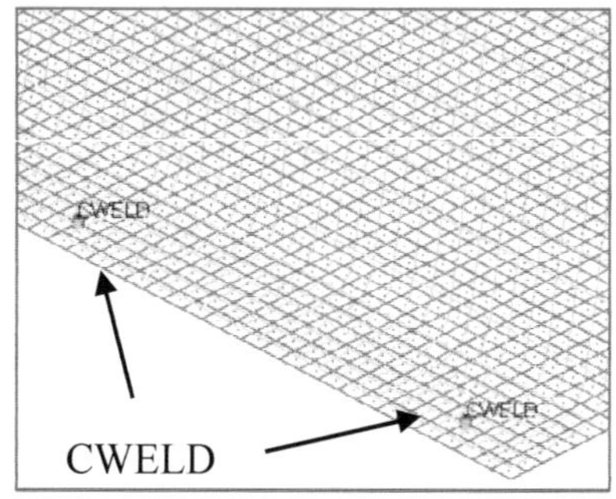

Figure 5. CWELD element connectors

In this research, normal mode analysis was run using MSC NASTRAN with SOL 103 solver that was used to identify the natural frequencies and mode shapes of the finite element model. The equation of motion that has been used to discretise the system to a finite element model, also known as 2nd order differential equation is given as

$$\mathbf{M}\ddot{x}(t) + \mathbf{C}\dot{x}(t) + \mathbf{K}x(t) = f(t) \quad (1)$$

where **M**, **C** and **K** are symmetric matrices of mass, damping and stiffness. Meanwhile $\ddot{x}$, $\dot{x}$ and x represent the vector of accelerations, velocities and displacement respectively and f(t) is vector of external forces. It has been found that, the damping value of the structure can only be determined experimentally. In this study, the experimental frequency response function (FRF) data as shown in Figure 6 indicates that the laser spot welded structure tends to be undamped. The value of damping is less than one percent and it can be considered lightly damped [20]. Therefore, the effect of damping can be theoretically neglected in finite element modelling. As a result, for the undamped free vibration analysis, the equation (1) can be simplified as

$$\mathbf{M}\ddot{x}(t) + \mathbf{K}x(t) = 0 \quad (2)$$

The equation (2) can be solved by assuming the harmonic solution in the form of

$$x = \phi sin\omega t \quad (3)$$

where ω and ϕ are the mode shape and natural frequency of the system. If the differentiation of the assumed harmonic solution is performed and substituted in equation (2), the equation of motion yields and simplified to the following

$$(\mathbf{K} - \omega^2\mathbf{M})\phi = 0 \quad (4)$$

The natural frequencies and mode shapes of the laser spot welded structure can be predicted by solving the equation (4) using finite element commercial software such as MSC NASTRAN.

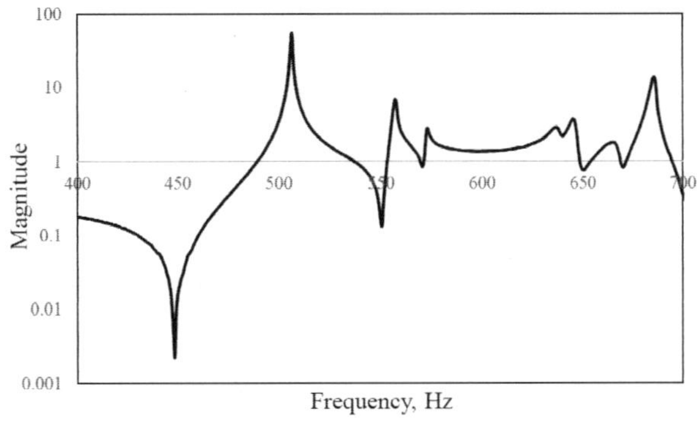

Figure 6. Frequency response function (FRF) of laser spot welded structure

4. Response Surface Methodology for Finite Element Model Updating

Finite element model updating based on response surface methodology is an approach to acquire the global approximations of the structural response that contains objectives and constraints established from functional evaluations at various points in the design space. Basically, finite element model updating based on response surface methodology often involves a combination of experimental strategies, mathematical, probability and statistical methods and much useful for the engineers to optimising structures based on the research interest [22].

4.1. Latin hypercube sampling method

In order to develop a response surface that will serve as a meta-model for the finite element model, one of the basic process is to calculate the initial predicted response features at various points in the parameter space by performing experiment at each of the points. The values that are featured in the experiment ran across the parameter domain fit with a response surface. The term of experimentation herein refers to either physical experiments or computer experiments that will serve as sampling. Latin hypercube sampling (LHS) is a method of sampling that can be used to sample the design space by bounding the upper lower limits of each of the design variable [23]. However, this method is performed by generating random sample points and ensuring that all portions of the design space are represented in the same probability trend. This method operates by subdividing the sample space into smaller regions and evenly distributed sampling points by ensuring a good coverage of the random parameter [24].

4.2. Response surface regression

It is essential to construct a response surface form that can represent structural response accurately. The selected response surface form should be capable of attaining surfaces that meet specific smoothness requirements of an application. In the case of finite element model updating in structural dynamics, polynomials are popular forms representing a response surface because the calculations are simple and the resulting function is closed-form algebraic expression [25]. The first step in response surface methodology is to find a suitable approximation for the true functional relationship between y and the set of its independent variables of x. Usually, a low-order polynomial in some regions of the independent variable is employed. If the response is well modelled by a linear function of the independent variables, the approximating function is the first order model represented as:

$$y = \beta_0 + \sum_{i=1}^{n} \beta_i x_i \tag{5}$$

Since the real world problems are usually very complicated, linear estimation may not perform well in providing a good representation of the objective function. If a curvature appears in the system, then quadratic model will be:

$$y = \beta_0 + \sum_{i=1}^{n} \beta_i x_i + \sum_{i=1}^{n} \beta_{ii} x_i^2 + \sum_{i=1}^{n} \sum_{j=1+1}^{n} \beta_{ij} x_i^2 x_j + \sum_{i=1}^{n=2} \sum_{j=i+1}^{n=1} \beta_{ijk} x_i x_j x_k \quad (6)$$

Equation (5) and equation (6), y is the response variable, β_0 is the constant term, β_i is the coefficient of the linear term, β_{ii} is the coefficient of quadratic single term, β_{ij} is the coefficient of the quadratic cross product term, β_{iii} is the coefficient of the cubic single term, β_{iij} is the coefficient of the cubic two cross product terms, and β_{ijk} is the coefficient of the cubic three cross product terms. The x_i, x_j, x_k terms represent the independent variables.

4.3. Finite element model updating

Model updating is an approach to improve the correlation of finite element model and the test structure by correcting the invalid assumptions of the model to an acceptable level of accuracy. In the structural dynamics, the structural response are often eigen-solutions related to such as natural frequencies and mode shapes. In this research, natural frequencies are employed as objective response. Therefore, the optimised objective function is formulated in terms of the residuals between analytical and measured natural frequencies and can be expressed as

$$J = \sum_{i=1}^{n} W_i \left(\frac{\lambda_i^{\text{fe}}}{\lambda_i^{\text{exp}}} - 1 \right)^2 \quad (7)$$

where, λ_i^{exp} is the *i*-th experimental eigenvalue and λ_i^{fe} is the *i*-th predicted eigenvalue from the finite element model and n is the number of eigenvalues involved in the updating procedure.

5. Results and Discussion

In this study, the dynamic characteristic of interest which is natural frequencies of the laser spot welded structure were successfully obtained using the finite element method and experimental modal analysis. CQUAD4 shell elements were used to model the structure, and CWELD element connectors were implemented to represent as laser spot welds. Initially, MSC NASTRAN SOL 103 was used to calculating the predicted dynamic characteristic of the structure. Meanwhile, SOL 200 was used for updating the initial finite element model using response surface method in light of the experiment results. Tables 2 shows the results of natural frequencies obtained using the above-mentioned methods.

Table 2 shows the comparison of results between the initial finite element model and experiment modal analysis of the first six vibration modes. An enormous total error of 21.38 percent was identified in the initial predicted model with 2nd mode was contributed to the largest error. The other noticeable modes that have a high contribution to the total error were observed in the 3th and 6th modes with 3.69 percent and 3.67 percent respectively. It has been found that, these relative errors occurred due to the inability of the initial finite element model to mimic the actual structure accurately. This is because, the finite element model was constructed based on engineering assumptions on geometry, material properties, and spot weld joints [26]. Therefore, some improvement on the assumptions in the initial finite element model must be done to accurately represent the structure.

Finite element model updating using response surface method was successfully applied to the initial finite element model with the minimisation of the natural frequencies in a light of measured natural frequencies as stated in the objective function. As tabulated in Table 2, a great improvement was achieved with the total error showing a huge decrement from 21.38 percent to 14.54 percent. The results in Table 2 also show that the updating has led to a reduction in the individual error of every vibration mode particularly for the 5th mode, from 3.17 percent to 1.86 percent.

Table 2. Comparison between the measured, predicted and updated results for the structure

Mode	I. EMA (Hz)	II. FE (Hz)	III. Error between I & II (%)	IV. RSM (Hz)	V. Error between I & IV (%)
1	503.83	515.57	2.33	512.32	1.69
2	555.53	583.48	5.03	575.58	3.61
3	572.48	593.58	3.69	589.36	2.95
4	632.32	654.40	3.49	645.51	2.09
5	643.03	663.44	3.17	654.99	1.86
6	664.89	689.26	3.67	680.53	2.35
	Total Error		21.38		14.54

6. Conclusion

In this paper, an application of the finite element model updating based on response surface methodology of the laser spot welded structure is presented. The response surface using quadratic polynomial was constructed using the finite element so as to improve the initial prediction response such as natural frequencies. On top of that, the Latin hypercube was successfully used to develop the replacement model of the initial finite element model of the structure. In the part of finite element modelling, CQUAD shell element and CWELD element connectors were implemented by assuming the material properties of mild steel to represent as hat-flat shape plate and laser spot welds respectively. After that, the initial prediction was optimising using finite element model updating based on response surface methodology. By replacing the original finite element model to the response surface form, the model updating process becomes efficient and capable to reduce the discrepancies of the initial finite element model.

Acknowledgments

The authors wish to acknowledge the Malaysia Ministry of Higher Education (MOHE), Research Management Institute (RMI) and Institute of Graduate Studies of Universiti Teknologi MARA (UiTM) for providing financial support for this research through the fundamental research grant scheme (FRGS) 600-IRMI/FRGS 5/3 (172/2019). They would also like to extend their sincere gratitude for the support and help given by Prof. Hadariah Bahron, all SDAV members and Assoc Prof Dr Hjh Siti Akmar Abu Samah.

References

[1] S. E. Abdel Raheem, A. K. Abdel Zaher, and A. M. Taha, "Finite element modeling assumptions impact on seismic response demands of MRF-buildings," *Earthq. Eng. Eng. Vib.*, vol. 17, no. 4, pp. 821–834, 2018.

[2] M. A. Yunus, M. A. S. Aziz Shah, M. N. Abdul Rani, W. I. I. Wan Iskandar Mirza, and S. Kushairi, "Experimental and Finite Element Analysis of Laser Stitch Welded Structure," *J. Mech. Eng.*, vol. 4, no. 3, pp. 155–162, 2017.

[3] M. Zhan, Q. Guo, L. Yue, and B. Zhang, "Modeling and Stochastic Model Updating of Bolt-Jointed Structure," *Shock Vib.*, vol. 2018, pp. 1–12, 2018.

[4] M. A. S. Aziz Shah, M. A. Yunus, M. N. Abdul Rani, N. A. Tormidi, A. A. Mat Isa, and W. I. I. Wan Iskandar Mirza, "Closed Mode Separation of The Symmetrical Structure," *J. Mech. Eng.*, 2018.

[5] S. Daouk, F. Louf, O. Dorival, L. Champaney, and S. Audebert, "Uncertainties in structural dynamics: overview and comparative analysis of methods," *Mech. Ind.*, vol. 16, no. 4, p. 404, 2015.

[6] M. A. Yunus, M. N. A. Rani, M. S. M. Sani, and M. A. S. A. Shah, "Finite element model updating of riveted joints of simplified model aircraft structure," *AIP Conf. Proc.*, vol. 1952, no.

1, p. 20013, 2018.
[7] M. Friswell and J. E. Mottershead, *Finite Element Model Updating in Structural Dynamics*. Springer Netherlands, 1995.
[8] W. X. Ren and H. B. Chen, "Finite element model updating in structural dynamics by using the response surface method," *Eng. Struct.*, vol. 32, no. 8, pp. 2455–2465, 2010.
[9] J. E. Mottershead and M. I. Friswell, "Model Updating In Structural Dynamics: A Survey," *Journal of Sound and Vibration*, vol. 167, no. 2. pp. 347–375, 1993.
[10] M. S. Mohd Zin, M. N. Abdul Rani, M. A. Yunus, W. I. I. Wan Iskandar Mirza, A. A. Mat Isa, and Z. Mohamed, "Modal and FRF Based Updating Methods for the Investigation of the Dynamic Behaviour of a Plate," *J. Mech. Eng.*, vol. 4, no. 3, pp. 175–189, 2017.
[11] Y. Yuan and H. Liu, "An iterative method for solving finite element model updating problems," *Appl. Math. Model.*, vol. 35, no. 2, pp. 848–858, 2011.
[12] J. E. Mottershead, M. Link, and M. I. Friswell, "The sensitivity method in finite element model updating: A tutorial," *Mech. Syst. Signal Process.*, vol. 25, no. 7, pp. 2275–2296, 2011.
[13] M. A. Yunus, H. Ouyang, M. N. A. Rani, and A. A. M. Isa, "Modal test and model updating for a welded structure made from thin steel sheets," *20th Int. Congr. Sound Vib. 2013, ICSV 2013*, no. 7, pp. 7–11, 2013.
[14] M. N. Abdul Rani, S. Kasolang, M. H. Othman, M. A. Yunus, W. I. I. Wan Iskandar Mirza, and H. Ouyang, "Finite Element Modelling and Modal Based Updating for the Dynamic Behaviour of a Laser Spot Welded Structure," *ICSV 2016 - 23rd Int. Congr. Sound Vib. From Anc. to Mod. Acoust.*, pp. 1–8, 2016.
[15] G. Boscato, S. Russo, R. Ceravolo, and L. Zanotti Fragonara, "Global Sensitivity-Based Model Updating for Heritage Structures," *Comput. Civ. Infrastruct. Eng.*, vol. 30, 2015.
[16] L. Deng and C. S. Cai, "Bridge Model Updating Using Response Surface Method and Genetic Algorithm," *J. Bridg. Eng.*, vol. 15, no. 5, pp. 553–564, 2009.
[17] T. Marwala, *Finite-element-model updating using computional intelligence techniques: Applications to structural dynamics*. 2010.
[18] S. Sehgal and H. Kumar, "Structural Dynamic Model Updating Techniques: A State of the Art Review," *Arch. Comput. Methods Eng.*, vol. 23, no. 3, pp. 515–533, 2016.
[19] M. Palmonella, M. I. Friswell, J. E. Mottershead, and A. W. Lees, "Guidelines for the implementation of the CWELD and ACM2 spot weld models in structural dynamics," *Finite Elem. Anal. Des.*, vol. 41, no. 2, pp. 193–210, 2004.
[20] R. M. N. . Abd Halim. N.I, Yunus.M.A, Kasolang.S, "Finite Element Model of the Spot Welded Joints of Door in White (DIW)," *Eng. Appl. Sci.*, vol. 11, no. 11, pp. 2474–2478, 2016.
[21] M. S. M. Zin, M. N. A. Rani, M. A. Yunus, M. S. M. Sani, W. I. I. Wan Iskandar Mirza, and A. A. Mat Isa, "Frequency response function (FRF) based updating of a laser spot welded structure," *AIP Conf. Proc.*, vol. 020055, p. 020055, 2018.
[22] M. A. Yunus, S. Nazri, M. N. Abdul Rani, A. Tormodi, and S. Kasolang, "Response surface reconciliation method of bolted joints structure," *MATEC Web Conf. AiGEV 2016*, vol. 90, p. 01013, 2017.
[23] G. G. Wang, "Adaptive Response Surface Method Using Inherited Latin Hypercube Design Points," *J. Mech. Des.*, vol. 125, no. 2, p. 210, 2003.
[24] S. W. Doebling, F. M. Hemez, J. F. Schultze, and A. L. Cundy, "A Metamodel-Based Approach to Model Validation for Nonlinear Finite Element Simulations Model Validation – Beyond Test / Analysis Correlation," *Simulation*, 2002.
[25] C. Mares, J. E. Mottershead, and M. I. Friswell, "Stochastic model updating of a spot welded structure," *Int. Conf. Noise Vib. Eng.*, vol. 1–8, pp. 1885–1898, 2005.
[26] M. S. M. Zin, M. N. A. Rani, M. A. Yunus, M. A. Ayub, M. S. M. Sani, and M. A. S. A. Shah, "Modal based updating for the dynamic behaviour of a car trunk lid," in *AIP Conference Proceedings*, 2019, p. 020001.

Comparison of crash behaviour prediction of a car crash box using initial and updated finite element model

N.A.Z. Abdullah[1], M.S.M. Sani[1,2*], M.S. Salwani[2]
[1]Advanced Structural Integrity & Vibration Research (ASIVR), Faculty of Mechanical and Manufacturing Engineering, Universiti Malaysia Pahang, 26600 Pekan, Pahang, Malaysia
[2]Automotive Engineering Centre, Universiti Malaysia Pahang, 26600 Pekan, Pahang, Malaysia

[*]Corresponding author: mshahrir@ump.edu.my

Abstract. Topic of crashworthiness is gaining popularity as vehicle safety is becoming main concern to the vehicle manufacturers worldwide. Due to this factor, energy absorbing members on vehicle such as crash box is researched by many which lead to various strategies of design optimization has been carried out in order to seek a good energy absorber on vehicle. Among the popular method of determining the crash performance of a structure is by using dynamic analysis in computational finite element model (FEM). However, since finite element analysis (FEA) is only an approximation method which in many cases can be less reliable, model updating method is often suggested to improve its reliability. This study seeks to investigate the crash behaviour of initial and updated crash box model. The velocity of impactor, the deformation, collapse distance and energy transferred to the two models of crash box are compared. The findings show that the initial and updated model of the crash box in this study, do have different crash behaviour. However, the difference level is relatively small. The investigation on the method of validation for the updated model in terms of crash behaviour is highly recommended.

1. Introduction

In this day and age, number of vehicle has grown with the continuous development of the world economy. With the increasing number of automobile, rate of traffic accidents also raised considerably. Therefore, crashworthiness has become one of the most vital research topic when it comes to the concern of producing safer vehicles for consumers. Car crash box for instance, which is considered as an important energy absorbing members in vehicle, is researched by many [1–4]. There are numerous studies on crash box to study their crashworthiness performance [5]. Design optimization with numerous strategies have been proposed in order to produce an enhanced energy absorbing member on vehicles [6–8]. Mostly the computational analyses on these new designs or any enhancement done on crash box have to be validated in crash experiment [6,9-10]. This computational analysis which is mostly based on finite element analysis is widely used in the field of crashworthiness as it helps engineers and researchers to have the knowledge on the structural behaviour and its performance during impact or collision.

However, not all the finite element analysis simulated the exact result as obtained through experiment [10]. This is due to the fact that finite element analysis is only an approximate technique and the level of accuracy of the displayed results could vary from the experimental data. In fact, most of the previous studies have shown the concern of discrepancies in crashworthiness analysis. Therefore, it is essential for researchers to provide the solution to overcome the discrepancies problem

As researchers cannot totally rely on data from finite element analysis only, many studies has demonstrated that the finite element results are often need to be correlated and validated using experimental data [3,4,6,11–14]. There are many cases where discrepancies between finite element analysis data and the experimental data are unavoidable [13,15–17]. This discrepancy may be due to the wrong assignment of boundary conditions, material properties, assumption of localized effects like welding, bolt torque or any other reasons. In order to overcome the problems of discrepancies between these two sets of data, a method called model updating is introduced.

Model updating of a structural system is one of important topic with application in different problems especially in the field of structural engineering design and analysis. With the aid of model updating method, prediction capability of finite element model and the technique of modelling can be improved, as well as damage identification tool can be provided as shown in many previous study [14,18–23]. These studies explain that the model updating method normally employs numerically simulated data and experimental data from modal analysis to calibrate the model in study to better correlate the actual structure [23–30].

The aim of the present work is to investigate the crash behaviour of a crash box using initial and updated FE models. It is expected to have the different results of crash behaviour because both modes are developed using different model properties.

2. Finite element modelling and model updating

In order to facilitate this study, a top hat structure is selected as the representative structure of crash box (see figure 1). The configuration of the top hat used in this study are designed based on vehicle front substructures that are normally used in a car body which consists of dissimilar materials that are joined through spot welds. Two different types of steel used to make the top hat are joined together using resistance spot welding (RSW). Both materials are of different thickness; 1.0mm for the hat structure and 1.7mm for the flat surface. The geometrical dimensions of the structure are as shown in the figure 2.

The structure was modelled as surface where shell elements in quad shape are assigned to the model. In order to determine the appropriate element size, mesh convergence tests were carried out. The meshing properties for the top hat are shown in table 1. Material input for the structure was assigned as tabulated in table 2 below. CWELD element connectors were used to model the RSW joints between the flat and the top hat plate. CWELD element connectors were extensively utilized in modelling the spot weld joints due to the ability to replicate the stiffness of spot weld joints in the actual structure. The Young's modulus and the diameter of the spot weld are declared to be 1000 GPa and 0.001m respectively. 20 CWELD elements are used in the FE model of the top hat (refer to figure 3).

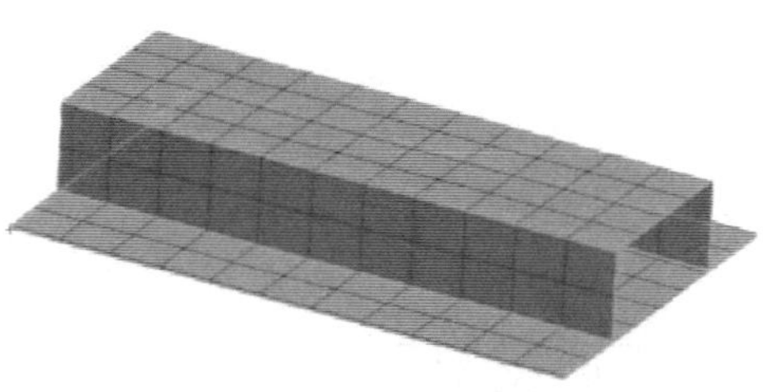

Figure 1. FE model of crash box structure.

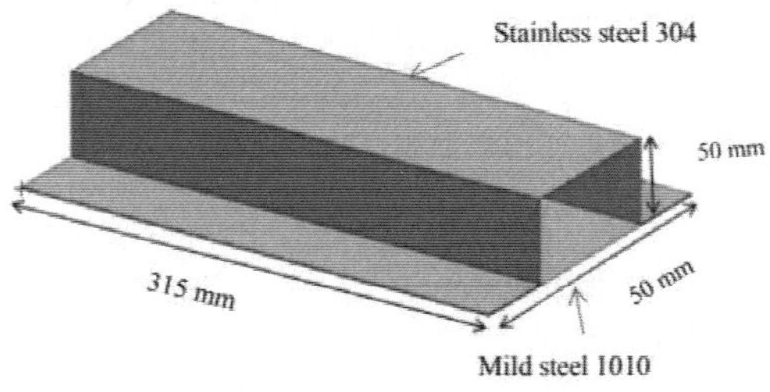

Figure 2. Dimensional value of crash box model.

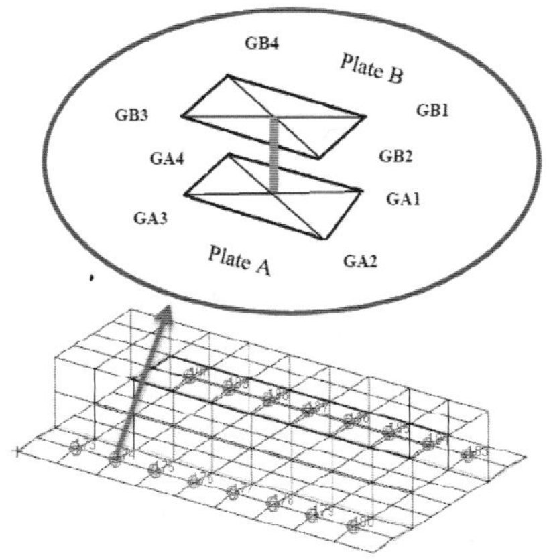

Figure 3. Location of spot weld joint.

Table 1. Meshing properties for top hat plate structure

Properties of meshing	
Element type	CQUAD4
Element number	171
Nodes number	266
DOF number	1330

Table 2. Input properties of material.

Material	Mild steel 1010	Stainless steel 304
Young's modulus (GPa)	200	190
Density (kg/m^3)	8000	8030
Poisson ratio	0.33	0.30
Thickness (m)	0.0017	0.001

Model updating was performed on the finite element model of the crash box. Updating parameters were determined by conducting sensitivity analysis on all existing parameters of the crash box such as material and physical properties used for the model. These properties include the Young's modulus and density of mild steel and stainless steel, plates thickness, diameter of CWELD elements and Young's modulus of the CWELD elements. At the end, the sensitivity analysis had determined four parameters, which only involve the material properties of the model, as the updating parameters. Table 3 shows the updated values of each the updating parameters selected during sensitivity analysis. The value shown for updated model will be the properties used in the updated model.

Table 3. Updated values of updating parameters.

Material	Initial value	Updated value
Young's modulus of mild steel (GPa)	200	192.74
Young's modulus of stainless steel (GPa)	190	161.50
Density of mild steel (kg/m^3)	8000	9200.00
Density of stainless steel (kg/m^3)	8030	8142.42

3. Crashworthiness analysis

Crash simulation was performed on both initial and updated model in order to study the crashworthiness performance of both the initial and updated model. The rear part of the crash box is constraint to be fixed and the impactor, which is modelled as a box, is modelled as rigid (refer to figure 4). The velocity of the impactor is set to be 10 m/s. Hard-contact interaction was defined between the impactor box and the end of the box that is not declared as rigid with additional self-contact interaction between the two structures during the process of deformation. The friction coefficient is set as 0.2 and the time step is set as 10 ms.

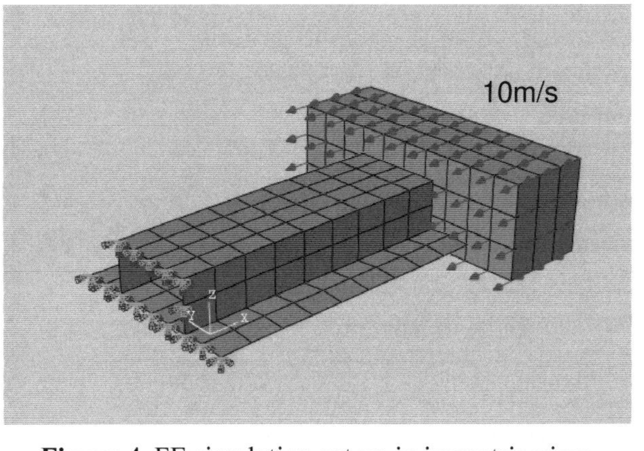

Figure 4. FE simulation set up in isometric view.

4. Crashworthiness of initial and updated crash box model

Several crashworthiness output are analysed in evaluating the crash behaviour of the initial model and updated model of the crash box. Figure 5 compares the velocity of the impactor in the direction of crash for both of the models, while figure 6 shows the resultant displacement of the impactor in both simulation of the different model. It appears that from the both figures, the updated crash box model is able to decelerate the impactor slightly faster than the initial model. Also, based on the displacement

IOP Conf. Series: Journal of Physics: Conf. Series **1262** (2019) 012003 doi:10.1088/1742-6596/1262/1/012003

curve in figure 6, the crush distance of the updated crash box model is slightly shorter than the initial one which suggests that the updated model is capable of stopping crash faster that initial model [10].

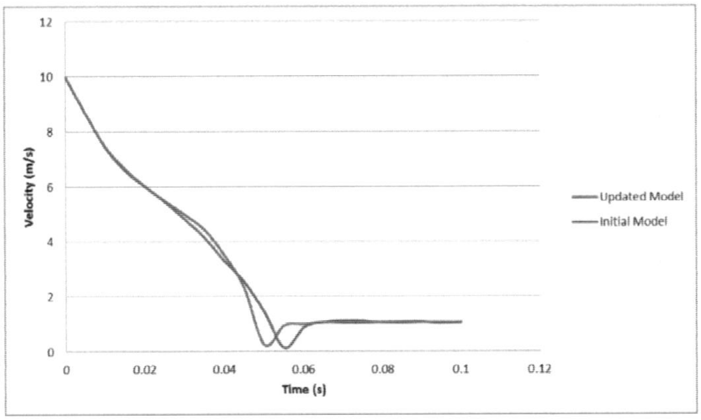

Figure 5. Velocity curve of impactor.

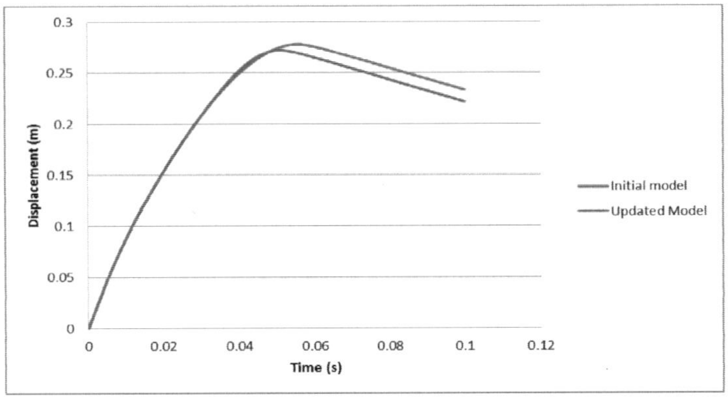

Figure 6. Resultant displacement of impactor.

During the time of impact by the rigid impactor, the impactor's kinetic energy was mainly dissipated by internal energy due to the crash box's deformation. The dissipated energy is considered as the energy absorbed by the crash box in order to prevent further crash on itself. The energy absorption ability is always considered as the key content of crashworthy structure. Figure 8 shows the change of energy between the impactor and crash box. It could be seen that both models absorb impact kinetic energy in almost similar behaviour. However, the kinetic energy is absorbed slightly faster than the initial model and therefore suggests that the updated model has slightly greater potential in enhancing the energy absorption capacity. This is consistent with the findings of a past studies that stated that energy absorption capacity is better when energy absorption process is completed faster [4,31].

IOP Conf. Series: Journal of Physics: Conf. Series **1262** (2019) 012003 doi:10.1088/1742-6596/1262/1/012003

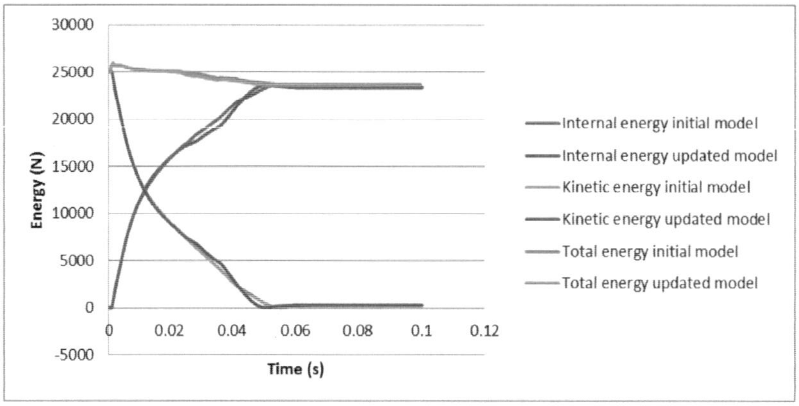

Figure 7. Energy balance for the whole model.

Table 4 shows the qualitative comparison of deformation for the initial and updated model. Comparison of crash box deformation during the impact is presented through sequential deformation in each time frame in the simulation. Although the pattern of deformation is almost the same, the stress distribution in each element is different in both models. In addition, the pattern of deformation started to behave a bit different at step time 15 ms onwards.

Table 4. Sequential crush behaviour of crash box in stress distribution.

Step time (s)	Initial model	**Updated model**
0.005	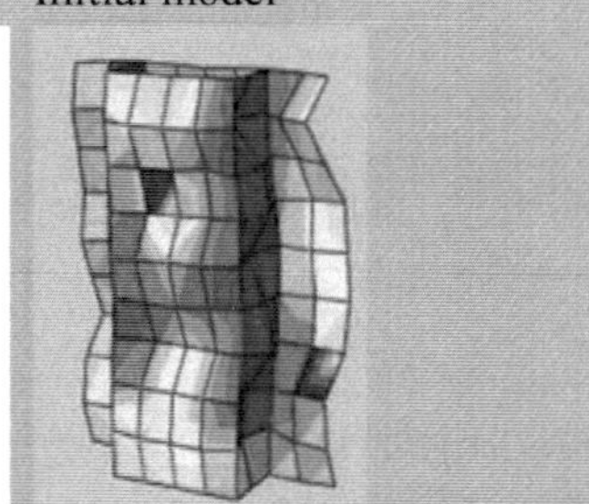	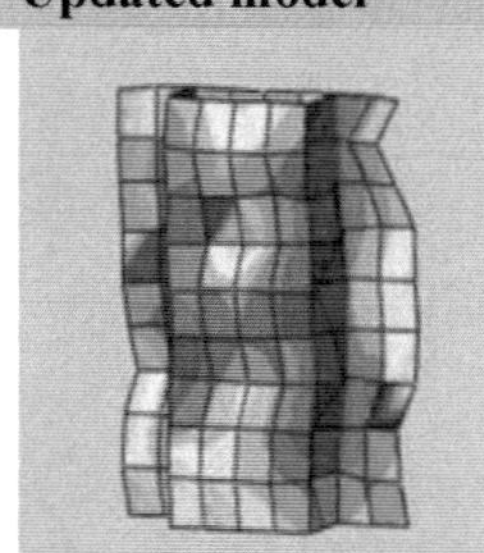
0.010	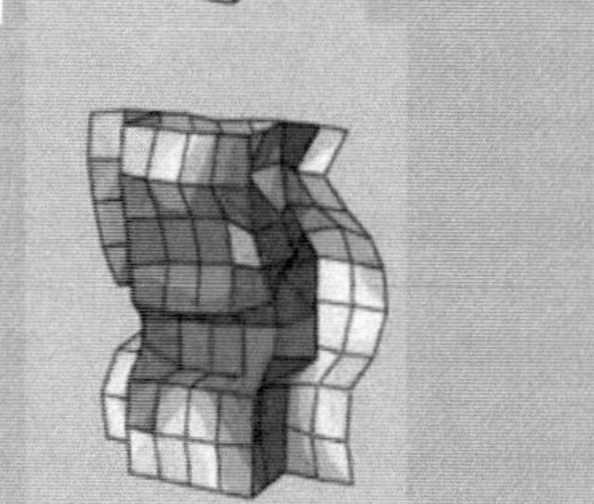	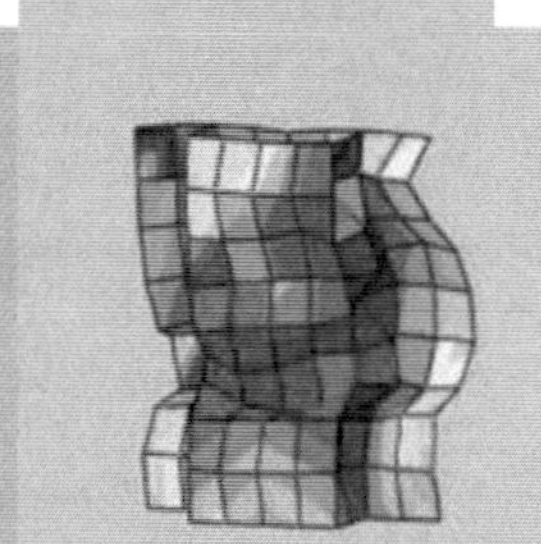
0.015		

0.020		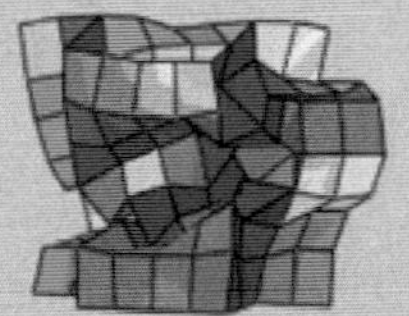
0.025	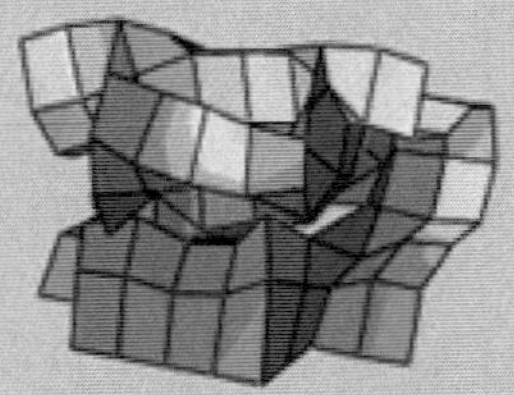	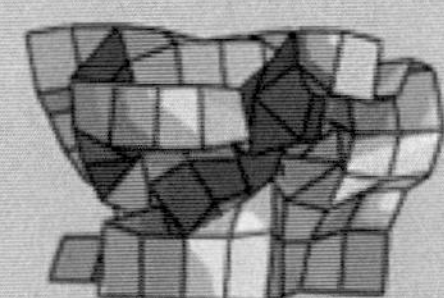
0.030	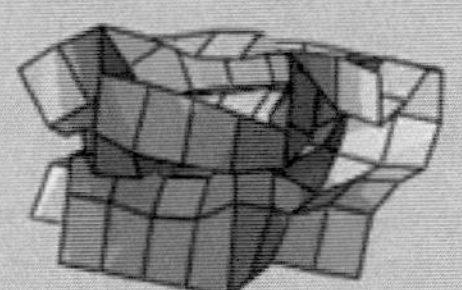	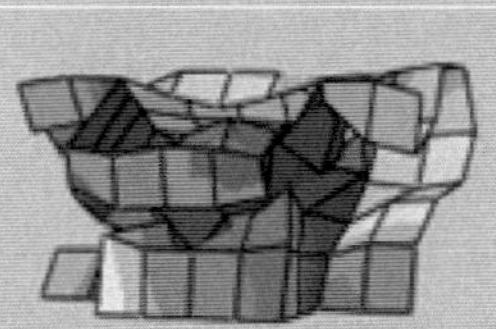
0.035	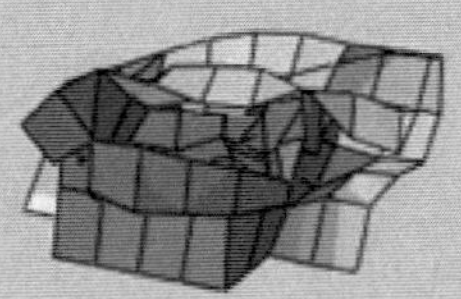	
0.040	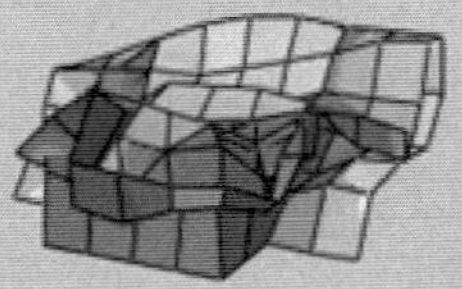	
0.045	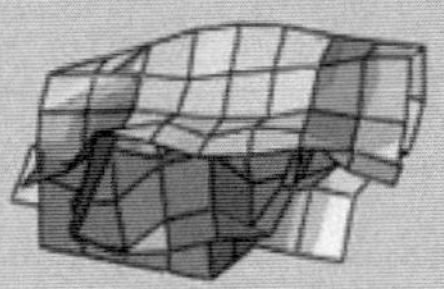	
0.050	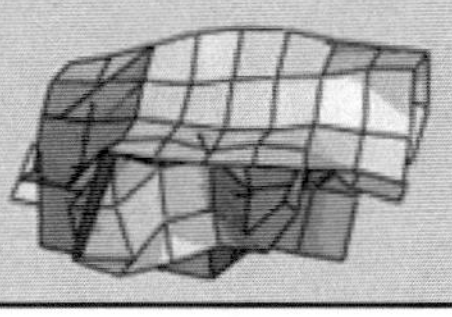	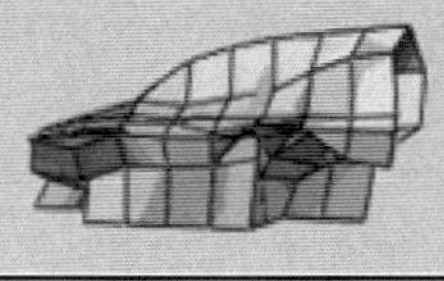

5. Conclusion

This study has been set out to investigate and compare the crash behaviour of a crash box design using two different finite element models, which is the original and the updated model. Some important observations are noted, which show that the initial and updated model of the crash box in this study, do have different crash behaviour. However, the percentage of the difference is relatively small. In the terms of crashworthiness performance, even though the updated model shows slightly better crash characteristics, it would be less significant to stress out its performance since the model only outshine the other model in just a slight manner. It is more significant if the crash outputs of both models are correlated with the output of crash experiment. This research will serve as a base for further studies of investigating the performance of updated finite element model in simulating the crash behaviour of its actual structure. Further work needs to be done where the method of validation for the updated model in terms of crash behaviour is highly needed in order to determine the level of accuracy of the information provided by the updated model in terms of crashworthiness.

Acknowledgement
The authors of this paper would like to acknowledge a great support and encouragement by focus group of Advanced Structural Integrity of Vibration Research (ASIVR), Universiti Malaysia Pahang (UMP) for providing all the equipment used for this work Fundamental Research Grant Scheme (FRGS/1/2017/TK03/UMP/02-19) – RDU 170123.

References
[1] Kaczyński P 2018 Progressive crushing as a new mechanism of energy absorption. The crushing study of magnesium alloy crash-boxes *Int. J. Impact Eng.*
[2] Ataollahi S, Taher S T, Eshkoor R A, Ariffin A K and Azhari C H 2012 Energy absorption and failure response of silk/epoxy composite square tubes: Experimental *Compos. Part B Eng.* 43 542–8
[3] Hussain N N, Regalla S P and Rao Y V D 2017 Comparative Study of Trigger Configuration for Enhancement of Crashworthiness of Automobile Crash Box Subjected to Axial Impact Loading *Procedia Eng.* 173 1390–8
[4] Wang C Y, Li Y, Zhao W Z, Zou S C, Zhou G and Wang Y L 2018 Structure design and multi-objective optimization of a novel crash box based on biomimetic structure *Int. J. Mech. Sci.* 138–139 489–501
[5] Baroutaji A, Sajjia M and Olabi A G 2017 On the crashworthiness performance of thin-walled energy absorbers: Recent advances and future developments *Thin-Walled Struct.* 118 137–63
[6] Gao Q, Zhao X, Wang C, Wang L and Ma Z 2018 Multi-objective crashworthiness optimization for an auxetic cylindrical structure under axial impact loading *Mater. Des.* 143 120–30
[7] Du X and Zhu F 2018 A new data-driven design methodology for mechanical systems with high dimensional design variables *Adv. Eng. Softw.* 117 18–28
[8] Arjomandi Rad M and Khalkhali A 2018 Crashworthiness multi-objective optimization of the thin-walled tubes under probabilistic 3D oblique load *Mater. Des.* 156 538–57
[9] Xu X, Zhang Y, Wang J, Jiang F and Wang C H 2018 Crashworthiness design of novel hierarchical hexagonal columns *Compos. Struct.* 194 36–48
[10] Li Y, Hu X, Liu X, Lu R, Liu W, Gao W, Hu X, Liu W, Li Y and Liu X 2018 Crushing analysis and crashworthiness optimization of tailor rolled tubes with variation of thickness and material properties *Int. J. Mech. Sci.* 136 67–84
[11] Nasir Hussain N, Prakash Regalla S and Daseswara Rao Y V. 2017 Low velocity Impact Characterization of Glass Fiber Reinforced Plastics for Application of Crash Box *Mater. Today Proc.* 4 3252–62
[12] Gao Q, Ge C, Zhuang W, Wang L and Ma Z 2019 Crashworthiness analysis of double-arrowed auxetic structure under axial impact loading *Mater. Des.* 161 22–34
[13] Sani M S M, Nazri N A, Zahari S N, Abdullah N A Z and Priyandoko G 2016 Dynamic Study of Bicycle Frame Structure *IOP Conf. Ser. Mater. Sci. Eng.* 160 012009

[14] Zahari S N, Sani M S M, Husain N A, Ishak M and Zaman I 2016 Dynamic analysis of friction stir welding joints in dissimilar material plate structure *J. Teknol.* 78
[15] Sani M S M, Abdullah N A Z, Zahari S N, Siregar J P and Rahman M M 2016 Finite element model updating of natural fibre reinforced composite structure in structural dynamics *MATEC Web of Conferences* 83 (03007)
[16] Jelani K M, Razip M H M, Nazri N A, Sani M S M and Yasar M 2018 Dynamics investigation on motorcycle chassis based on Finite Element (FE) modelling and updating *IOP Conf. Ser. Mater. Sci. Eng.* 469 012103
[17] Sani M S M and Ouyang H 2016 Identification of Nonlinearities in Joints of a Wing Structure *MATEC Web of Conference* 83 (03006)
[18] Husain N and Snaylam A 2009 FE Model Updating for Damage Detection–Application to a Welded Structure *Key Eng. ...* 414
[19] Bakir P G, Reynders E and De Roeck G 2007 Sensitivity-based finite element model updating using constrained optimization with a trust region algorithm *J. Sound Vib.* 305 211–25
[20] Magalhães F, Cunha A and Caetano E 2012 Vibration based structural health monitoring of an arch bridge: From automated OMA to damage detection *Mech. Syst. Signal Process.* 28 212–28
[21] Astroza R, Ebrahimian H and Conte J P 2019 Performance comparison of Kalman–based filters for nonlinear structural finite element model updating *J. Sound Vib.* 438 520–42
[22] Abu Husain N, Haddad Khodaparast H and Ouyang H 2012 Parameter selection and stochastic model updating using perturbation methods with parameter weighting matrix assignment *Mech. Syst. Signal Process.* 32 135–52
[23] Mottershead J E, Link M and Friswell M I 2011 The sensitivity method in finite element model updating: A tutorial *Mech. Syst. Signal Process.* 25 2275–96
[24] Khodaparast H H, Mottershead J E and Friswell M I 2008 Perturbation methods for the estimation of parameter variability in stochastic model updating *Mech. Syst. Signal Process.* 22 1751–73
[25] Mthembu L, Marwala T, Friswell M I and Adhikari S 2011 Model selection in finite element model updating using the Bayesian evidence statistic *Mech. Syst. Signal Process.* 25 2399–412
[26] Omar R, Rani M N A, Wan Iskandar Mirza W I I, Yunus M A and Othman M H 2017 Finite Element Modelling and Updating for Bolted Lap Joints *Journal of Mechanical Engineering* 4 202–22
[27] Abdullah N A Z, Sani M S M, Hussain N A, Rahman M M and Zaman I 2017 Dynamics properties of a Go-kart chassis structure and its prediction improvement using model updating approach *International Journal of Automotive and Mechanical Engineering (IJAME)* 14 3887-97
[28] Mohd Zin M S, Rani M N A, Yunus M A, Wan Iskandar Mirza W I I, Mat Isa A A and Mohamed Z 2017 Modal and FRF Based Updating Methods for the Investigation of the Dynamic Behaviour of a Plate *Journal of Mechanical Engineering* 4 175–89
[29] Wan Iskandar Mirza W I I, Rani M N A, Othman M H, Kasolang S and Yunus M A 2016 Reduced order model for model updating of a jointed structure *Journal of Engineering and Applied Sciences* 11 2383–6
[30] Fouzi M S M, Jelani K M, Nazri N A and Sani M S M 2018 Finite Element Modelling and Updating of Welded Thin-Walled Beam *International Journal of Automotive and Mechanical Engineering (IJAME)* 15(4) 5874-89
[31] Wang C Y, Wang W W, Zhao W Z, Wang Y and Zhou G 2018 Structure design and multi-objective optimization of a novel NPR bumper system *Compos. Part B Eng.* 153 78–96

A Comparative Study of CWELD and ACM2 Element Connectors on the Dynamic Behaviour of a Laser Spot Welded Hat-Plate Structure under Initial Stress Influence

C Peter[1], M N Abdul Rani[1*], H Hashim[1], M A Yunus[1], M.N Machmud[2] and M S Mohd Zin[1]

[1] Structural Dynamics Analysis and Validation, Faculty of Mechanical Engineering, Universiti Teknologi MARA (UiTM), 40000 Shah Alam, Selangor, Malaysia

[2] Laboratory of Structural Dynamics, Department of Mechanical and Industrial Engineering, Universitas Syiah Kuala, JL. Syech Abdurrauf No.7 Bandar Aceh 23111, Indonesia.

*Corresponding author: mnarani@salam.uitm.edu.my

Abstract. Analytical modelling of the dynamic behaviour of a laser spot welded structure can be developed using two element connectors which are CWELD and ACM2. In the presence of initial stress in the structure, the performance of these two connectors can be different. In order to properly investigate the dynamic behaviour of the structure, the difference of these two element connectors in representing the laser spot welded in the structure should be evaluated. The finite element method was used to develop the welded structure comprising two structural components. The element connectors were then used to represent the laser spot welds in the structure. Normal modes analysis was performed on the FE model of the welded structure to calculate the modal parameters. Experimental modal analysis was carried out to measure the modal parameters of the physical welded structure. The modal parameters calculated from the two different FE models were evaluated by the experimental counterparts in terms of natural frequencies, mode shapes and MAC value. The accumulated error registered from all 6 modes from the CWELD element connector FE model is 10.59%. Meanwhile, the accumulated error calculated from all 6 modes of the ACM2 element connectors based FE model is 26.67%. It was found that the FE model with CWELD element connectors has shown a good correlation with the physical welded structure. The findings can have an important effect on how analytical modelling of the dynamic behaviour of the welded structure under initial stress can be effectively and accurately carried out.

1. Introduction

Thin metal sheets are found to be widely used in the automotive and aerospace industries because of their light in weight and high structural integrity. For example, a car body in white, comprising a large number of structural components made from thin metal sheets. In practice, the structural components are usually assembled together by joints such as rivets, bolts and welds [1]. In recent years, the use of laser welding for joining the structural components has received much attention in the industries [2]. This is because, the welded joints do not only provide strong connections between structural components, but they also significantly contribute to the dynamic behaviour of the structures [3].

Laser spot welding offers higher process speed and excellent flexibility for automation [4]. However, laser spot welded joints are found to be complex and have many uncertainties when it comes to analytical modelling [5] due to the presence of local effect in the welded areas (Palmonella et al., 2005). Although a large number of studies on spot welded joints have been carried out, very little work has been found on the attempt to establish a proper procedure for designing and verifying laser spot welded structures via the finite element method [6]. For example, Nurul et. al [7] used CWELD element connectors to represent laser spot welds in the structure under study. In the study, modal based updating was performed on the hat-plate structure to minimise the differences between the measured and FE results. However, the differences were not successfully reduced. Recently, the frequency response function (FRF) based updating was adopted by Syazwan et. al [8] to improve the predicted results of the laser spot welded structure. It was found that the method adopted has led to a strong correlation between the predicted and measured frequency response function, particularly for lower modes. The powerful and successful model updating in improving the predicted results, however, remains to be very challenging and have many fundamental issues to be handled properly. This is because the use of the common updating properties of model and joints in updating the FE models is found to be no longer successful, especially for updating an assembled structure with different types of joints. Therefore, other potential updating properties can be fundamentally explored for a better correlation between the physical test structure and analytical model. One of the most potential updating candidates is initial stress resulting from welding and assembly of structural components.

Initial stress may arise from the way structural components are assembled using welded joints to form an assembled structure, particularly, with a large thin span surface. Initial stress is also connected with residual stress and the fatigue behaviour of a structure, in connection of those two subjects is extremely significant for researchers to learn more about this topic [9]. The attempt to understand the effect of initial stress was discussed by [10] in which the fatigue behaviour between resistance spot welds and laser spot welds of a structure with dissimilar materials was evaluated. The findings of the attempt showed that there were still noticeable errors. This result may be explained by the fact that there is the possibility of the presence of initial stress in the welded structure. Thus, the effects of initial stress on structural dynamics are still completely unknown and required further investigation [11].

This paper presents and discusses finite element modelling of the dynamic behaviour of a laser spot welded structure under initial stress using two different types of element connectors which are CWELD and ACM2 in representing the laser spot welded joints in the physical test hat-plate structure.

2. Description of the Welded Hat-Plate Structure and Experimental Set up

In this study, the structure under investigation is a laser spot welded hat-plate made from steel sheets of a uniform thickness of 1.5mm, 564mm long and 80mm wide. The structure consists of two components welded together with a total of 20 laser spot welds as shown in Figure 1. The test structure was set up in free-free boundary conditions by suspending the structure using strings and rubber bands as shown in Figure 1. The dynamic characteristics of the test structure were obtained using impact testing with two accelerometers. An impact hammer was used to excite the structure with one accelerometer used as a reference by fixing to the predefined excitation point. The other accelerometer was roved around to measure 60 points throughout the structure. LMS SCADAS system was used to process the input and output responses of the structure under the test.

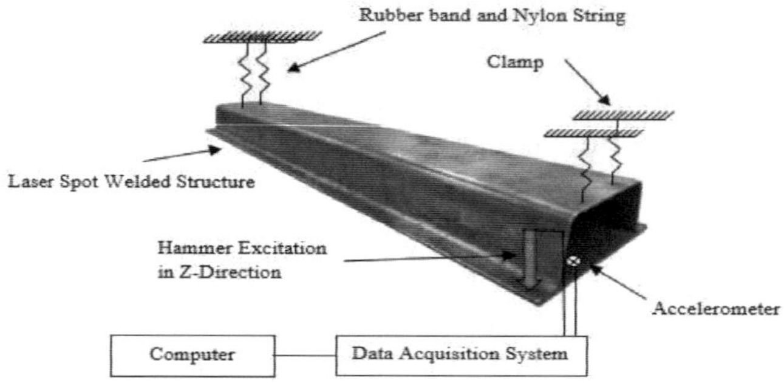

Figure 1: Experimental set-up of the test structure.

3. Finite Element Modelling and Analysis

The finite element model of the welded hat-plate structure was developed using HyperMesh software. MSC NASTRAN SOL 103 and SOL 200 were used for the analytical analysis of the dynamic behaviour of the welded hat-plate. Specifically, the former was used for normal modes analysis and the latter for model updating. Full details of the methods used in this study are available in [12-18]. Figures 2 and 3 show two different types of the finite element model of the welded hat-plate developed in this study. The finite element models of the hat and plate were developed based on CQUAD4 shell elements with 17673 elements and 18146 nodes. In Figure 2, CWELD element connectors were used to represent the laser spot welds in the test hat-plate structure. Meanwhile, Figure 3 shows the same FE model but ACM2 element connectors were used instead. Beforehand, the actual dimensions of the test structure were measured using a coordinate measuring machine (CMM). The purpose was to reduce the uncertainties in the FE model developed. Detailed information on the model properties used in the FE model of the welded hat-plate structure is given in Table 1 and Table 2. In this study, the main area of initial stress which was largely found on the bottom plate of the test welded hat-plate structure was modelled in the FE model by discretising the plate into 56 collectors. In order to investigate the effect of initial stress on the natural frequencies of the welded hat-plate structure, the initial value of initial stress modelling was set to be 1 which is the same setting that was found in [4]. The frequency of interest used in the normal modes analysis was from 0Hz to 1000Hz. Only the first six frequencies and mode shapes were investigated because the confidence level in the modes is greater than the other modes.

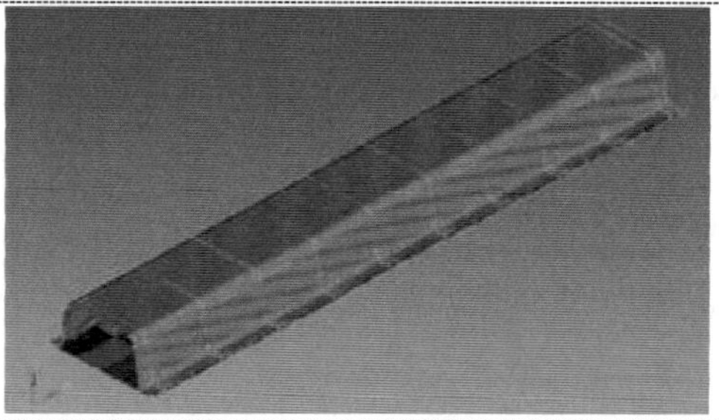

Figure 2: Hat-plate with CWELD element connectors

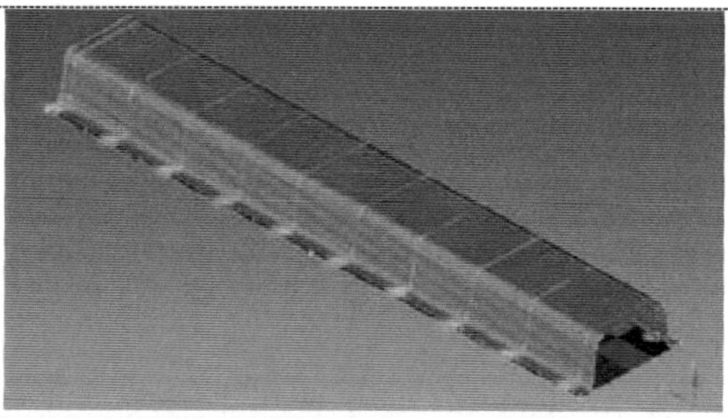

Figure 3: Hat-plate with ACM2 Element connectors

IOP Conf. Series: Journal of Physics: Conf. Series **1262** (2019) 012004 doi:10.1088/1742-6596/1262/1/012004

Table 1: Model properties of the FE model of the welded hat-plate structure

The name of property	Nominal Value
Young's Modulus	210 GPa
Poisson's Ratio	0.3
Density	7500 kg/m^3

Table 2: Model properties of CWELD and ACM2 element connectors used in the FE model of the welded hat-plate structure

The name of property	Nominal Value
Young's Modulus	210 GPa
Poisson's Ratio	0.3
Density	7500 kg/m^3

3.1 Modal Assurance Criterion

In a brief historical view, Modal Assurance Criteria originated from the necessity for a quality assurance indicator for experimental modal vectors that are predicted from measured frequency response functions. Thus, it leads to the discovery of Modal Assurance Criterion which can be defined as a scalar constant relating the degree of consistency (linearity) between one modal and another reference modal vector [19], as shown in Equation 1, the equation for MAC. This method is widely used to validate the results between experimental modal analysis and finite element analysis because the modal assurance criterion takes on values from zero, representative a no consistent correspondence, to one, representing a consistent correspondence. In this manner, if the modal vectors under consideration accurately display a consistent, linear relationship, the modal assurance criterion should approach unity and the value of the modal scale factor can be considered reasonable [20]. However, unlike the orthogonality calculations, the modal assurance criterion is normalized by the magnitude of the vectors and, thus, is bounded between zero and one. Thus, imaginary value from the experimental modal analysis and finite element analysis can be validated by using this method on top of the degree of the accuracy of natural frequencies and mode shapes.

$$\text{MAC} = \Phi_m \Phi_a = \frac{\left|\Phi_m^T \Phi_a\right|^2}{(\Phi_a^T \Phi_a)(\Phi_m^T \Phi_m)} \quad (1)$$

Where:

Φ_a= Modal vector for test mode

Φ_m=Modal vector for FE mode

Φ_a^T=Transpose of test modal vector

Φ_m^T=Transpose of FE modal vector

4. Results and Discussion

In this study, the predicted and measured modal parameters, particularly natural frequencies and mode shapes of the welded hat-plate structure were obtained from the finite element method and experimental modal analysis respectively. Two different types of element connectors which are CWELD and ACM2 were used in representing the laser spot welds. The influence of initial stress on the physical welded hat-plate was introduced in the finite element model as described in the section of finite element modelling and analysis. The level of correlation between the predicted and measured results of both CWELD and ACM2 model was evaluated by frequency deviation and MAC value.

Table 3 shows the comparison of natural frequencies obtained from three different sources which are the experimental modal analysis (column II), the finite element model with CWELD (column III) and the finite element model with ACM2 (column V). Therefore, from a direct comparison, particularly, between the model with CWELD element connector (column III) and EMA (column II), in column IV it was found that the largest percentage error has been recorded from the 2nd mode with 2.97%, while the lowest error which is 1.22% has been registered from the 1st mode. The accumulated error registered from all 6 modes is 10.59%. The next comparison of the natural frequencies which are tabulated in column VII, is between the finite element modal with ACM2 element connector (column VI) and EMA (column I). From column VII, it is shown that the largest error which is 6.71%, has been registered from the 2nd mode and the lowest error of 2.97% has been recorded from the 1st mode. Meanwhile, the accumulated error calculated from column VII is 26.67%.

MAC values of corresponding mode shapes (mode pairing) between EMA and FEM were used to quantify the mode shapes calculated from both types of the finite element models. Figure 4 shows MAC analysis on the FE model with CWELD element connectors model and the findings are tabulated in Table 3 (column V). While Figure 5 shows the MAC analysis on the FE model with ACM2 element connectors; the calculated MAC values are tabulated in Table 3 (column VIII). In general, any mode with MAC value more than 0.7 is regarded as a well correlated mode shape while MAC value is less than 0.7 is considered as a poor correlated mode shape. In Table 3, the lowest MAC value has been recorded from the 1st mode of both types of the FE models.

It is worthwhile to note that the comparisons of the MAC values for both types of the FE models reveal that the FE model with CWELD element connectors has recorded the highest MAC values for all 6 mode shapes in comparison with that of the FE model with the ACM2 element connectors. However, there are three modes with MAC values of less than 0.7 recorded from the former. This may be due to the low quality of the calculated mode shapes as a result of the incapability of the FE model to accurately represent the physical laser welded structure, especially the local effects of the laser welding and the presence of initial stress. One way to increase confidence in the quality of the FE model is the use of model updating methods. Previous studies [21-23] revealed that initial stress has a profound impact on structural dynamic behaviour which is crucial to the performance of engineering products.

Table 3: Comparison of the natural frequencies between the EMA, FEA and MAC values of both types of FE models with different element connectors (CWELD and ACM2)

Modes (I)	EMA (Hz) (II)	CWELD (Hz) (III)	Error % (IV)	CWELD MAC (V)	ACM2 (Hz) (VI)	Error % (VII)	ACM2 MAC (VIII)
1	503.8	509.95	1.22	0.38	514.24	2.97	0.37
2	555.5	572	2.97	0.88	592.79	6.71	0.87
3	572.5	586.39	2.43	0.43	594.07	3.77	0.42
4	634.43	640.93	1.36	0.46	664.2	5.05	0.42
5	643	651.03	1.25	0.88	670.37	4.26	0.88
6	664.9	673.93	1.36	0.71	696.91	4.81	0.66
		Total error	10.59		Total error	26.67	

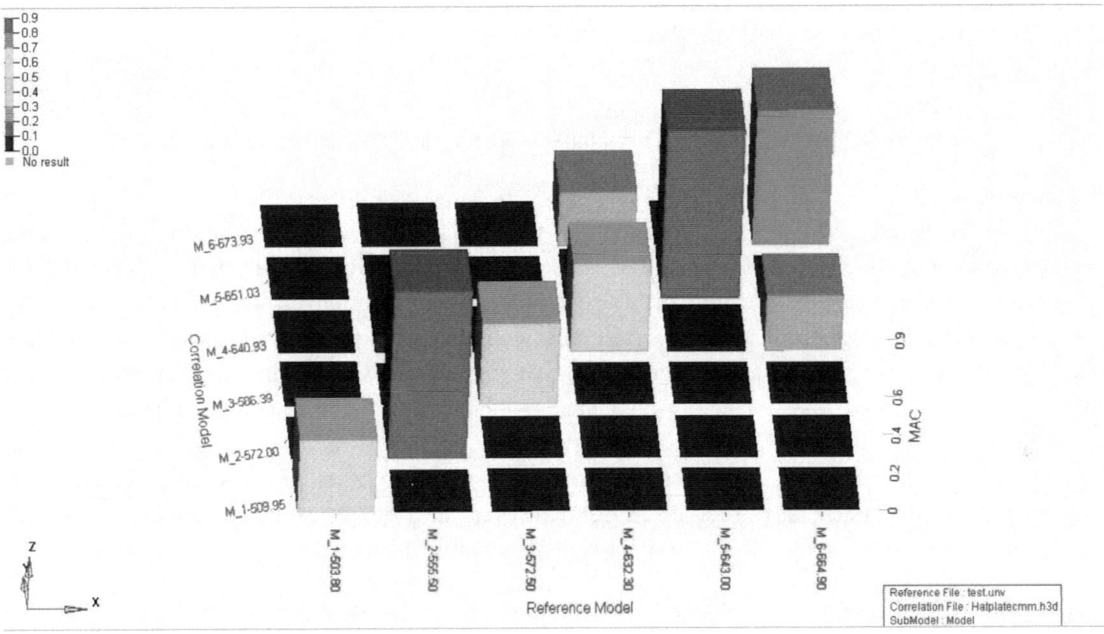

Figure 4: MAC values of the hat-plate model with CWELD element connectors

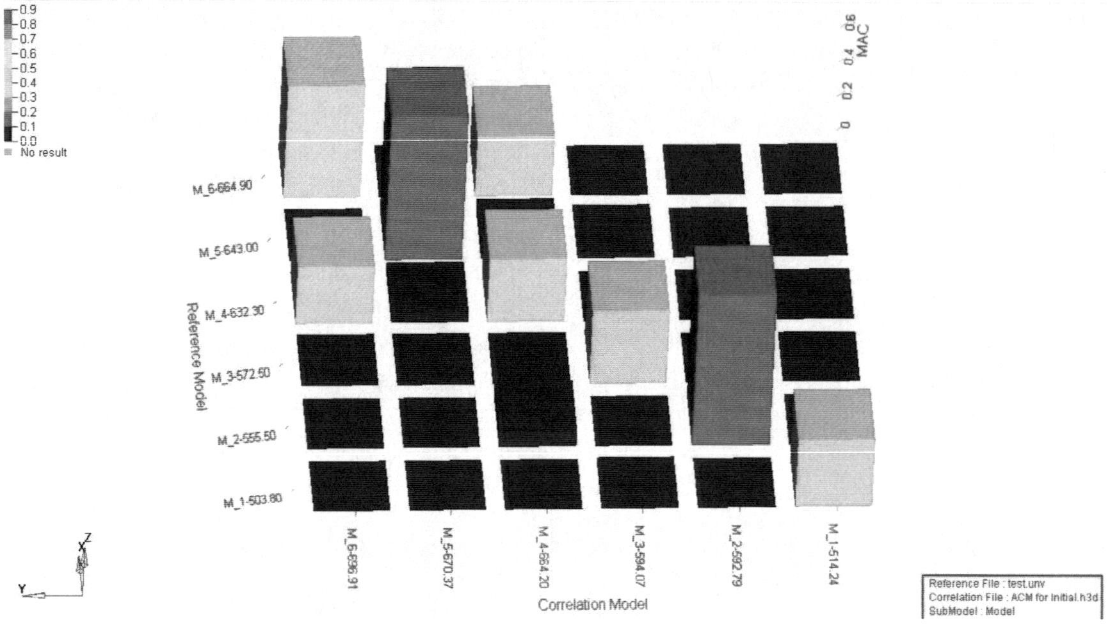

Figure 5: MAC values of the hat-plate model with ACM2 element connectors

5.Conclusions

This study gives an account of a comparative study of modelling and analysing the dynamic characteristics of the laser spot welded hat-plate structure using two different types of element connectors. As mentioned earlier in the introduction, the purpose of this study was to investigate the best candidate of element connectors in representing the dynamic behaviour of the laser spot welded hat-plate structure under initial stress using the finite element method and experimental modal analysis. The results of the investigation show that the FE model with CWELD element connectors has demonstrated the best capability in representing the dynamic behaviour of the laser welded structure under the influence of initial stress. The findings of the study also disclose that the inclusion of initial stress to the FE model has improved the degree of accuracy of the FE model. However, the accuracy of the correlation between the measured and predicted data can be further improved by adopting the modal based updating method.

Acknowledgments

The authors gratefully acknowledge the Malaysia Ministry of Higher Education (MOHE), Research Management Institute (RMI) and Institute of Graduate Studies of Universiti Teknologi MARA (UiTM) for providing financial support for this study through the fundamental research grant scheme (FRGS) (600-RMI/FRGS 5/3 (164/2019)). They would like to extend their sincere thanks for the support and help given by Dr. Hafizan Bin Hashim and all SDAV members.

References

[1] Omar R, M N A, Wan Iskandar Mirza W I I, Yunus M A and Othman M H 2017 Finite Element Modelling and Updating for Bolted Lap Joints Journal of Mechanical Engineering 4 202–22

[2] Husain N A, Khodaparast H H, Snaylam A, James S, Sharp M, Dearden G & Ouyang H 2009 Modal testing and finite element model updating of laser spot welds. *Journal of Physics: Conference Series, 181*, 012024. doi:10.1088/1742-6596/181/1/012024

[3] Mottershead J E *et al.* 1994 Experience in Mechanical Joint Model Updating *19th International Seminar on Modal Analysis* 481-495

[4] Rani M N A, Stancioiu D, Yunus M, Ouyang H, Deng H, & James S. 2011 Model Updating for a Welded Structure Made from Thin Steel Sheets. *Applied Mechanics and Materials, 70*, 117-122. doi: 10.4028/www.scientific.net/amm.70.117

[5] Florea R S, Hubbard, Camden, Solanki K, Bammann D, Whittington W R & Marin E 2012 Quantifying Residual Stresses in Resistance Spot Welding of 6061-T6 Aluminum Alloy Sheets Via Neutron Diffraction Measurements *Journal of Materials Processing Technology* 212. 2358–2370. 10.1016/j.jmatprotec.2012.06.024.

[6] Palmonella M, Friswell M I, Mottershead J E and Lees A W 2005 Finite element models of spot welds in structural clynamics: Review and updating *Computer and Structure* 83: 648-661.

[7] Husain N A, Khodaparast H H, Snaylam A, James S., Dearden G, & Ouyang H 2010 Finite-element modelling and updating of laser spot weld joints in a top-hat structure for dynamic analysis. *Proceedings of the Institution of Mechanical Engineers, Part C: Journal of Mechanical Engineering Science,224*(4), 851-861. doi:10.1243/09544062jmes1787

[8] Zin M S, Rani M N, Yunus M A, Sani M S M, Wan Iskandar Mirza W I I & Isa A A 2018 Frequency response function (FRF) based updating of a laser spot welded structure *AIP Proceeding* 1952 020055 doi:10.1063/1.5032017

[9] Martinson P, Daneshpour S, Koçak M, Riekehr S & Staron P 2009 Residual stress analysis of laser spot welding of steel sheets. *Materials & Design 30*(9) 3351-3359 doi: 10.1016/j.matdes.2009.03.041

[10] Daneshpour, S., Riekehr, S., Koçak, M., & Gerritsen, C. H. (2009). Mechanical and fatigue behaviour of laser and resistance spot welds in advanced high strength steels. *Science and Technology of Welding and Joining, 14*(1), 20-25. doi:10.1179/136217108x336298

[11] De Faria R and De Almeida S F M 2006 The maximization of fundamental frequency of structures under arbitrary initial stress states *International Journal for Numerical Methods in Engineering* 65 445-460

[12] Labeas G, Tsirkas S, Diamantakos J. and Kermanidis A 2005 Effect of Residual Stresses due to Laser Welding on the Stress Intensity Factors of Adjacent Crack *Proc. of 11th International Conference on Fracture*

[13] Leissa W and Kadi A S 1971 Curvature Effects on Shallow Shell Vibrations *Journal of Sound and Vibration* 16 173

[14] Fong K L J 2003 A Study of Curvature Effects on Guided Elastic Wave, PhD, Imperial College London, University of London.

[15] Liu X H, Paurobally R, and Pan J 2008 The effect of prestress on vibration behaviour of structures in impedance method damage detection *Structural Integrity and Failure* 41-42 401-406.

[16] Abdullah N A Z, Sani M S M, Hussain N A, Rahman M M and Zaman I 2017 Dynamics properties of a Go-kart chassis structure and its prediction improvement using model updating approach International Journal of Automotive and Mechanical Engineering (IJAME) 14 3887-97

[17] Zahari S N, Sani M S M, Husain N A, Ishak M and Zaman I 2016 Dynamic analysis of friction stir welding joints in dissimilar material plate structure *J. Teknol.* 78

[18] Sani M S M and Ouyang H 2016 Identification of Nonlinearities in Joints of a Wing Structure *MATEC Web of Conference* 83 (03006)

[19] Allemang R J and Phillips A W 2014 Un-weighted and Weighted Versions of the Modal Assurance Criterion (MAC) for Evaluation of Modal Vector Contamination. *Topics in Modal Analysis I, Volume 7 Conference Proceedings of the Society for Experimental Mechanics Series,* 173-180. doi:10.1007/978-3-319-04753-9_17

[20] Phillips A W & Allemang R J 2014 Normalization of Experimental Modal Vectors to Remove Modal Vector Contamination *Topics in Modal Analysis II, Volume 8 Conference Proceedings of the Society for Experimental Mechanics Series* 29-41 doi:10.1007/978-3- 319-04774-4_4

[21] MSC.NASTRAN 2004. MSC.Nastran 2004 Release Guide.

[22] Mohd Zin M S, Rani M N A, Yunus M A, Wan Iskandar Mirza W I I, Mat Isa A A and Mohamed Z 2017 Modal and FRF Based Updating Methods for the Investigation of the Dynamic Behaviour of a Plate Journal of Mechanical Engineering 4 175–89

[23] Rani M N A, Kasolang S, Othman M H, Yunus M A and Wan Iskandar Mirza W I I 2016 Finite Element Modelling and Modal Based Up- Dating for The Dynamic Behaviour of a Laser Spot Welded Structure ICSV 2016 - 23rd International Congress on Sound and Vibration: From Ancient to Modern Acoustics pp 1–8

IOP Conf. Series: Journal of Physics: Conf. Series **1262** (2019) 012005 doi:10.1088/1742-6596/1262/1/012005

Finite Element Modelling and Updating of a Bolted Structure Using Elements Representing the Stiffness Members, Bolts and Affected Areas of the Joints

R Omar[1,2], M N Abdul Rani[1*] and M A Yunus[1]

[1] Structural Dynamics Analysis and Validation, Faculty of Mechanical Engineering, Universiti Teknologi MARA (UiTM), 40000 Shah Alam, Selangor, Malaysia
[2] Machinery Technology Centre, SIRIM Industrial Research, 44200 Hulu Selangor, Selangor, Malaysia

[*]Corresponding author: mnarani@salam.uitm.edu.my

Abstract. Efficient and accurate analytical representation of bolted joints in an assembled structure for the prediction of the dynamic behaviour of the structure has been a crucial unsolved issue in the finite element (FE) community. 3D modelling of bolted joints, possessing a large number of degrees of freedom, leads to very high computational time. This paper proposes an efficient representation of bolted joints for the prediction of the dynamic behaviour of a bolted structure. The FE model of the bolted joints was developed using CBUSH element connectors for bolt shanks and CELAS element connectors for the stiffness of the members at the affected areas of the bolted joints. The results of the normal modes analysis of the FE model of the bolted structure were in a very good correlation with experimental modal analysis (EMA) counterparts quantified by a range of MAC values between 0.91 to 0.99 for all modes. Finite element model updating was used to improve the accuracy of the FE model of the bolted joints. The updated FE model of the assembled structure was validated with the physical tested structure. It was found that the comparison of the results has shown good agreement. In addition, the natural frequencies and mode shapes predicted from the proposed model have been found to be more accurate than the results obtained from the previous analytical models reported in the literature on bolted joint modelling of dynamic behaviour investigation. The proposed model can be used for a large and complex structure.

1. Introduction

Bolted joints are extensively used as the main joining method for structures in various industrial sectors due to the low cost and quick assembling work. Furthermore, it also plays a substantial role in obtaining an optimum dynamic response of a bolted structure due to the higher energy dissipation and higher damping ratio [1]–[3]. However, modelling of bolted joints with an efficient and accurate methods for the investigation of structural dynamic behaviour has always been found to be very difficult [4], [5].

It has been found that the stiffness member, bolt modelling and joint affected areas [5]–[9] are the main factors that lead to the development of an efficient FE modelling of the bolted structure. However, very little evidence of research work is available, especially on incorporating all the main factors in the development of an efficient FE model of the bolted structure and on updating the developed FE model using model updating methods.

Normally, a bolted structure comprises several structural members jointed with a series of bolts and nuts. Therefore, a structure with multiple members should have the overall stiffness members represented in an FE model by several springs connected in series [10], [11]. In this study, CELAS element connectors are used to represent the overall stiffness members in the FE model of the bolted structure.

There are also various methods for bolt modelling used by researchers [12]–[17]. Elements such as CBAR, CELAS, CBEAM and CBUSH were employed to represent the bolts [7], [18]–[20]. In this work, CBUSH elements are used to represent the bolts. The flexibility of the bolts is calculated based on the Swift's formula from which the shear stiffness and rotational stiffness components of the CBUSH element are determined [16], [21]–[25].

There are several cases demonstrated in previous studies. The interfaces of the bolted structure were modelled with special elements such as beam-bar elements, contact zone elements, partitioned thin layer elements and connective layer elements [5], [26]–[28]. However, in this work, the contact bodies are defined as sub-parts of the affected areas of the joints [29]. Therefore, the inclusion of the parameters of the joints' affected areas such as the Young's modulus, mass density and thickness may lead to an improvement in the prediction of the dynamic behaviour of the bolted structure.

This paper put forward a proposal of an efficient and economical modelling method for the bolted structure focusing specifically on the stiffness members, bolt representation and the joints' affected areas. Normal modes analysis is performed, and the predicted natural frequencies and mode shapes are compared with the experimental modal analysis (EMA) counterparts. FE model updating is used to update the initial FE model of the bolted structure in the light of experimental data in order to improve the accuracy and reliability of the FE model of the bolted structure.

2. FE Modelling of the Bolted Structure

A bolted structure was developed in this study to investigate the dynamic behaviour of the structure with bolted joints. Figure 1 shows the 3D CAD model of the bolted structure which was developed using NX 11 design software. The bolted structure consists of two equal dimensional plates which are made of steel and jointed together by using stainless-steel bolts and nuts. Each of the plates is termed as Plate A and Plate B respectively. The dimensions of each plate are 380 mm in length, 45 mm in width and 6 mm thick. In addition, there are two cylindrical features in each plate with the diameter of 18 mm and the thickness of 2 mm as shown in Figure 2. The purpose of having the specially designed features is to reduce the contact zones between the assembled plates and they are named as the affected area of the plate [9].

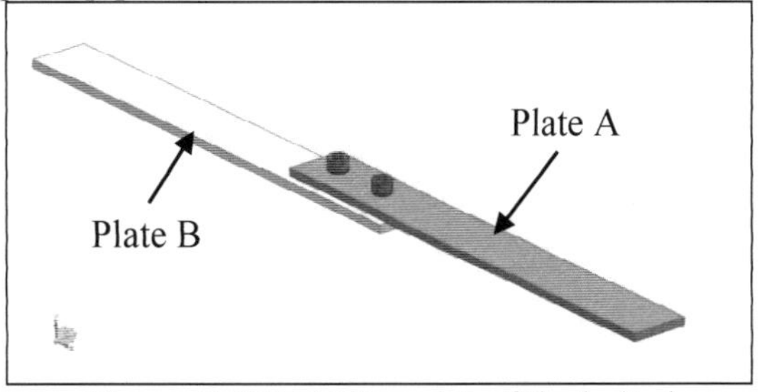

Figure 1. 3D CAD model of the bolted structure.

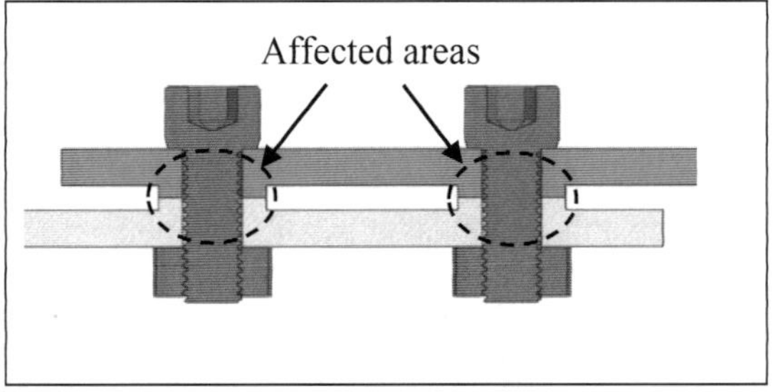

Figure 2. Cross-section of the bolted structure shows the affected areas of the bolted joints.

The material properties of Plate A and Plate B are shown in Table 1. The tabulated values slightly differ from that of the textbook [30]. The values of the material properties used for the plates were systematically adjusted in the previous study using FE model updating [7], [31]–[33]. The use of the FE model updating technique produced accurate modelling of Plate A and Plate B prior to assembling both plates to form a bolted structure. On the other hand, the material for the M10 bolts and nuts are Stainless Steel and the material properties are also tabulated in Table 1 [30].

Table 1. Material properties for Plate A, Plate B and, bolts and nuts.

Property	Plate A	Plate B	Bolts and nuts
Young's modulus (N/mm^2)	206,400	209,400	193,000
Poisson's ratio	0.302	0.316	0.27
Mass density (kg/mm^3)	7.458×10^{-6}	7.45×10^{-6}	7.86×10^{-6}

The mid-surface of the solid geometry of Plate A and Plate B was extracted. The mid-surface which is in the form of 2D thin shell was discretised into finite elements. The FE model of plate A and Plate B was created using 1288 CQUAD8 elements with each element is 5 mm in size. The determination of the element size was selected based on several convergent tests [34]. Figure 3 shows the developed FE model of the bolted structure after the FE models of the plates A and B were assembled together. In the construction of the FE models of the bolted structure, CELAS elements were used to represent the jointed stiffness members. On top of that, the bolt's shank was represented by CBUSH elements, while the bolt's heads and nuts were represented by RBE3 elements. The modelling of the full FE model of the bolted structure is illustrated in Figure 4.

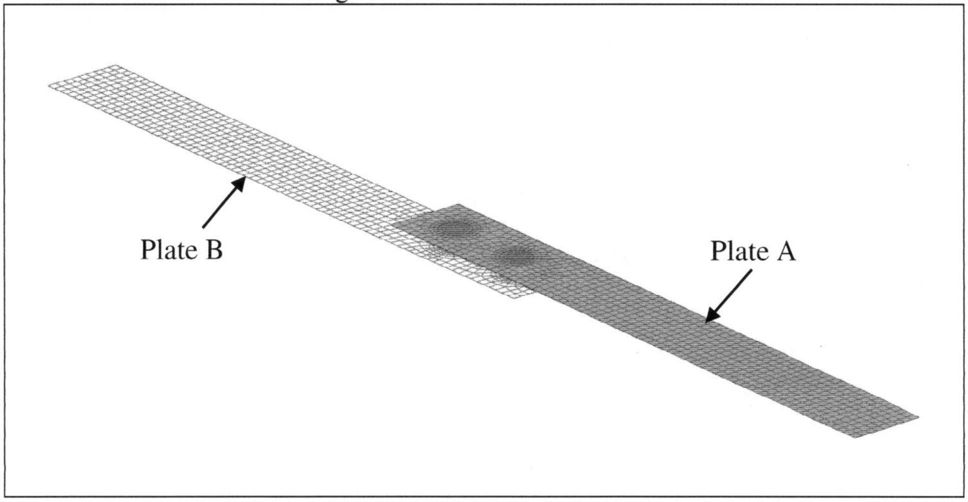

Figure 3. FE model of the bolted structure.

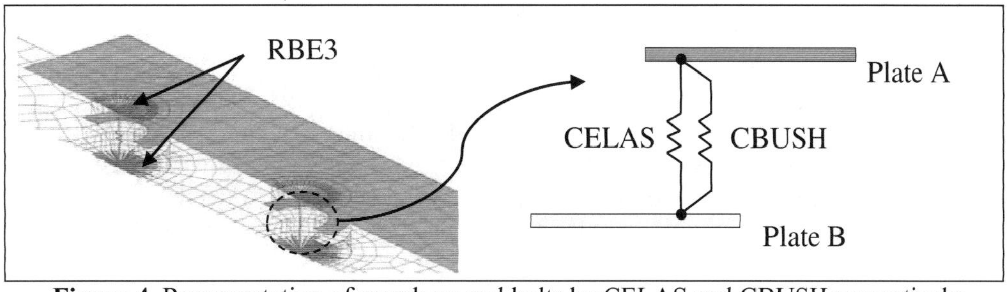

Figure 4. Representation of members and bolts by CELAS and CBUSH respectively.

In this work, the stiffness members was calculated by assuming a uniform pressure with conical envelope under the bolt's head or nut at the affected area as shown in Figure 5 [10], [11], [35], [36]. The 1st stiffness member and 2nd stiffness member were calculated using equation (1) and equation (2) respectively. The variables of the stiffness member used are tabulated in Table 2. The stiffness values of k_1 and k_2 were used to calculate the overall stiffness members using equation (3). The overall stiffness members were then used as the component of translational stiffness of the CELAS element in the FE modelling of the bolted structure.

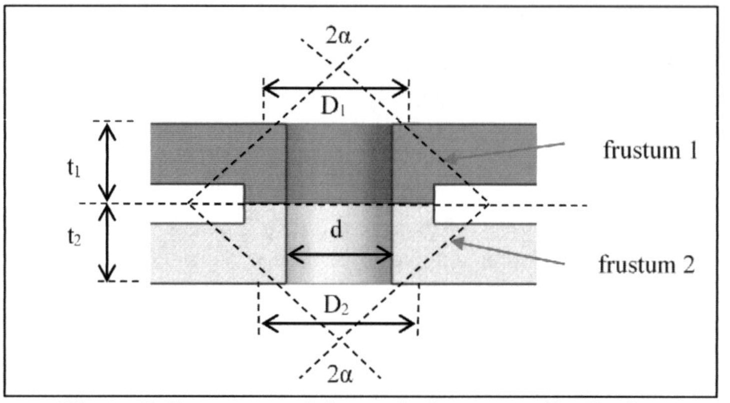

Figure 5. Frustums and the conical pressure distribution.

$$k_1 = \frac{0.577\pi E_1 d}{ln\frac{(1.15t_1 + D_1 - d)(D_1 + d)}{(1.15t_1 + D_1 + d)(D_1 - d)}} = 6.642 \times 10^6\ N/mm \tag{1}$$

$$k_2 = \frac{0.577\pi E_2 d}{ln\frac{(1.15t_2 + D_2 - d)(D_2 + d)}{(1.15t_2 + D_2 + d)(D_2 - d)}} = 6.954 \times 10^6\ N/mm \tag{2}$$

$$k_m = 1/\left(\frac{1}{k_1} + \frac{1}{k_2}\right) = 3.397 \times 10^6\ N/mm \tag{3}$$

Table 2. Nomenclatures for the calculation of member stiffness.

Variable	Description
t_1	Thickness of Plate A
t_2	Thickness of Plate B
E_1	Young's modulus of Plate A
E_2	Young's modulus of Plate B
d	Diameter of the bolt
D_1	Diameter of bolt's head
D_2	Diameter of nut

In this work, bolt's shank was represented by CBUSH elements. The components of stiffness of a CBUSH element have six degrees of freedom as illustrated in Figure 6. All stiffness components are named as axial stiffness (K_1), shear stiffness (K_2, K_3) and rotational stiffness (K_4, K_5, K_6). The value of each component is defined using the formulae as stated in equation (4) to equation (7). The Swift's formula [16], [21]–[25] as shown in equation (6) was used to calculate the flexibility of the bolt. Furthermore, the rotational stiffness around the axis of the bolt, K_4, is fixed at 100 N.mm. Table 3 shows the description of the variables used in the equations while Table 4 is the calculated values for the stiffness components.

Normal modes analysis was performed on the FE model of the bolted structure. The solution type used is the SOL 103 Real Eigenvalues of the NX11 Simcenter software. The analysis was carried out to predict the first ten natural frequencies and mode shapes of the bolted structure. The predicted results were used as the initial FE results. The initial FE results were then compared with the measured results of the plates obtained from EMA.

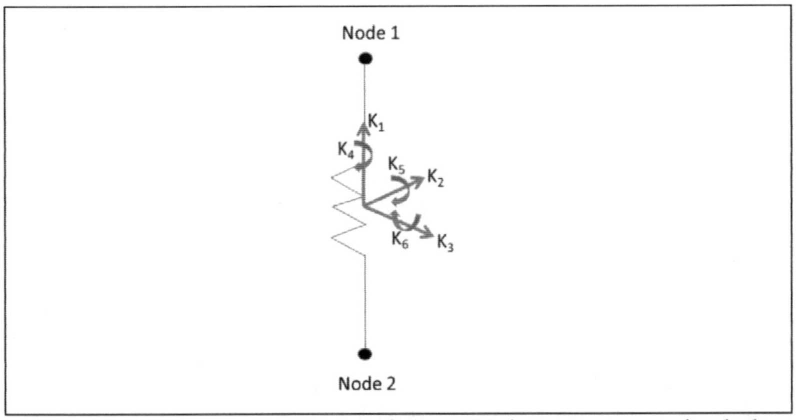

Figure 6. Stiffness components of CBUSH element representing bolt.

$$K_1 = \frac{E_b \times S}{L} \tag{4}$$

$$K_2 = K_3 = \frac{1}{C} \tag{5}$$

$$C = \frac{1}{E_1 d}\left(A + Bd\left(\frac{1}{t_1} + \frac{1}{t_2}\right)\right) \tag{6}$$

$$K_{5,6} = \frac{1}{\varepsilon} \times \left(\mathrm{Max}(K_2, K_3) \times \frac{(L_{CBUSH})^2}{4}\right) \tag{7}$$

where $1/\varepsilon$ is fixed at 100, $\mathrm{Max}(K_2, K_3)$ is the maximum value taken from the calculation of K_2 or K_3 and L_{CBUSH} is the length of the CBUSH element.

Table 3. Nomenclatures for the calculation of stiffness of CBUSH element.

Variable	Description
E_1	Young's modulus of Plate A
E_b	Young's modulus of the bolt
S	Section area of the bolt
L	Length of the bolt
D	Diameter of the bolt
A	Constant value used for steel bolt
B	Constant value used for steel bolt
L_{CBUSH}	Length of the CBUSH element

Table 4. Value of stiffness components of CBUSH element.

Stiffness components	Value
Axial stiffness, K_1	9.474×10^5 N/mm
Shear Stiffness, $K_{2,3}$	5.403×10^5 N/mm
Rotational Stiffness, K_4	1.000×10^2 N.mm
Rotational Stiffness, $K_{5,6}$	3.458×10^6 N.mm

3. Experimental Modal Analysis (EMA)

EMA is a method to extract the structural dynamic characteristics, which are the natural frequencies and mode shapes, of the actual bolted structure [37]. In this work, the bolted structure was constructed by joining two identically fabricated steel plates using two sets of M10 stainless steel bolts and nuts. The bolted structure was suspended from a test rig by using rubber bands to simulate free-free boundary conditions. The experimental setup of the bolted structure is shown in Figure 7.

The items of test equipment used to measure the dynamic characteristics of the bolted structure are the LMS system with the LMS Test.Lab software, triaxial accelerometers and impact hammer. The impact hammer was used to excite the bolted structure at a fixed point while the accelerometers were roved to 42 pre-determined points of the bolted structure to measure the responses. The excitation needs to be done in Z-direction and Y-direction to successfully extract the first 10 modes. Figure 8 shows the frequency response function of the bolted structure.

Figure 7. Experimental configuration of the bolted structure.

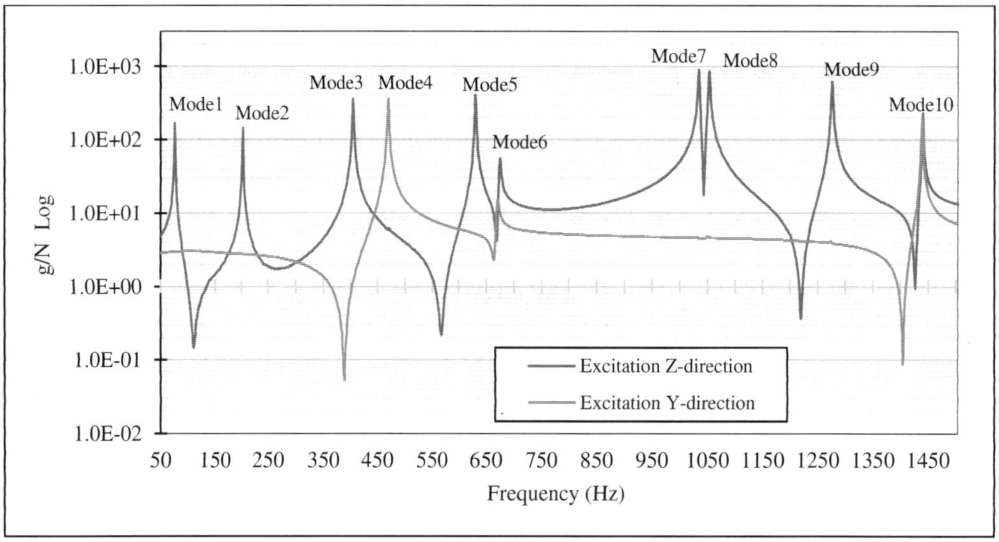

Figure 8. Frequency response functions of bolted structure based on the excitation in the directions of Y and Z.

4. FE Model Updating

FE model updating is an analytical method from which the accuracy and reliability of a FE model can be improved by systematically adjusting the parameters of the FE model within an identified range [38], [39]. The method was employed in this work in order to reduce the uncertainties introduced in the initial

IOP Conf. Series: Journal of Physics: Conf. Series **1262** (2019) 012005 doi:10.1088/1742-6596/1262/1/012005

FE models as a result of invalid assumptions about model properties. The selection of the parameters used for the FE model updating procedure is based on the result of the sensitivity analysis [40].

In this work, the SOL 200 Model Update of NX 11 Simcenter 3D software was utilised in the effort to update the FE models of the bolted structure. The first ten natural frequencies were incorporated in the objective function to improve the correlation between the predicted and measured natural frequencies. The objective function used in this work is as in equation (8) [32], [41]:

$$min \sum_{i=1}^{m} W_i \left(\frac{\omega_i^n}{\omega_i^e} - 1 \right)^2 \qquad (8)$$

where ω_i^n is the i-th predicted frequency, ω_i^e is the i-th measured frequency, the weightage, W_i is set to unity and $i = 1,\ldots,10$.

5. Results and Discussion

In this work, the natural frequencies and mode shapes of the bolted structure were obtained experimentally and numerically using the EMA and FE method respectively. Differences in the results between EMA and FE method were reduced using the FE model updating method through which the selected parameters of the initial FE model are systematically adjusted to match the EMA results as close as possible.

The 10 modes of the natural frequencies and mode shapes obtained from EMA and the initial FE model of the bolted structure are tabulated in Table 5. The comparison of the natural frequencies obtained from EMA and the initial FE shows that the total error is 14.59%. Meanwhile, the quantification of mode shapes through the Modal Assurance Criterion (MAC) shows a range of values between 0.91 to 0.99 for all modes. This is an indication of a very good correlation between EMA and initial FE results.

Table 5. Comparison of EMA and initial FE of the bolted structure.

Mode	I. EMA (Hz)	II. Initial FE (Hz)	III. Error between I & II (%)	IV. MAC
1	75.08	75.29	0.28	0.98
2	200.77	202.44	0.83	0.94
3	404.93	401.63	0.81	0.99
4	469.31	446.94	4.77	0.95
5	630.78	632.63	0.29	0.97
6	674.39	661.27	1.95	0.96
7	1034.64	1019.16	1.50	0.94
8	1053.23	1067.47	1.35	0.91
9	1276.95	1278.41	0.11	0.96
10	1435.09	1473.76	2.69	0.94
	Total Error		14.59	

A list of the potential updating parameters is tabulated in Table 6. Based on the sensitivity analysis of the potential updating parameters, the natural frequencies of the bolted structure are most sensitive to the shear stiffness in x-axis and y-axis of the CBUSH element representing the bolt's shank. On top of that, the thickness, Young's modulus and mass density of the affected areas of Plate A and Plate B may also affect the natural frequencies. The results of the sensitivity analysis of the potential updating parameters are shown in Figure 9.

Table 6. Description of the potential updating parameters.

Parameter ID	Description	Unit
$K_{1\ CELAS}$	Stiffness of joint members	N/mm
t_{AA}	Thickness of affected area of Plate A	mm
t_{AB}	Thickness of affected area of Plate B	mm
E_{AA}	Young's modulus of affected area of Plate A	MPa
E_{AB}	Young's modulus of affected area of Plate B	MPa
ρ_{AA}	Mass density of affected area of Plate A	kg/mm^3
ρ_{AB}	Mass density of affected area of Plate B	kg/mm^3
v_{AA}	Poisson's ratio of affected area of Plate A	unitless
v_{AB}	Poisson's ratio of affected area of Plate B	unitless
$K_{1\ CBUSH}$	Axial stiffness of bolt	N/mm
$K_{2\ CBUSH}$	Shear stiffness (x-axis) of bolt	N/mm
$K_{3\ CBUSH}$	Shear stiffness (y-axis) of bolt	N/mm
$K_{4\ CBUSH}$	Rotational stiffness around the axial axis of bolt	N.mm
$K_{5\ CBUSH}$	Rotational stiffness around x-axis of bolt	N.mm
$K_{6\ CBUSH}$	Rotational stiffness around y-axis of bolt	N.mm

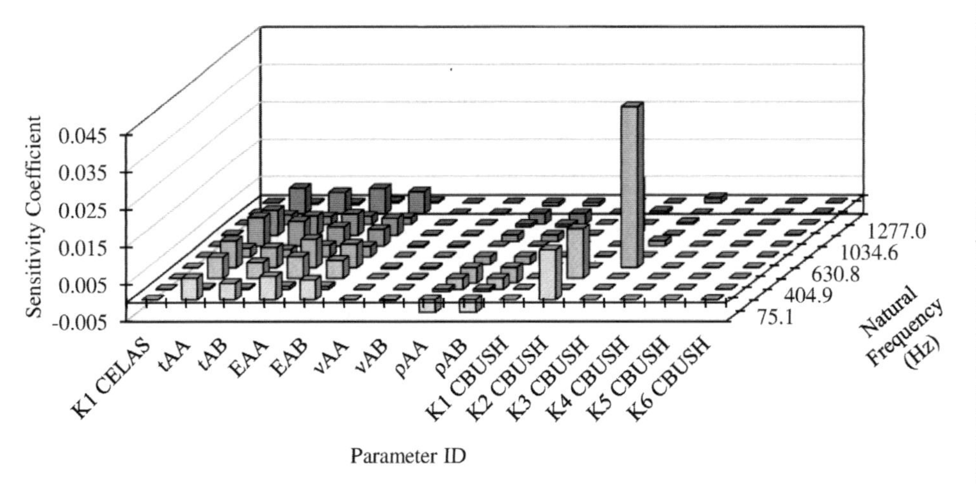

Figure 9. Sensitivity matrix of potential updating parameters.

For the FE model updating of the bolted structure, genetic algorithms were used as the optimizer. The comparison of results of EMA and the updated FE models is tabulated in Table 7 with several sets of updating parameters. The recorded total error has reduced to 14.39% using the shear stiffnesses $K_{2\ CBUSH}$ and $K_{3\ CBUSH}$ as the updating parameters. The total error has decreased to 14.24% with the Young's modulus of the affected areas of Plate A and Plate B. Further reduction in total error to 14.10% can be achieved with all selected updating parameters were included in the updating procedure. However, it was found that the lowest total error of 14.08% can be achieved when the mass densities of the affected areas of Plate A and Plate B were excluded. This comparison process to select appropriate updating parameters was also demonstrated in previous researches [5], [41].

The updated parameters resulted from the FE model updating procedure which gave the lowest total error are shown in Table 8. It was discovered that there is a substantial increase in the shear stiffnesses $K_{2\ CBUSH}$ and $K_{3\ CBUSH}$ of the bolt's CBUSH element. There is also significant increase in Young's modulus and small increment of thickness of the affected areas of Plate A and Plate B. These results showed that the bolts and nuts used to assemble the plates contributed significantly to the dynamic behaviour of the assembled structure as well as the affected areas of the Plate A and Plate B.

Table 7. Comparison of EMA and updated FE results.

Mode	I.	II.	III.	IV.	V.	VI.	VII.	VIII.	IX.
		After updating							
	EMA (Hz)	$K_{2\ CBUSH}$ & $K_{3\ CBUSH}$		$K_{2\ CBUSH}$, $K_{3\ CBUSH}$, E_{AA} & E_{AB}		$K_{2\ CBUSH}$, $K_{3\ CBUSH}$, E_{AA}, E_{AB}, ρ_{AA}, ρ_{AB}, t_{AA} & t_{AB}		$K_{2\ CBUSH}$, $K_{3\ CBUSH}$, E_{AA}, E_{AB}, t_{AA} & t_{AB}	
		FE (Hz)	Error between I & II (%)	FE (Hz)	Error between I & IV (%)	FE (Hz)	Error between I & VI (%)	FE (Hz)	Error between I & VIII (%)
1	75.08	75.29	0.28	75.31	0.31	75.36	0.37	75.34	0.35
2	200.77	202.44	0.83	202.45	0.84	202.46	0.84	202.45	0.84
3	404.93	401.66	0.81	401.75	0.79	402.00	0.72	401.89	0.75
4	469.31	447.80	4.58	448.58	4.42	449.23	4.28	449.29	4.27
5	630.78	632.63	0.29	632.68	0.30	632.84	0.33	632.77	0.32
6	674.39	661.31	1.94	661.44	1.92	661.83	1.86	661.72	1.88
7	1034.64	1019.23	1.49	1019.46	1.47	1020.13	1.40	1019.86	1.43
8	1053.23	1067.50	1.35	1067.58	1.36	1067.81	1.38	1067.74	1.38
9	1276.95	1278.41	0.11	1278.66	0.13	1279.43	0.19	1279.12	0.17
10	1435.09	1473.85	2.70	1473.93	2.71	1473.97	2.71	1473.97	2.71
	Total Error		14.39		14.24		14.10		14.08

Table 8. Updated parameters for the assembled structure with bolted joints.

Parameter ID	Initial Value	Updated Value	Unit
$K_{2\ CBUSH}$	540,314	556,125	N/mm
$K_{3\ CBUSH}$	540,314	661,007	N/mm
E_{AA}	206,400	216,196	MPa
E_{AB}	209,400	217,519	MPa
t_{AA}	2	2.11	mm
t_{AB}	2	2.08	mm

6. Conclusions

The work of developing and investigating an efficient, economic and reliable FE model of a bolted structure using the FE model updating method in light of the EMA results has been presented. It was found that the updated parameters of the physical and material properties of the FE model which are the stiffness members, bolts and joints' affected areas, have a significant influence towards the accuracy of the prediction of the dynamic behaviour of the bolted structure. Therefore, an accurate FE model can be confidently used for any subsequent analysis.

Acknowledgments

The authors would like to acknowledge the Ministry of Education of Malaysia and UiTM for providing financial support for this study. They would like to express their appreciation of helpful comments, suggestions and technical support given by Structural Dynamics Analysis & Validation (SDAV) and Machinery Technology Centre (MTC) members.

References

[1] Ibrahim R A and Pettit C L 2005 *J. Sound Vib.* **279** 857
[2] Ahmadian H and Jalali H 2007 *Mech. Syst. Signal Process.* **21** 1041

[3] Zaman I, Khalid A, Manshoor B, Araby S and Ghazali M I 2013 *IOP Conf. Series: Materials Science and Engineering* **50** 012018
[4] Daouk S, Louf F, Cluzel C, Dorival O, Champaney L and Audebert S 2017 *J. Sound Vib.* **392** 307
[5] Shokrollahi S and Adel F 2016 *J. Theor. Appl. Vib. Acoust.* **2** 65
[6] Kim J, Yoon J-C and Kang B-S 2007 *Appl. Math. Model.* **31** 895
[7] Omar R, Abdul Rani M N, Wan Iskandar Mirza W I I, Yunus M A and Othman M H 2017 *J. Mech. Eng.* **SI 4** 202
[8] Ahmadian H, Mottershead J E and Friswell M I 2001 *J. Sound Vib.* **247** 755
[9] Omar R, Abdul Rani M N, Yunus M A, Wan Iskandar Mirza W I I and Mohd Zin M S 2018 *AIP Conf. Proc.* **1952** 020082
[10] Haidar N, Obeed S and Jawad M 2011 *J. Mech. Sci. Technol.* **25** 2827
[11] Nassar S A and Abboud A 2009 *J. Mech. Des.* **131** 121001
[12] Montgomery J 2002 *ANSYS User's Conference*
[13] Adel F, Shokrollahi S, Jamal-Omidi M and Ahmadian H 2017 *J. Sound Vib.* **396** 172
[14] Grzejda R 2017 *J. Mech. Transp. Eng.* **69** 29
[15] Francavilla A B, Latour M, Piluso V and Rizzano G 2015 *Eng. Struct.* **100** 656
[16] Uysal I, Suer M and Ozkol I 2011 *Proc. of 5th Int. Conf. on Recent Adv. in Space Tech.* 68
[17] Tanlak N, Sonmez F O and Talay E 2011 *J. Strain Anal. Eng. Des.* **46** 213
[18] Rutman A, Boshers C, Pearce L and Parady J 2007 *Americas Virtual Product Development Conf.*
[19] Yunus M A, Nazri S, Abdul Rani M N, Tormodi A and Kasolang S 2016 *MATEC Web of Conferences AiGEV 2016* **90** 01013
[20] Wan Iskandar Mirza W I I, Abdul Rani M N, Yunus M A, Roslan L and Ahmad Basri A B 2017 *MATEC Web of Conferences AiGEV 2016* **90** 01012
[21] Huth H 1986 *Fatigue in Mechanically Fastened Composite and Metallic Joints, ASTM STP 927*, (Philadelphia: American Society for Testing and Materials) p 221
[22] Jarfall L 1986 *Int. J. Veh. Des.* **7** 337
[23] Whitman Z L 2012 University of Utah
[24] Gunbring F 2008 Linkoping University Institute of Technology
[25] Chandregowda S and Ganga Reddy C R 2018 *AIP Conf. Proc.* **1943** 020001
[26] Saeed S, Ahmad N and Farhad A 2014 *21st Int. Congress on Sound and Vibration 2014* pp 1
[27] Yao X, Wang J and Zhai X 2016 *Proc. Inst. Mech. Eng. Part G J. Aerosp. Eng.* **231** 823
[28] Shokrollahi S, Ahmadian H and Adel F 2016 *Modares Mech. Eng.* **16** 35
[29] McCarthy M A, McCarthy C T, Lawlor V P and Stanley W F 2005 *Compos. Struct.* **71** 140
[30] Hibbeler R C 2011 *Mechanics of Materials* (Prentice Hall)
[31] Abdul Rani M N, Ouyang H, Yunus M A and Aminudin B A 2013 *20th International Congress on Sound and Vibration 2013 pp* 1
[32] Abdul Rani M N, Kasolang S, Othman M H, Yunus M A, Wan Iskandar Mirza W I I and Ouyang H 2016 *23rd International Congress on Sound and Vibration* pp 315
[33] Omar R, Abdul Rani M N, Yunus M A, Wan Iskandar Mirza W I I and Mohd Zin M S 2018 *25th Internatinal Congress on Sound and Vibration* pp 4624
[34] Omar R, Abdul Rani M N, Yunus M A, Mat Isa A A, Wan Iskandar Mirza W I I and Mohd Zin M S and Roslan L 2018 *Int. J. Automot. Mech. Eng.* **15** 5695
[35] Lehnhoff T F, Il Ko K and McKay M L 1994 *J. Mech. Des.* **116** 550
[36] Shigley J E and Mischke C R 2006 *Mech. Eng. Des.* (New York, USA: McGraw-Hill)
[37] Packard H 1997 *Methods* 1
[38] Mottershead J E and Friswell M I 1993 *J. Sound Vib.* **167** 347
[39] Sehgal S and Kumar H 2016 *Arch. Comput. Methods Eng.* **23** 515
[40] Mottershead J E, Link M and Friswell M I 2011 *Mech. Syst. Signal Process.* **25** 2275
[41] Ahmadian H, Mottershead J E, James S, Friswell M I and Reece C A 2006 *Mech. Syst. Signal Process.* **20** 868

IOP Conf. Series: Journal of Physics: Conf. Series **1262** (2019) 012006 doi:10.1088/1742-6596/1262/1/012006

Identification dynamic behaviour of disc brake based on finite element model updating approach

M. S. M. Fouzi[1,3], M. S. M. Sani[1,2*] and M. F. R. Ishak[1]
[1]Advanced Structural Integrity and Vibration (ASiVR), Faculty of Mechanical & Manufacturing Engineering, Universiti Malaysia Pahang, 26600 Pekan, Pahang, Malaysia.
[2]Automotive Engineering Centre, Universiti Malaysia Pahang, 26600 Pekan, Pahang, Malaysia.
[3]Jabatan Pendidikan Politeknik & Kolej Komuniti.

*Corresponding author: mshahrir@ump.edu.my

Abstract. Braking system is an essential mechanism that designed to provide a good handling in order to slow down or to stop a vehicle depends on situation and driver's capability. In order to slow down or to stop the vehicle, brake pad will be pushed and make contact with the disc/rotor to form the friction. Resultant from braking event, it will produce vibration on disc/rotor structure. Thus, this study presents numerical prediction approach to identify the dynamic behaviour of disc brake structure using normal mode analysis in FEA software. Since the result of FEA is a prediction, the fidelity of FE model might be questioned. Therefore, measured data from EMA used to validate the predicted result of FEA through correlation process. As FE model used in this study made from assumption and simplification apart of real structure, it drive to discrepancy between FEA result and EMA data during correlation process. Hence, FE model updating approach implemented in order to reduce the discrepancy and increase the trustworthiness of FE model and it is become the main goal in the present study. The updating process has successfully reduced the discrepancy between FEA and EMA from 9.00 % of error to 4.05 % of error from its original value. As conclusion, the use of model updating process is important in preparing a reliable FE model of test structure no matters how simple or complex it is before being used for further analysis.

1. Introduction

Predominantly, each of vehicles has been designed with braking system to control the speed and give better handlings to the driver. Brake system is built from several components such as disc/rotor, piston, pad, calliper and knuckle as illustrated in Figure 1. In order to slow down the vehicle speed, the brake pad will be pushed by the piston to make a sliding friction with the disc/rotor. Due to this sliding friction, the structure tends to experience the vibration formed from dissipated kinetic energy of moving part. This friction-induced vibration has resulted the phenomenon called brake squeal [1]. In addition, [2] have mentioned most of the scientist and engineers have agreed that squeal noise in the disc brake is initiated by the instability due to friction forces, contributing to self-excited vibrations. Hence, the identification of dynamic behaviour of the disc brake structure subjected to the induced-vibration becomes a major interest in this study using modal analysis.

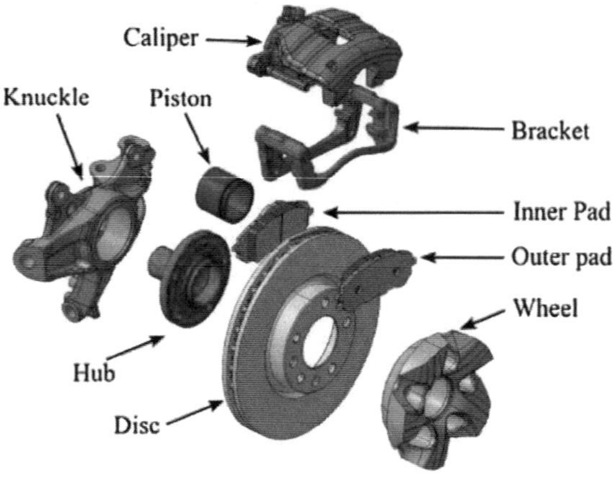

Figure 1. Exploded view of brake system [3].

Regarding to [4], modal analysis was conducted numerically and experimentally to determine the structural dynamic characteristics, in terms of modal frequencies, damping ratio, and mode shape. It is vital to identify the dynamic behaviour of the structure in the early development stage to avoid the resonance. Resonance effects are activated when the frequency content of the loading gets closer to the natural frequencies of the structure [5]. This was agreed by [6], it is paramount importance to periodically monitor the technical condition of components in order to avoid resonance or overload problems that may cause undesirable behaviour or even lead to critical failure. Thus, a reliable finite element (FE) model is required to monitor the structure periodically rather than make a repeated measurement of structure response that required highly expenses on expensive equipments and more efforts.

In order to prepare the reliable FE model, the predicted result computed in finite element analysis (FEA) needs to validate with measured data of frequency response functions (FRFs) that executed from experimental modal analysis (EMA). This kind of validation of FE model was reported by [7] using correlation process which compared the modal parameters (natural frequency and mode shape) of a body-in-white (BIW) obtained from EMA as benchmark and analyzed in FEA as prediction. However, the discrepancy possibly occurred between EMA data and FEA result in correlation process. This happen due to inaccuracies of FE model; poorly known boundary conditions of structure, the unknown material properties of the structure or because of the simplification in the modelling of very complex structural system [8]. In encounter this problem, an approach called FE model updating will be used to reduce the discrepancy and reconcile the predicted result with its measured counterpart by adjusting certain parameters of the FE model.

As mentioned by [9], FE model updating techniques are used to update the finite element model of a structure in order to improve its correlation with the experimental dynamic test data. Regarding [10], model updating techniques can be broadly classified as the direct methods and iterative methods. Direct methods are one-step procedures that seek to make a minimum change in the structural matrices so that the measured natural frequencies and mode shapes are reproduced by the updated model [10]. Despite of direct method, iterative methods are based on minimizing an objective function that is generally a non-linear function of selected updating parameters [11]. Practically, the iterative method seems more useful for larger and more complex structure in model updating technique. In addition, the sensitivity method probably the most successful of the many approaches to the problem of updating FE models of engineering structures based on vibration test data [12]. In the sensitivity method, the eigensolutions are often used to construct the objective function and the physical parameters of analytical FE model adjusted iteratively to minimize the discrepancy of dynamic properties between the analytical FE model and the measured counterpart [13]. This kind of iteratively updating based on sensitivity method successfully reported in their works [7, 14, 15].

Due to outstanding capability of iterative model updating technique reported from the previous researchers, this technique has been adopted in preparing the reliable FE model to replicate disc brake structure and it has become the main goal for the present study. Last but not least, this updating technique managed to shrink the discrepancy between predicted dynamic properties with the measured counterpart. This supported by the previous published works [7, 16, 17] that reported their updating process successfully reduce the discrepancies between FEA and EMA.

2. Finite element analysis (FEA) of disc brake

Disc/rotor is major structure in brake system and directly exposed to friction induced-vibration. Hence, numerical approach using FEA is adopted to identify the dynamic behaviour of the structure using normal mode analysis SOL103 in MSC Nastran/Patran. This normal mode analysis used to extract natural frequency and mode shape of undamped system [18]. The entire process involved during this analysis has been summarized in Figure 2.

Initially, the disc/rotor structure was modelled using computational aided design (CAD) software, SolidWork. Then, it was imported into FEA software, MSC Patran in *"parasolid"* format for pre-processing stage. The similar approach has been reported in the previous work, where the test structure was modelled using CATIA V-5 software and been imported into Hypermesh for pre-processing in FEA and iterated by MSC Nastran solver [19].

During pre-processing stage, the FE model was meshed using solid element CTETRA10 topology that produced 3712 number of elements with 7243 number of nodes. Then, the FE model was assigned with the following nominal value of material properties; Modulus of Young, $E = 130$ GPa, density, $\rho = 7100$ kg/m^3, and Poisson's Ratio, $v = 0.26$. This nominal value was considered in the midst of the study for material properties effect on disc brake squeal by [2].

Once pre-processing stage done, the mathematical model generated for normal mode analysis in bulk data file (*bdf.*) format that will be used in iteration process by MSC Nastran solver to extract the modal parameter (natural frequency and mode shape) of disc brake. Regarding [7], neither boundary condition nor external forces were applied to the model as the model was let to be free-free boundary condition for calculation of modal properties using normal mode analysis SOL103 in MSC Nastran/Patran.

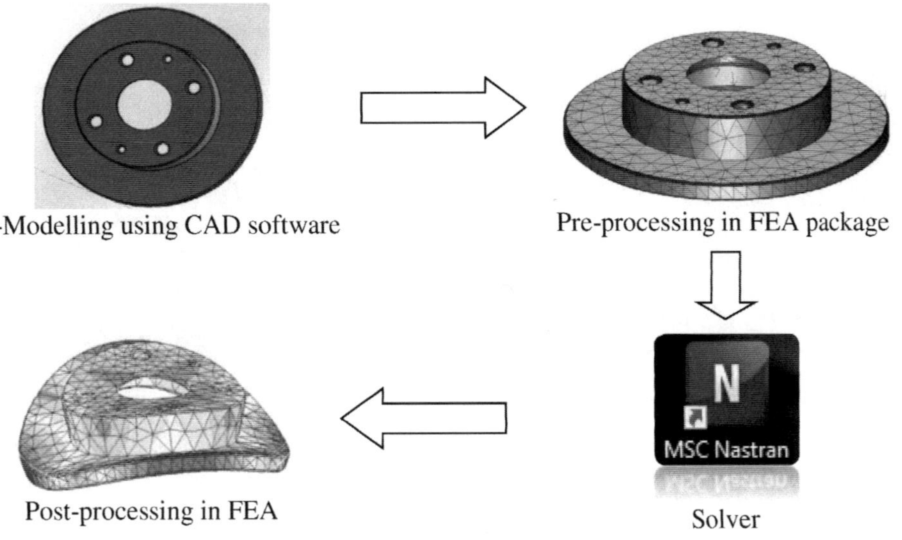

Figure 2. Flow for normal mode analysis of exhaust structure.

IOP Conf. Series: Journal of Physics: Conf. Series **1262** (2019) 012006 doi:10.1088/1742-6596/1262/1/012006

3. Modal testing procedure of disc brake structure

In order to validate the predicted result computed in FEA, the measured dynamic properties of disc brake was carried out using modal testing. Modal testing or familiar as experimental modal analysis (EMA) was conducted to measure the structural response. As reported by [20], one of preferable technique to measure the structural response is impact hammer excitation owing to its fast, convenient, and fast diagnostic. This technique was implemented to measure the structural response of disc brake in the present study.

In the beginning of experimental procedure, the configuration of equipment and test structure were setup as depicted in Figure 3. The disc brake was suspended by elastic cord such as Figure 4 to represent free-free boundary condition and this approach was reported by [21] in their work. The measurements were made using EMA software with several equipment such as 4 channels National Instruments (NI) data acquisition system (DAQ), impact hammer with transducer sensitivity 2.25 mV/g, tri-axial accelerometer with transducer sensitivity 100 mV/g for x-axis, 104 mV/g for y-axis, and 102 mV/g for z-axis. Roving accelerometer technique was employed in this setting same as practiced by[7]. In this technique, the disturbance will be created by impact hammer at fixed point while the accelerometer will be roved around 48 measurements point that assigned on the structure. The response will specifically relate to each point sketched in wire-frame structure such as Figure 5 in EMA software to simulate the mode shape once the measurement completed.

During the testing, the output response measured by the accelerometer is in the form of Fast Fourier Transform (FFT). In order to extract the modal parameters, DAQ is used as FFT analyzer which converter the response in the form of FFT into frequency response functions (FRFs). Due to measured FRFs, modal parameters can be extracted using curve-fitting method in EMA software as illustrated in Figure 6. As supported by [4], EMA techniques have been dramatically improved within the past two decade with accurate modal post processing curve-fitting technique (modal extraction).

The extracted modal parameters (natural frequency and mode shape) then were used to validate the predicted result from FEA in correlation process.

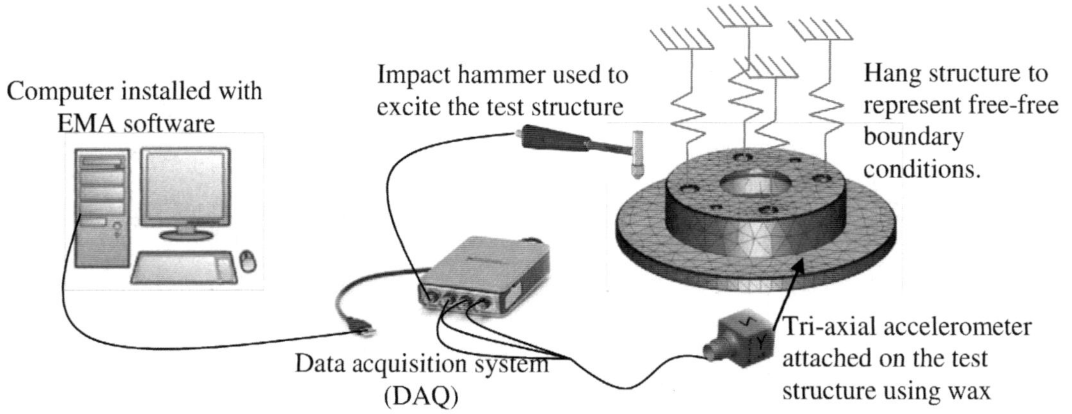

Figure 3. Configuration of equipments and disc brake in experimental modal analysis (EMA).

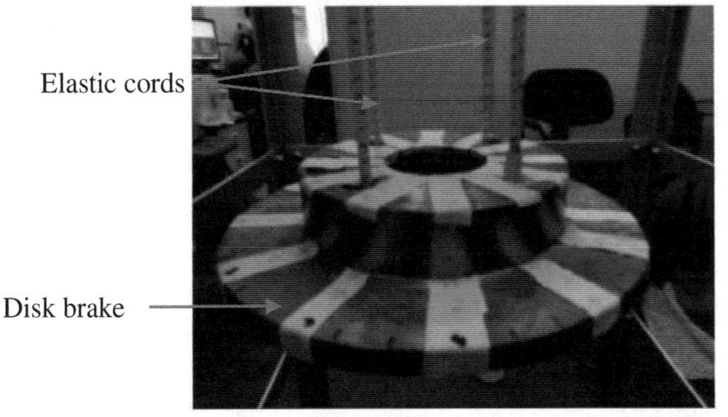

Figure 4. Test structure hanged to replicate free-free boundary condition.

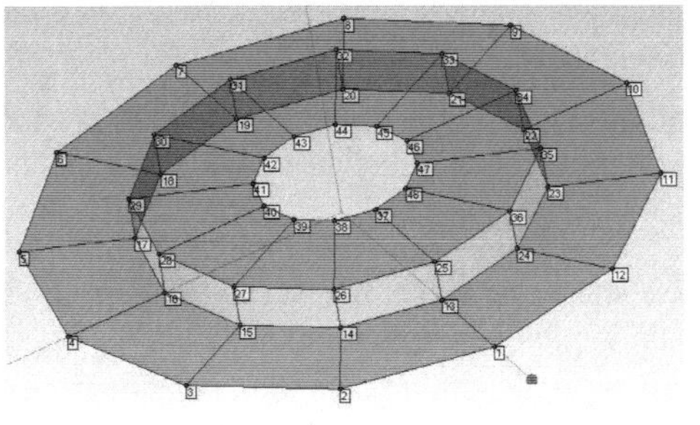

Figure 5. Wire frame structure sketched in EMA software.

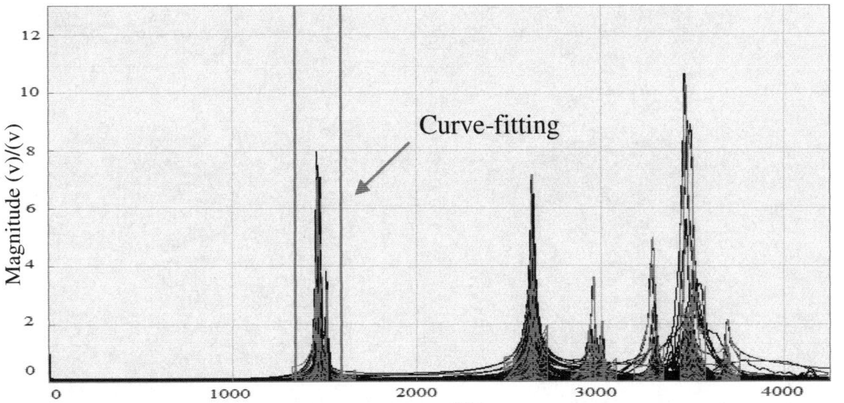

Figure 6. Overlay traces of FRFs.

4. Correlation

In verifying and validate the predicted result from FEA, correlation process is implemented in this study. Regarding [22], correlation is a process where data from the experiment are compared with theoretical results. In addition, this correlation process also implemented by [7] where numerical analysis for dynamic behaviour of a body-in-white (BIW) structure are compared with modal testing results.

Comparison of eigenvalue or natural frequency is tabulated in Table 1 while in Table 2 is comparison of eigenvector or mode shape between predicted result and measured counterpart. In equation (1), the percentage of error is calculated to demonstrate how far the predicted of FEA agreed with measured data in EMA. The bigger value of percentage of error indicates huge disparity while the small error

IOP Conf. Series: Journal of Physics: Conf. Series **1262** (2019) 012006 doi:10.1088/1742-6596/1262/1/012006

shows the agreement between EMA and FEA. Hence, the smaller error means the FE model is reliable to represent actual structure. The calculated percentage of error showed there is 9.00 % of error between EMA and FEA in Table 1. The discrepancies are able to be minimized using FE model updating approach that be discussed in section 5.

$$Percentage\ of\ error = \left|\frac{f^{predicted} - f^{measured}}{f^{measured}}\right| x\ 100 \tag{1}$$

Table 1. Comparison between predicted results from FEA with measured data in EMA.

Mode	Natural frequency (Hz)		
	EMA	FEA	Error (%)
1	1480.00	1659.50	12.13
2	2640.00	3061.40	15.96
3	3480.00	3455.60	0.70
4	3520.00	3873.00	10.03
5	3700.00	3929.30	6.20
Total average error			9.00

Table 2. Comparison of mode shape between measured FRF in EMA and numerical prediction eigenvector by FEA for the first 5 mode of interests.

Mode	EMA	FEA
1	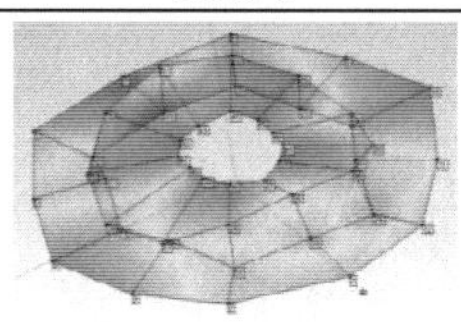 1480.00 Hz	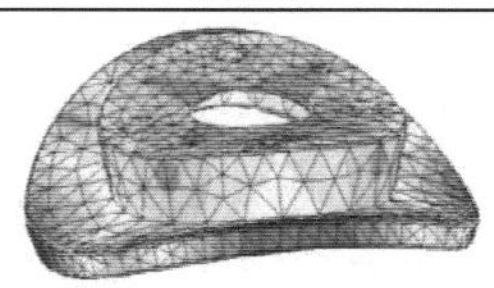 1659.50 Hz
2	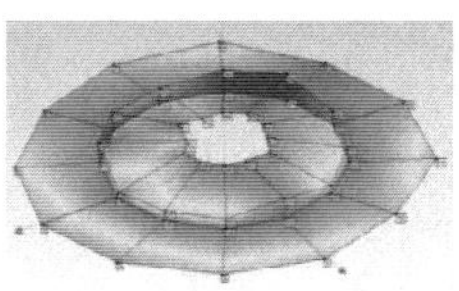 2640.00 Hz	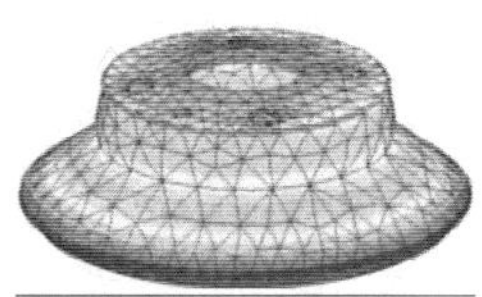 3061.40 Hz
3	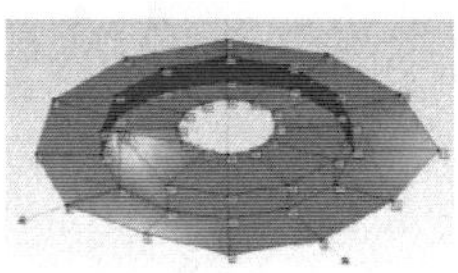 3480.00 Hz	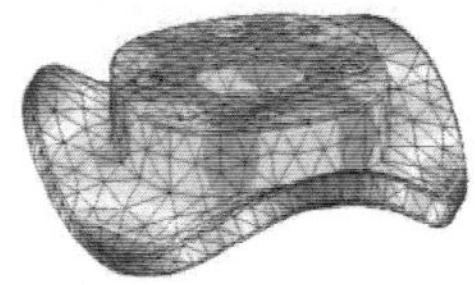 3455.60 Hz
4	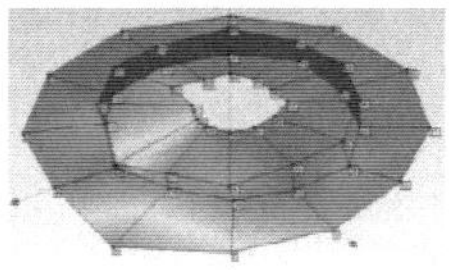 3520.00 Hz	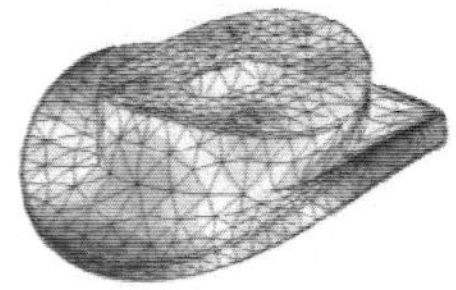 3873.00 Hz

5	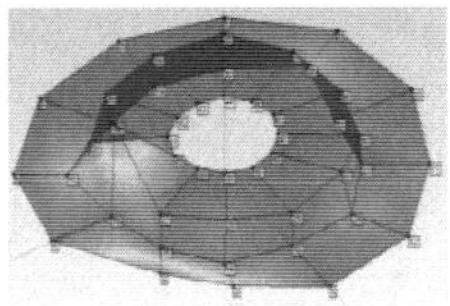 3700.00 Hz	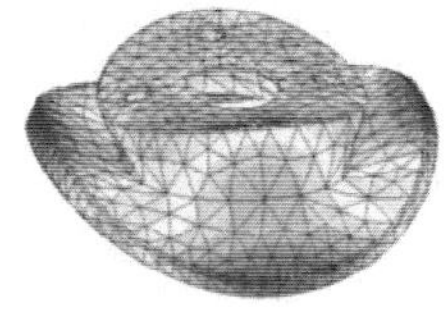 3929.30 Hz

5. FE model updating

Since FEA is numerical prediction method, it is possible that disparity will occurred when correlate with the measured counterpart from EMA. The difference occurred due to inaccurate FE model since simplification of structure details for complex structure and lack of knowledge of exact material properties [10]. FE model updating adopted in this study to overcome this obstacle. Model updating is essentially a process of adjusting certain parameters of FE model [12]. The following equation (2) is sensitivity analysis used to identify the most sensitive parameter in the analysis.

$$\mathbf{S} = \boldsymbol{\phi}_i^{\mathrm{T}} \left[\frac{\partial \mathbf{K}}{\partial \theta_j} - \lambda_i \frac{\partial \mathbf{M}}{\partial \theta_j} \right] \boldsymbol{\phi}_i \quad (2)$$

where **S** indicates the sensitivity matrix, **K** and **M** are the stiffness and mass matrices respectively, while ϕ, λ and θ represent eigenvector, eigenvalue and parameter respectively. Furthermore, i indicate the i-th eigenvalue and j for the j-th parameter [23-25].

Meanwhile, the minimisation of the discrepancies between the measured and predicted values was carried out via a residual based objective function in the form of equation (3)

$$min \sum_{j=1}^{m} W_j \left(\frac{\omega_i^n}{\omega_i^e} - 1 \right)^2 \quad (3)$$

Where ω_i^n is the i-th predicted frequency and ω_i^e represent the i-th measured frequency; W_j indicates the weighting coefficient through which certain modes that required particular attention are assigned [14]. Both equation (2) and (3) are embedded in the optimization algorithm SOL200 that been running into MSC Nastran. The outputs of optimization accessible via F06 file such as updated eigenvalue and the sensitivity of parameters.

As depicted in Figure 7 and tabulated in Table 3, three parameters which are Modulus Young (E), Density (RHO) and Poisson's Ratio (Nu) are considered in the sensitivity analysis for five mode of interests. It showed Modulus Young and density are sensitive parameters while Poisson's Ratio is light sensitive. There are only 4 iterations are taken to converge in this analysis such as showed in Figure 8. Regarding to [12], the updating procedure are significantly be made only to erroneous assumptions such material parameters not on model-structure errors. Hence, only three parameters are available to update in this study as mentioned earlier. This was supported in the published work by [2], there are only three parameters (Modulus Young, density and Poisson's Ratio) were considered in identifying material properties effect on disc brake squeal.

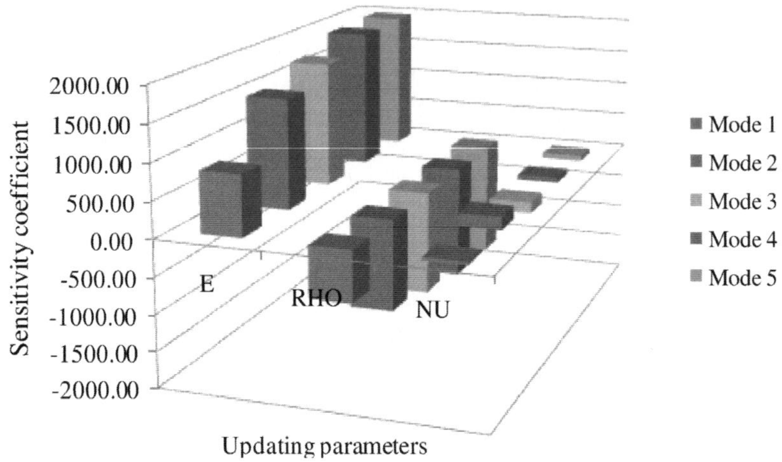

Figure 7. Graph of sensitivity matrix analysis for the present study.

The updated natural frequency of FE model accessed from F06 file then compared with the initial value of natural frequency as tabulated in table 3. The discrepancy successfully reduced from initial value 9.00 % of total average error to 4.05 % of total average error. Updated value of parameters that be suggested from F06 is tabulated in Table 4 with the changes calculated from its original value.

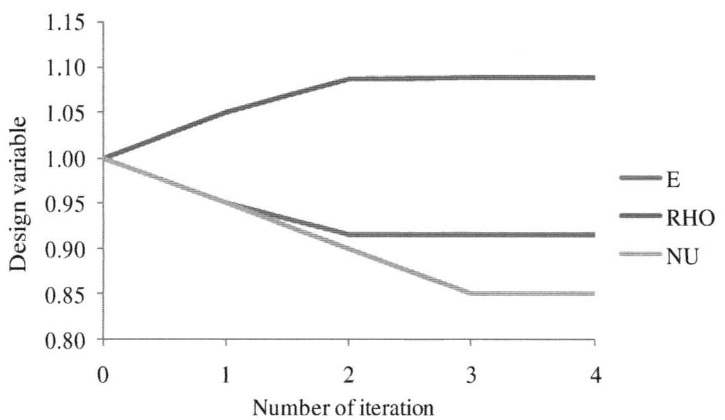

Figure 8. Graph of iteration.

Table 3. Comparison between initial FEA and updated FEA with measured data from EMA.

Mode	Natural frequency (Hz)				
	EMA	Initial FEA	Error (%)	Updated FEA	Error (%)
1	1480.00	1659.50	12.13	1527.45	3.21
2	2640.00	3061.40	15.96	2777.72	5.22
3	3480.00	3455.60	0.70	3190.52	8.32
4	3520.00	3873.00	10.03	3560.00	1.14
5	3700.00	3929.30	6.20	3611.93	2.38
Total average error			9.00		4.05

$$Changes\ value\ of\ parameters = \left|\frac{Updated\ value - Initial\ value}{Initial\ value}\right| \quad (4)$$

Table 4. Value of parameters based on design variable of model updating approach.

Parameter	Initial value	Updated value	Changes
Modulus of Young, E (GPa)	130	119	0.08
Density, ρ (kg/m^3)	7100	7728.35	0.09
Poisson's Ratio, v	0.26	0.22	0.15

6. Conclusion

The present work is identified the dynamic behaviour of disc brake rotor using modal analysis approach. The analysis has been carried out through numerical prediction method via finite element analysis (FEA) software. Since the accuracy issue of material properties used to assign in the modelling process made from assumption, the discrepancy occurred when compared with measured dynamic parameters. Modal testing was carried out to measure the structural response in order to verify the predicted result computed in FEA. In attempt to reduce the discrepancy between predicted and its measured counterpart, FE model updating technique applied in this study with respective updating parameters. After been updated iteratively, the discrepancy between EMA and FEA successfully minimized from 9.00 % of total average error to be 4.05 % of total average error. As conclusion, FE model updating approach is feasible to produce a reliable FE model in numerical analysis before been used for further analysis.

Acknowledgement

The authors of this paper would like to acknowledge a great support and encouragement by focus group of Advanced Structural Integrity of Vibration Research (ASIVR), Universiti Malaysia Pahang (UMP) for providing all the equipment used for this work Fundamental Research Grant Scheme (FRGS/1/2017/TK03/UMP/02-19) – RDU 170123. Special thanks for Ministry of Education (MOE) under Higher Education and Jabatan Pendidikan Politeknik and Kolej Komuniti (JPPKK) for financial assistance, support and study leave via Hadiah Latihan Persekutuan (HLP) scholarship.

References

[1] Ouyang H, Nack W, Yuan Y, and Chen F 2005 "Numerical analysis of automotive disc brake squeal: a review" *International Journal of Vehicle Noise Vibration* vol. 1 pp. 207-231

[2] Kharate N K and Chaudhari S S 2018 "Effect of Material Properties On Disc Brake Squeal And Performance Using FEM and EMA Approach" *Materials Today: Proceedings* vol. 5 pp. 4986-4994

[3] Nacivet S and Sinou J-J 2017 "Modal Amplitude Stability Analysis and its application to brake squeal" *Applied Acoustics* vol. 116 pp. 127-138

[4] Eslaminejad A, Ziejewski M, and Karami G 2019 "An experimental–numerical modal analysis for the study of shell-fluid interactions in a clamped hemispherical shell" *Applied Acoustics* vol. 152 pp. 110-117 doi:https://doi.org/10.1016/j.apacoust.2019.03.029

[5] Kihm F, Langelier A, and Munson K 2018 "Influence of the modal damping on the estimated fatigue life" *Procedia engineering* vol. 213 pp. 270-281 doi:10.1016/j.proeng.2018.02.028

[6] Huňady R, Pavelka P, and Lengvarský P 2019 "Vibration and modal analysis of a rotating disc using high-speed 3D digital image correlation" *Mechanical Systems and Signal Processing* vol. 121 pp. 201-214 doi:https://doi.org/10.1016/j.ymssp.2018.11.024

[7] Abdullah N A Z, Sani M S M, Rahman M M, and Zaman I 2017 "Correlation of numerical and experimental analysis for dynamic behaviour of a body-in-white (BIW) structure" *MATEC Web of Conferences* vol. 90 p. 01020 doi:10.1051/matecconf/20179001020

[8] Abdullah N A Z, Sani M S M, Rahman M M, and Zaman I 2015 "A review on model updating in structural dynamics" *IOP Conference Series: Materials Science and Engineering* vol. 100 pp. 1-6 doi:10.1088/1757-899X/100/1/012015

[9] Nehete D V, Modak S V, and Gupta K 2015 "Structural FE model updating of cavity systems incorporating vibro–acoustic coupling" *Mechanical Systems and Signal Processing* vol. 50-51 pp. 362-379 doi:http://dx.doi.org/10.1016/j.ymssp.2014.05.028

[10] Modak S V 2014 "Model updating using uncorrelated modes" *Journal of Sound and Vibration* vol. 333 pp. 2297-2322 doi:http://dx.doi.org/10.1016/j.jsv.2014.01.013

[11] Pradhan S and Modak S V 2012 "Normal response function method for mass and stiffness matrix updating using complex FRFs" *Mechanical Systems and Signal Processing* vol. 32 pp. 232-250 doi:http://dx.doi.org/10.1016/j.ymssp.2012.04.019

[12] Mottershead J E, Link M, and Friswell M I 2011 "The sensitivity method in finite element model updating: A tutorial" *Mechanical Systems and Signal Processing* vol. 25 pp. 2275-2296 doi:10.1016/j.ymssp.2010.10.012

[13] Zhu T, Tian W, Weng S, Ge H, Xia Y, and Wang C 2018 "Sensitivity-based finite element model updating using dynamic condensation approach" *International Journal of Structural Stability and Dynamics* vol. 18 pp. 1-23 doi:10.1142/S0219455418400047

[14] Mohd Zin M S, Rani M N A, Yunus M A, Wan Iskandar Mirza W I I, Mat Isa A A and Mohamed Z 2017 Modal and FRF Based Updating Methods for the Investigation of the Dynamic Behaviour of a Plate *Journal of Mechanical Engineering* 4 175–89

[15] Husain N A, Khodaparast H H, and Ouyang H 2010 "FE model updating of welded structures for identification of defects" *International Journal of Vehicle Noise and Vibration* vol. 6 pp. 163-175

[16] Jelani K M, Razip M H M, Nazri N A, Sani M S M, and Yasar M 2019 "Dynamics investigation on motorcycle chassis based on Finite Element (FE) modelling and updating" *IOP Conference Series: Materials Science and Engineering* vol. 469 pp. 1-10 doi:10.1088/1757-899X/469/1/012103

[17] Fouzi M S M, Sani M S M, and Muchlis Y 2019 "Finite Element Modelling and updating of welded joint for dynamic study of exhaust structure" *IOP Conference Series: Materials Science and Engineering* vol. 469 pp. 1-10 doi:doi:10.1088/1757-899X/469/1/012099

[18] AbuBakar A R and Ouyang H 2006 "Complex eigenvalue analysis and dynamic transient analysis in predicting disc brake squeal" *International Journal of Vehicle Noise Vibration* vol. 2 pp. 143-155

[19] Vinutha B S and Hemanth K 2017 "Study of static and frequency responsible analysis of hangers with exhaust system" *IJASRE* vol. 3

[20] Carne T G and Stasiunas E C 2006 "Lessons learned in modal testing—part 3: Transient excitation for modal testing, more than just hammer impacts" *Experimental Techniques* vol. 30 pp. 69-79 doi:10.1111/j.1747-1567.2006.00046.x

[21] Renault A, Massa F, Lallemand B, and Tison T 2016 "Experimental investigations for uncertainty quantification in brake squeal analysis" *Journal of Sound Vibration* vol. 367 pp. 37-55

[22] Rajadurai M S, Kavin R, Rejinjose, Prabhakaran, and Rajeshraman 2016 "A system approach to dynamic characteristics of hanger rod in exhaust system" *International Journal of Innovative Science, Engineering & Technology* vol. 3

[23] Fouzi M S M, Jelani K M, Nazri N A, and Sani M S M 2018 "Finite Element Modelling and Updating of Welded Thin-Walled Beam" *International Journal of Automotive and Mechanical Engineering* vol. 15 pp. 5874-5889 doi:10.15282/ijame.15.4.2018.12.0449

[24] Abdullah N A Z, Sani M S M, Hussain N A, Rahman M M and Zaman I 2017 Dynamics properties of a Go-kart chassis structure and its prediction improvement using model updating approach *International Journal of Automotive and Mechanical Engineering (IJAME)* 14 3887-97

[25] Sani M S M and Ouyang H 2016 Identification of Nonlinearities in Joints of a Wing Structure *MATEC Web of Conference* 83 (03006)

Identifying the Appropriate Frequency Response Function Driving Point of a Car Door Using Finite Element Analysis and Modal Testing

W I I Wan Iskandar Mirza[1], M N Abdul Rani[1*], M F Musa[1], M A Yunus[1], C Peter[1] and M A S Aziz Shah[1]

[1] Structural Dynamics Analysis and Validation, Faculty of Mechanical Engineering, Universiti Teknologi MARA (UiTM), 40000 Shah Alam, Selangor, Malaysia

*Corresponding author: mnarani@uitm.edu.my

Abstract. Identifying the appropriate frequency response function (FRF) driving point of complex structures is crucial to experimental modal analysis (EMA), but it is often found to be challenging and problematic. Traditionally, EMA analysts tend to place a reference sensor at several potential driving points and excite the points during testing. However, the approach is time-consuming. The aim of this work is to demonstrate an effective procedure for identifying the appropriate FRF driving point of a complex structure using finite element pre-test analysis and modal testing. The demonstration is performed on a car door structure comprising several geometrically complex structural components. The effective impedance method (EIM) is used to identify thirty potential driving points from the finite element model of the car door. The FRF data of the driving points is derived by using the FRF synthesis method and the derived data is compared with the EMA FRF data for validation purposes. Using EIM, three appropriate driving points covering all the modes required within the frequency of interest that is 0 to 100Hz has been successfully identified. The achievement suggests that the use of the effective impendence method for identifying the appropriate driving points is highly dependent on the accuracy and reliability of the finite element model.

1. Introduction

Identifying the appropriate frequency response function (FRF) driving points of a structure is a crucial to experimental modal analysists as the driving points are extremely important for excitations to be performed in modal testing [1,2]. However, when it comes to performing modal testing on a dynamically complex structure, identifying FRF driving points using modal testing has been seen to be more challenging and time-consuming [3–5]. Therefore, finite element pre-test analysis [6] is a good alternative approach for the identification..

The pre-test analysis can be performed using various methods as presented in [7]. One of the popular pre-test methods is the effective impedance method (EIM). Algorithms used in the method define a summation of all terms corresponding to particular DOF and yield vectors whose elements represent one DOF's contribution to all modes of interest [8] . The detailed information about the method can be found in [9].The method has been successfully used in identifying the optimal sensor placement for system identification and structural health monitoring as demonstrated in [10]. Another striking example as shown in [11] the optimal sensor placement for hyperbolic paraboloid fabric structure was successfully determined using EIM.

The pre-test analysis for a geometrically complex structure, however, requires a comprehensive procedure as the pre-test analysis is a case dependent [12]. This paper puts forward an idea of effectively identifying the appropriate FRF driving points of a complex structure using EIM pre-test analysis and experimental modal analysis. A geometrically complex car door consisting of several structural components is used for the identification. The accuracy of the pre-test analysis is evaluated by comparing the FE FRF with the measured counterparts obtained from modal testing

1.1. Description of the test structure

Figure 1 shows the car door that consists of several structural components. The structural components are assembled using spot welded joints. The structure is made of galvanized steel, and the thicknesses vary from 0.8mm to 1.5mm. The height and width of the structure are approximately 1000mm and 1200mm respectively. The mass of the structure is 13.6kg.

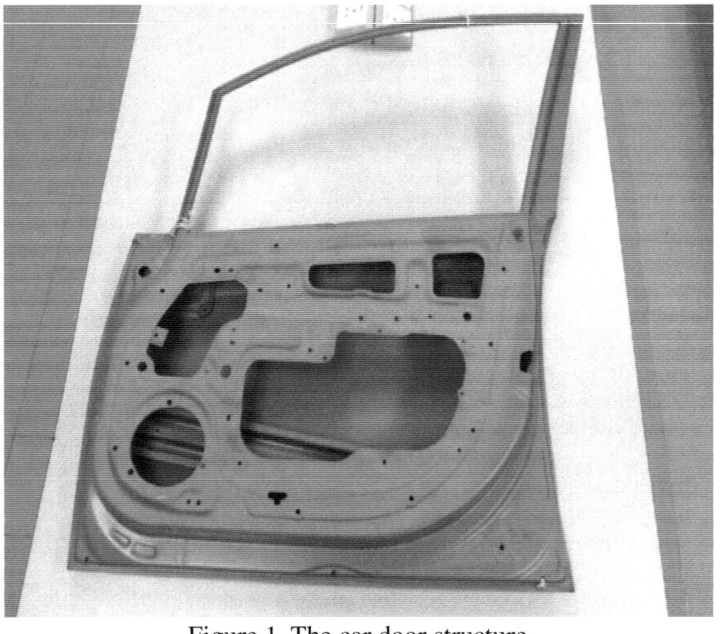

Figure 1. The car door structure

Initially, the dynamic behaviour of the structure is calculated using the finite element method. The FE model is then used to identify several appropriate measurement points. The test structure is investigated under free-free boundary conditions in accordance with previous studies [13,14]. The frequency of interest for the investigation is between 15Hz and 100Hz, which contains twelve modes. The rigid body modes are neglected during the investigation.

2. Finite Element Modelling

Figure 3 presents the finite element model of the car door. The model was developed approximately similar to the test car door in [15] . The model of the structure was discretised into 35000 2D shell elements, and the size of the elements varies from 7mm and 8mm. The standard material properties for galvanized steel [15,16] was defined to the elements as follows; density; 7.850 kg/, Poisson's ratio: 0.3, and Young's Modulus: 210,000 MPa. The natural frequencies and mode shapes of the finite element model were calculated using Nastran normal modes solution.

IOP Conf. Series: Journal of Physics: Conf. Series **1262** (2019) 012007 doi:10.1088/1742-6596/1262/1/012007

Figure 2. The finite element model of the car door

The effective impendence method (EIM) was used to identify and suggest the appropriate measurement points based on the modal data calculated from the FE model [9]. The identification was only carried out in a single direction that is Z-direction. The measurement points were determined with the aid of Siemens Virtual.Lab software. Figure 3 shows the suggested measurement points obtained from the FE model.

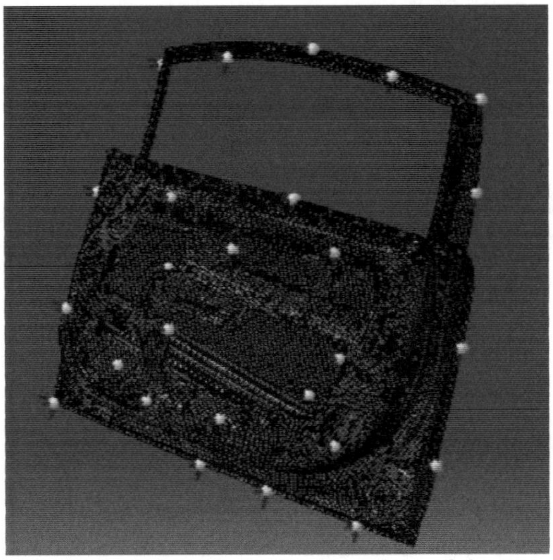

Figure 3. The suggested measurements points of the test structure

Since the FE model was developed based on assumptions, the FRF derived from the FE model is compared with the measured counterparts for validation purposes. The finite element FRF was derived using FRF synthesis method. For this method, the synthesized FRF matrix $H_{syn}(\omega_k)$ and mode shapes are expressed by:

$$H_{syn}(\omega_k) = \sum_{i=1}^{N} \frac{\{\emptyset\}_i\{\emptyset\}_i^T}{\left(\omega_{n_i}^2 - \omega_k^2\right) + j2\xi_i\omega_k\omega_{n_i}} \tag{1}$$

IOP Conf. Series: Journal of Physics: Conf. Series **1262** (2019) 012007 doi:10.1088/1742-6596/1262/1/012007

where N is the number of calculated modes, $\{\emptyset\}_i$ is the ith mass normalised mode shapes, ω_{n_i} is ith natural frequency and ξ_i is the ith modal damping ratio.

3. Test Configuration for the Test Structure

In this work, thirty drive points FRF were measured using experimental modal analysis (EMA). The experimental configuration of the test car door is illustrated in Figure 4. The experimental work was carried out by suspending the test car door using two soft springs to simulate free-free boundary conditions. This experimental configuration was designed as to the FE model [17–20]. The test car door was excited by using an impact hammer, and the resulting response was measured using a uniaxial accelerometer. The sensitivity of the transducers used was 21.65mV/N and 10mV/g respectively. Leuven Measurement System (LMS) SCADAS was used to acquire the dynamic data. In this study, the frequency bandwidth was set between 0 to 100 Hz with the frequency resolution of 0.5 Hz.

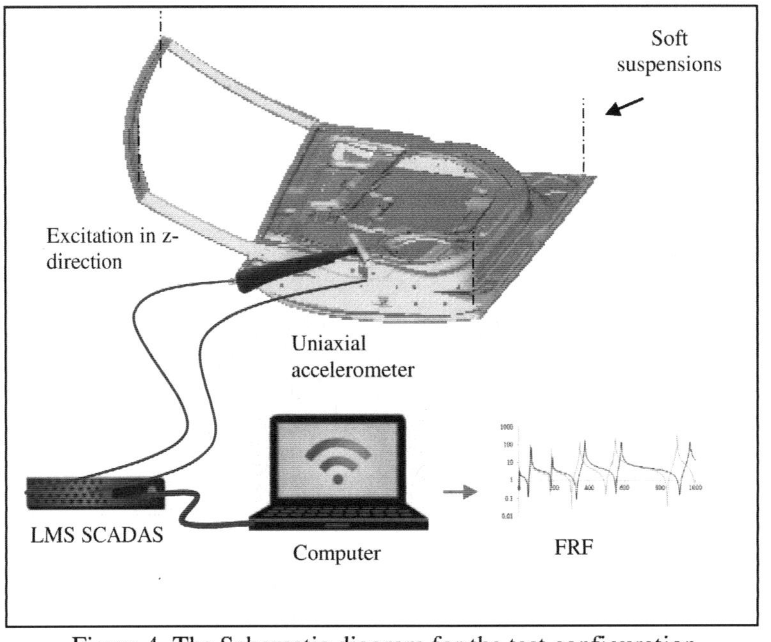

Figure 4. The Schematic diagram for the test configuration

4. Results and Discussion

The level of accuracy of the pre-test analysis is evaluated by comparing the FE FRF data with the measured counterparts. In this work, thirty sets of FRF data were acquired from the synthesised and modal testing. Based on the normal modes analysis carried out, twelve modes have been identified within the frequency of interest between 0 to 100 Hz and the data is presented in Table 1 in which the number of modes indicates the number of the resonance frequencies found in the FRF peaks.

Table 1. The modes of the test model within the frequency of interest

No. of Mode	Resonance Frequency, Hz
1	39.756
2	47.950
3	55.068
4	64.481
5	67.131
6	71.160
7	76.768
8	78.058
9	85.506
10	89.304
11	91.549
12	95.886

4.1. FRF data calculated from the FE model

The appropriate FRF driving point points are identified based on the visibility of the resonance peaks within the frequency of interest. Table 2 shows the patterns obtained from the comparison of data between modes and FRF driving points. Green colour indicates a good and clear peak, yellow shows a poor peak and red illustrates no peak at all. Meanwhile, blue colour confirms a good driving point covering all modes with a good quality of FRF. From Table 2, it can be seen that only 37 per cent (11 points) of the total of the driving points has yielded a good quality of FRF data with good and clear peaks. The rest have shown evidence of missing modes and no peaks at all. However, the 14th driving point has marked the lowest visibility of resonance peak.

Table 2. Pattern summary of the number of modes against a number of the FE FRF driving points

	FE Driving Points														
Mode	1	2	3	4	5	6	7	8	9	10	11	12	13	14	15
1															
2															
3															
4															
5															
6															
7															
8															
9															
10															
11															
12															

	FE Driving Points														
Mode	16	17	18	19	20	21	22	23	24	25	26	27	28	29	30
1															
2															
3															
4															
5															
6															
7															
8															
9															
10															
11															
12															

To reinforce the understanding of the pattern summary, two different driving points, which are points 4 and 14, were used in plotting FRFs as shown in Figure 5. It clearly shows that the FRF plotted from point 4, covering all resonance peaks required in this study is far superior to that of point 14 with missing several resonance peaks. The missing resonance peaks are at 55.07 Hz and 85.51 Hz. Therefore, point 14 may be inappropriate to be used for a driving point.

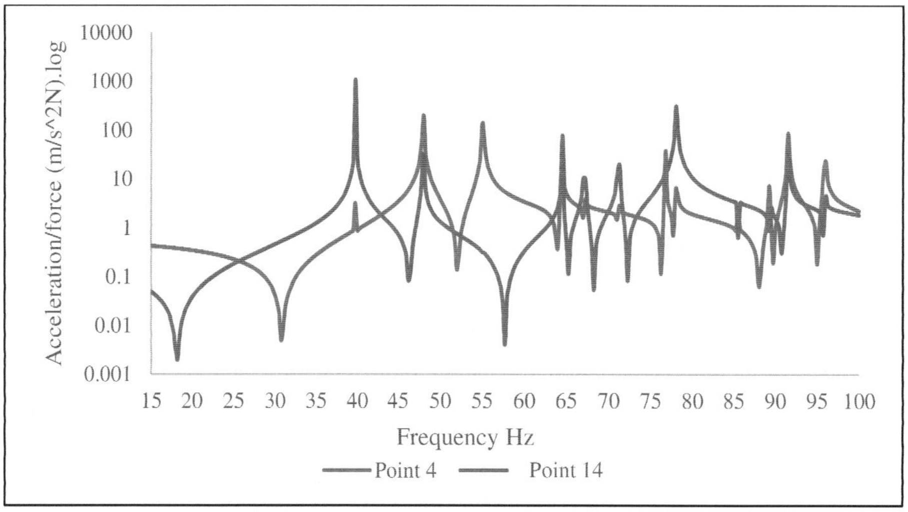

Figure 5. The FRF of point 4 and point 14

4.2. FRF data obtained from experimental modal analysis

As mentioned before, the FE model used was developed based on initial assumptions. The predicted results obtained from FE models sometimes differ from the measured counterparts. In this work, modal testing was performed on the car door to evaluate the accuracy of the numerically derived FRF driving points. Table 3 presents the summary of the number of modes related to the thirty points of measured driving points.

Table 3. Pattern summary of the number of modes against the number of the measured FRF driving points

	Measured Driving Points														
Mode	1	2	3	4	5	6	7	8	9	10	11	12	13	14	15
1															
2															
3															
4															
5															
6															
7															
8															
9															
10															
11															
12															

	Measured Driving Points														
Mode	16	17	18	19	20	21	22	23	24	25	26	27	28	29	30
1															
2															
3															
4															
5															
6															
7															
8															
9															
10															
11															
12															

From Table 3, it shows that most of the measured driving points have yielded a poor quality of results in FRF. However, only three driving points (points 16, 23 and 24) have evidence of good visibility of the resonance peaks, in contrast to the FE driving points, which it has gained 11 driving points. The comparisons of FRF data between the FE and measured driving points, particularly of points 16, 23 and 24 are depicted in Figures 6, 7 and 8.

IOP Conf. Series: Journal of Physics: Conf. Series **1262** (2019) 012007 doi:10.1088/1742-6596/1262/1/012007

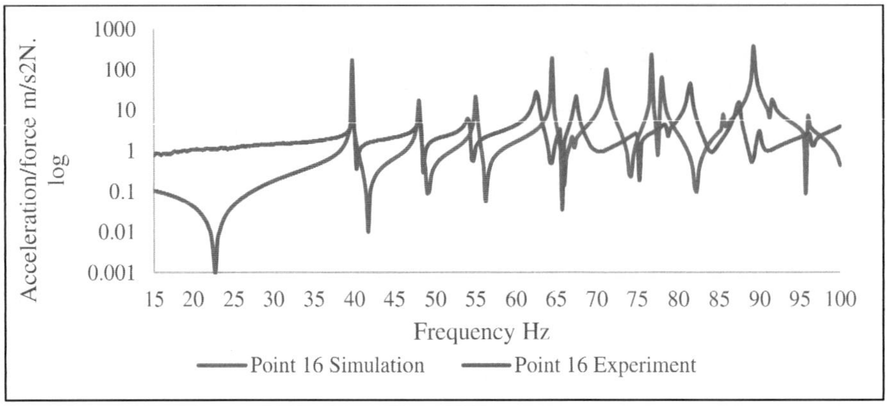

Figure 6. The comparison between the FE and measured FRF of point 16

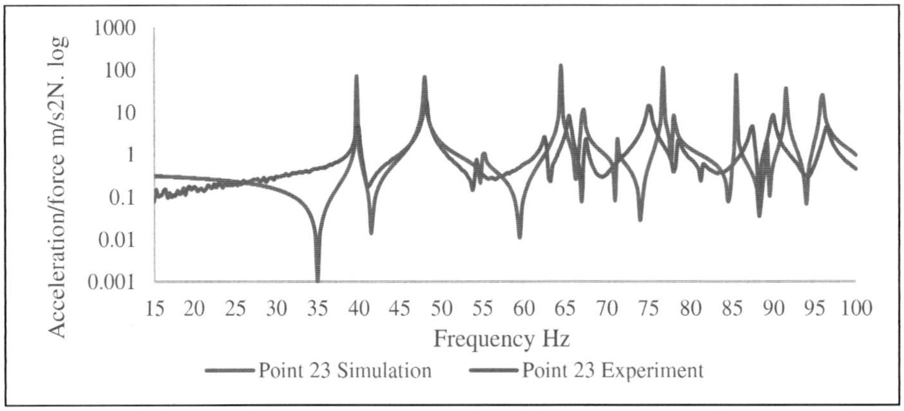

Figure 7. The comparison between the FE and measured FRF of point 23

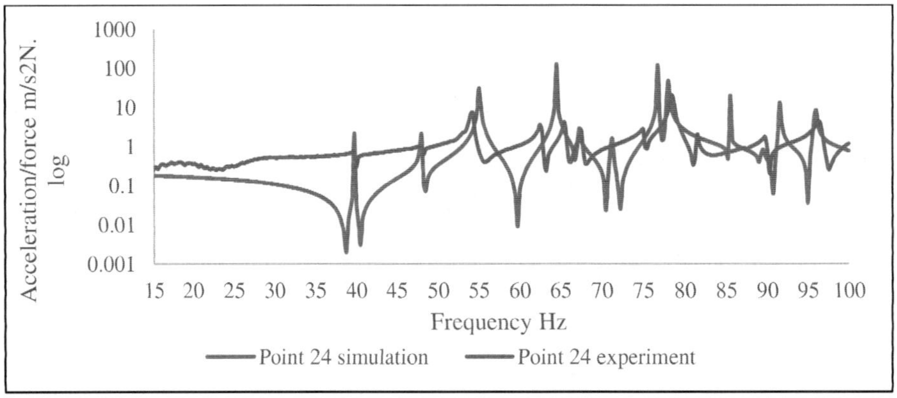

Figure 8. The comparison between the FE and measured FRF of point 24

From Figures 6, 7 and 8, the measured FRFs of the car door have low amplitude resonance peaks, indicating that the car door is a highly damped structure [21]. In addition, the FE FRF patterns presented in the figures are not in good agreement with the measured counterparts. This demonstrates that the FE model is incapable of accurately predicting the actual dynamic behaviour of the car door [22–24], whereas the FE model can be effectively used to identify the appropriate FRF driving points that are shown in Figure 9.

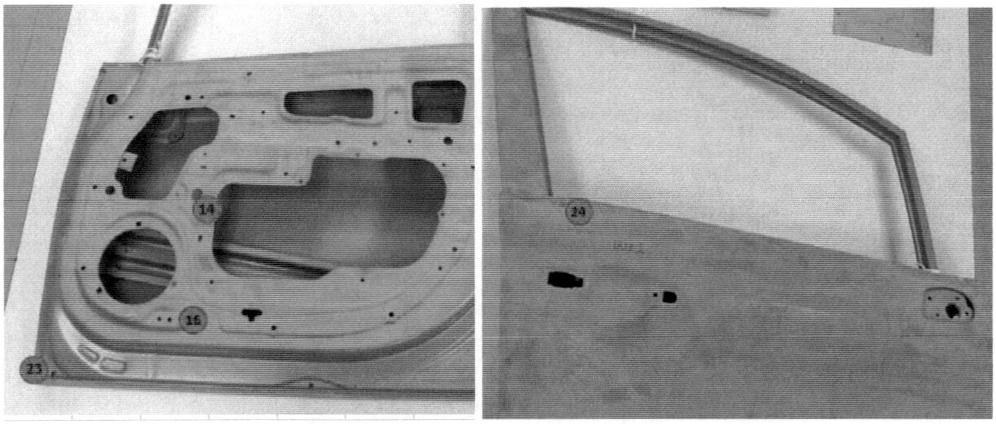

Figure 9. The location of the driving points 16, 23 and 14.

5. Conclusions and Recommendations

The procedure of identifying the appropriate FRF driving points of a car door using finite element and modal testing has been presented. The appropriate FRF driving points of the car door have been successfully identified using the effective impendence method, which is highly dependent on the accuracy and reliability of the FE model. The procedure demonstrated may be useful to the structural dynamics and experimental modal analysis community for confidently determining the dynamic behaviour of large, complex structures that are prevalent in the automotive and aerospace industries.

Acknowledgements

The authors are gratefully indebted to the Malaysian Ministry of Higher Education for providing financial support for this study through the GIP grant scheme (600-IRMI 5/3/GIP (012/2019). They would also like to express their gratitude for the help and support given by Prof. Ir. Dr. Hj. Abdul Rahman Omar, Prof. Dr. Hadariah Bahron, and SDAV group members.

References

[1] De Klerk D, Voormeeren S N and Petzsche T 2008 Enhanced FRF Determination using a 3D Impedance Head. *Proceedings of the Twentysixth International Modal Analysis Conference, Orlando, FL*

[2] Peter C, Rani M N A, Yunus M A, Zin M S M and Kalam A 2019 Inverse method for material properties identification of a kenaf reinforced composite Inverse Method for Material Properties Identification of a Kenaf Reinforced Composite *AIP Conference Proceedings* vol 020058

[3] Merkel R, C. G, B. K, Brown D and L. 1998 Important aspects of precise driving point FRF measurements using a mechanical impedance head sensor *Proceedings of the International Modal Analysis Conference - IMAC* **2** 795–9

[4] Lam H F, Yang J H and Hu Q 2011 How to install sensors for structural model updating? *Procedia Engineering* **14** 450–9

[5] Haapaniemi H, Luukkanen P, Nurkkala P, Rostedt J and Saarenheimo A 2003 Correlation Analysis of Modal Analysis Results from a Pipeline *International Modal Analysis Conference*

[6] Pickrel C R 1999 Practical approach to modal pretest design *Mechanical Systems and Signal Processing* **13** 271–95

[7] Riveros Jerez C A, García Aristizábal E F and Rivero Jerez J E 2013 A Comparative Study of Sensor Placement Techniques for Structural Damage Detection. *Revista EIA* **10** 23–37

[8] Zhang J, Maes K, De Roeck G, Reynders E, Papadimitriou C and Lombaert G 2017 Optimal sensor placement for multi-setup modal analysis of structures *Journal of Sound and Vibration* **401** 214–32

[9] Kammer D C 1991 Kammer, Daniel C. "Sensor placement for on-orbit modal identification and

correlation of large space structures." Journal of Guidance, Control, and Dynamics 14.2 (1991): 251-259. **14**

[10] Rao A R M, Lakshmi K and Krishnakumar S 2014 A generalized optimal sensor placement technique for structural health monitoring and system identification *Procedia Engineering* **86** 529–38

[11] Castro-triguero R, Macías E G, Gallego R, Sánchez Sánchez J and Vázquez Vicente E 2014 Optimal sensor placement methodology for operational modal system identification of a hyperbolic paraboloidal fabric *EWSHM - 7th European Workshop on Structural Health Monitoring* 1608–15

[12] Castro-Triguero R, Murugan S, Gallego R and Friswell M I 2013 Robustness of optimal sensor placement under parametric uncertainty *Mechanical Systems and Signal Processing* **41** 268–87

[13] Wan Iskandar Mirza W I ., Abdul Rani M N, Yunus M A, Ayub M A, Sani M S M and Mohd Zin M S 2019 Frequency Based Substructuringfor Structure with Double Bolted Joints: A Case Study *International Journal of Automotive and Mechanical Engineering* **16** 6188–99

[14] Omar R, Abdul Rani M N, Yunus M A, Mat Isa A A, Wan Iskandar Mirza W I ., Mohd Zin M S and Roslan L 2018 Investigation of Mesh Size Effect on Dynamic Behaviour of an Assembled Structure with Bolted Jointsusing Finite Element Method *International Journal of Automotive and Mechanical Engineering* **15** 5695–708

[15] Halim N I A, Yunus M A, Kasolang S and Rani M N A 2016 Finite element model of the spot welded joints of Door in White (DIW) *Journal of Engineering and Applied Sciences* **11** 2474–8

[16] Zin M S M, Rani M N A, Yunus M A, Ayub M A, Sani M S M and Shah M A S A 2019 Modal based updating for the dynamic behaviour of a car trunk lid *AIP Conference Proceedings* **2059**

[17] Mirza W I I W I, Rani M N A and Starbuck D P 2019 Using single-axis multipoint connection approach for coupling test and finite element model in the frequency based substructuring method *AIP Conference Proceedings* **020034**

[18] Zahari S N, Sani M S M, Husain N A, Ishak M and Zaman I 2016 Dynamic analysis of friction stir welding joints in dissimilar material plate structure *Jurnal Teknologi* **78** 57–65

[19] Wan Iskandar Mirza W I ., Abdul Rani M ., Othman M H, Kasolang S and Yunus M A 2016 Reduced order model for model updating of a jointed structure *Journal of Engineering and Applied Sciences* **11** 2383–6

[20] Omar R, Abdul Rani M N, Wan Iskandar Mirza W I ., Yunus M A and Othman M H 2017 Finite element modelling and updating for bolted lap joints *Journal of Mechanical Engineering* **SI 4** 202–22

[21] Bograd S, Schmidt A and Gaul L 2007 Modeling of damping in bolted structures *VDI Berichte* 97–100

[22] Yunus M A, Ouyang H, Abdul Rani M N and Misa A A 2013 Finite element modelling and updating of bolted joints in a thin sheet metal structure *20th International Congress on Sound and Vibration 2013, ICSV 2013* vol 2 pp 1146–52

[23] Yuen K 2012 Updating large models for mechanical systems using incomplete modal measurement *Mechanical Systems and Signal Processing* **28** 297–308

[24] Husain N A, Khodaparast H H, Snaylam A, James S, Dearden G and Ouyang H 2010 Finite-element modelling and updating of laser spot weld joints in a top-hat structure for dynamic analysis *Proceedings of the Institution of Mechanical Engineers, Part C: Journal of Mechanical Engineering Science* **224** 851–61

IOP Conf. Series: Journal of Physics: Conf. Series **1262** (2019) 012008 doi:10.1088/1742-6596/1262/1/012008

Noise Source Identification of Automotive Radiator by Using Sound Intensity Mapping Method

N F D Razak[1], M S M Sani[1,2*], W H Azmi[1,2] and B Zhang[3]
[1]Advanced Structural Integrity and Vibration Research (ASIVR), Faculty of Mechanical and Manufacturing Engineering, Universiti Malaysia Pahang, 26600 Pekan, Pahang, Malaysia
[2]Automotive Engineering Centre (AEC), Universiti Malaysia Pahang, 26600 Pekan, Pahang, Malaysia
[3]School of Mechanical Engineering, Ningxia University, 750021 Yinchuan, China

*Corresponding author: mshahrir@ump.edu.my

Abstract. Automotive radiator played an essential role as a heat exchanger that exchange the circulated heated coolant inside the engine compartment to return into its ambient temperature. Radiator cooling fan that attached at the back of radiator, is a device that work to regulate the coolant temperature when the heated coolant is passed inside the radiator tubes. It produces external flow of air that passes through the fins and tubes of the radiator. However, this external flow contributes high noise level into the system. In consequence of this risk, a detailed experimental study on noise analysis to identify the location of maximum sound pressure level (SPL) generated by variations of cooling fan speed by using sound intensity mapping method is presented. Ethylene Glycol (EG)-water based was used as working fluid operate with engine temperature range from 80 to 90 °C. The cooling fan speed of radiator cooling fan were varied from 750 to 1250 rpm and the water flowrates were 8.0, 11.0, and 14.0 l/min. The results indicate that the noise level is increasing with the fan speed. However, there is no significance difference of change in coolant flow rate and type of coolant fluids to the noise level. The location of maximum sound pressure level is identified and the higher SPL value most located on the blade trailing edge of the suction surface with highest SPL value of 81 db for 1250 rpm fan speed. The project delivers the reliable input for the engineering practice to reduce automotive radiator noise level.

1. Introduction

Noise is usually generated by the vibrating body which the vibrations is propagate through the medium to the listeners. It is also such a common part of everyday life [1]. When sound produced, energy have transferred from sound source to the surrounding air molecules takes place. The energy transfer is called Sound Power that have unit of sound power is W (Watt). The audible range of sound power is from 10^{-9} W to 1000 x 10^{-9} W is the lowest level which can heard by people that are closed to the source and 1000 W can immediately create hearing damage. Lower level of sound power also can create hearing damage but in long period of time.

Kang concluded that sound have affected human in positive and negative way. In positive way, sound can be defined as pleasant and relaxing [2]. In other way, the unpleasant and unwanted sound is called noise which are occurred in high decibel (dB) frequency. Noise affected people in everyday life in different ways depending on the type of noise, noise level, frequency characteristics, time of day and variance over time.

Beranek and Ver discovered that when a source produces sound power (P) it will created a certain sound intensity (I) at distance away from the source [3]. The intensity is a measure for the amount of power through a certain area of the distance. Sound intensity is the rate of acoustic energy flow per unit area. It is also a current of acoustic energy through medium that have magnitude and direction. Sound intensity mapping is by integrated over enclosing surface of an object and its sound power. Sound intensity measurement is able to determine the sound directional and localise noise hot spot. Mapping software have provided a contour map of noise level that are provided intuitive documentation of noise problems.

According to Kim and Thompson, noise measuring devices are typically used a sensor such microphones to receive noise signals from source [4]. However, sensor not only detect noise from the source, but it is also detected any ambient background noise. Sound measurement for machine is running, it is important to make the background noise in lower level to ensure any influence on the result. Therefore, the anechoic chamber is needed to measure the source of noise.

Hanif and Rasid stated that sound intensity probe involves two closely pressure microphones where the both sound pressure and pressure gradients can be measure between the microphones. By sound intensity analyzer, it is necessary to convert these signals into sound intensity values [5]. Ginn and Haddad emphasized that Noise source identification (NSI) techniques are used to optimise the noise emission from a wide range of products including vehicles, household goods and wind turbines [6]. Noise source have two categories which are vibration-induced noise and flow-induced noise. First category has included noise from rotating machinery, noise due to structural resonance, impact noise, and others. Then second category included fan noise, pump noise, jet noise, etc.

Other than that, NSI is sound field visualization techniques and find applications in non-destructive evaluation, underwater imaging and machine diagnosis. NSI techniques is useful in estimating the sources position and sound strength. Sani et al. [7] stated that other than conduct the experiment in an anechoic chamber room, this method can be applied on-site as long as the background noises are stationary. NSI is also to identify the major sources of noise radiation. The best reduction in noise can be achieved by reducing the noise that are radiated from the noisiest sources. The reduction noise is required not only the treatment of many different sources but also called for different approaches in each particular case depending upon the degree of the contribution of the various individual sources to the total noise.

Davy [8] presented that there will be energy flow in certain directions but not in others because sound intensity gives a measure of direction. Thus, sound intensity is vector quantity as it has both magnitude and direction whereas pressure is a scalar quantity as it has magnitude only. It is important to measure the intensity in normal direction which is at 90° to specified unit area through the sound energy flowing. Fahy [9] analyzed that it can be simply to measure the normal spatial averaged since Intensity, I is the power per area. This is according to intensity is over an area that encloses the source and then multiply it by the area follows the inverse square law for free field propagation to find the sound power intensity and pressure. Most of researchers revealed that contour plots give a more detailed picture of the sound field generated by a source which explained the intensity field intersecting the hood of crawler dozer [10-13]. Several sources and sinks can then be identified with accuracy. A grid is set up to define a surface.

Overall, this work aims to develop guidelines that can be used for noise analysis and advancement by process engineers to implement satisfactory operating flow ranges for their existing equipment. Data was collected and processed with commercially available software and transducers hence to emphasize the role of experimental activity in identifying possible sound power level problems, which could affect overall vehicle harmony.

2. Experimental setup and methodology

The experimental methodology is starting from fabrication of automotive radiator test rig, then followed by preparation of coolants and experimental setup acoustic devices in Advance Automotive Liquid Laboratory (A2LL). Next, it is followed by sensors calibration of intensity mapping prop. Noise measuring devices are used a sensor which is a pair of ½ inch microphone to receive noise signals from source. Then, the measurement intensity mapping can be conducted. The total sound power is computed and all of the experimental data is analyzed and evaluate according to the standard of experiment.

2.1. Automotive radiator test rig setup

The experimental facility consists of a car radiator (test section), coolant tank with heater, water pump with control valve, cooling fan with motor speed regulator, flow meter, temperature signals data acquisition system, and temperature sensor. A complete automotive radiator test rig to study the noise measurement analysis of automobile radiator is shown in figure 1 below.

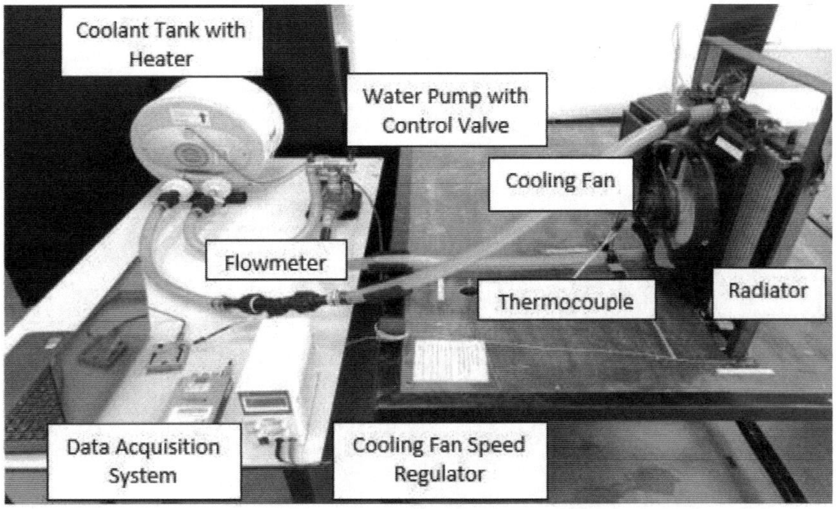

Figure 1. Automotive radiator test rig.

2.2. Test section (radiator system)

The test section is a conventional aluminium car radiator and its dimensions are 34.0×1.3×36.8 mm. It was consisted of 32 vertical aluminium tubes. The gap between the tubes is filled with thin perpendicular louvered fins as shown in the figure 2(a). Air was forced to flow through the fins of the radiator using a conventional forced draft fan with 5 blades in the range of 100 to 1250 rpm as shown in figure 2(b). The radiator cooling fan is made up of an electric motor with flange-mounted fan wheel. The cooling fan and its motor are mounted behind the radiator. Due to the air flow produced, the cooling fan withdraw heat from the coolant.

(a) (b)

Figure 2. (a)Automotive radiator, (b)Radiator cooling fan.

2.3. Preparation of acoustic device and measurement

Sound intensity mapping measurement is to determine the sound power level of noise that produced by radiator cooling fans. A special acoustical facility which is an anechoic chamber room is required to measure the accuracy of sound. However, the advantage of using this method is it can be applied on-site rather than in an anechoic chamber as long as the background noises are stationary. Fortunately, by using hand-held analyzer there's only noise from the source are detected. Sound measurement for machine is running, it is important to make the background noise in lower level to ensure any influence on the result.

2.4. Hand-held analyzer

The Bruel & Kjaer Hand-held analyzer Type 2270 as shown in figure 3, is an instrument that allow the study of sound intensity and sound power measurements to be perform. The FFT analysis of sound and vibration and sound intensity mapping results can be execute using this measurement channels.

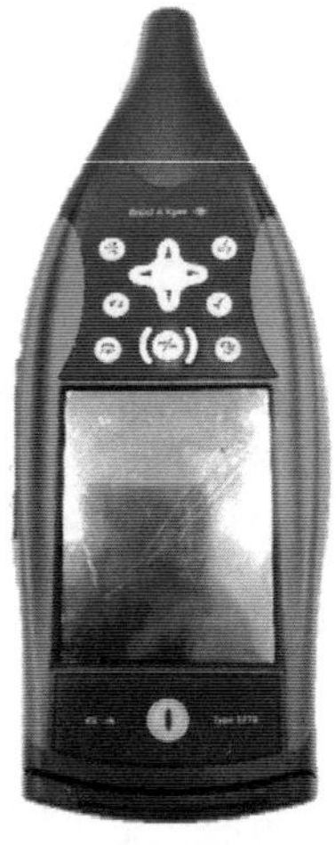

Figure 3. Hand-held analyzer type 2270.

2.5. Sound intensity microphone pair

Figure 4 shows the Sound intensity microphone pair Type 4197 is equipped with 8.5 mm, 12 mm and 50 mm spacers and 8.5 mm spacer was used in this study.

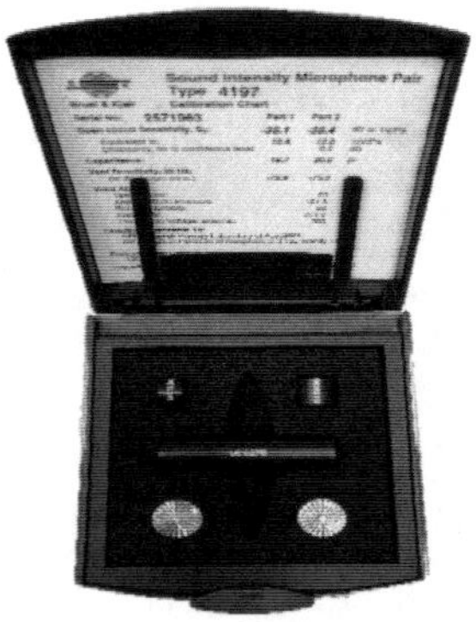

Figure 4. Microphone pair type 4197 and spacers.

The sensitivity of the transducer as well as the response of the electronic circuitry can vary slightly over time or could be affected by environmental conditions such as temperature and humidity. Therefore, before the experiment is conduct a calibration is important for an adjustment of hand-held analyzer to measure and display correct values. The calibration for microphone is followed calibration chart from Bruel & Kjaer type 4297 using Sound Calibrator Type 4231 as shown in figure 5.

Figure 5. Sound calibrator type 4231.

Figure 6 shows the microphone is calibrated until it reaches sound level around 94.0 dB. The calibration data provided included phase matching up to 1/3 octave centre frequency of 6.3 kHz, sensitivities at 250 Hz, actuator responses and individual free field frequency responses valid for the microphones mounted on a 1/4 preamplifier.

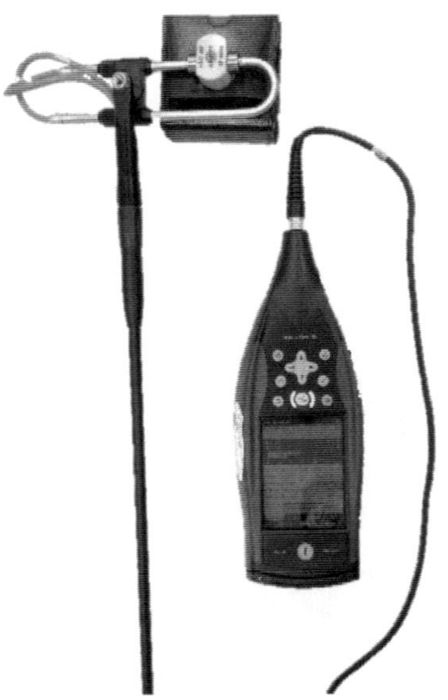

Figure 6. Calibration of hand-held analyzer Type 2270.

2.6. Experimental procedure of sound intensity mapping

The noise analysis of car radiator cooling fan in this study was set up properly in the Advance Auto Liquid Laboratory (A2LL) before the experiment is conduct. The devices and instruments used in this noise investigation are Bruel & Kjaer intensity probe, laptop, a pair of microphones, and hand held analyzer. First, before the experiment of noise source identification is conducted, a rectangular grid was directly set up by taking a photo of the measurement configuration first which is the car radiator by using the hand-held analyzer then 3x3 grid system is plotted to define its surface as shown in figure 7.

IOP Conf. Series: Journal of Physics: Conf. Series **1262** (2019) 012008 doi:10.1088/1742-6596/1262/1/012008

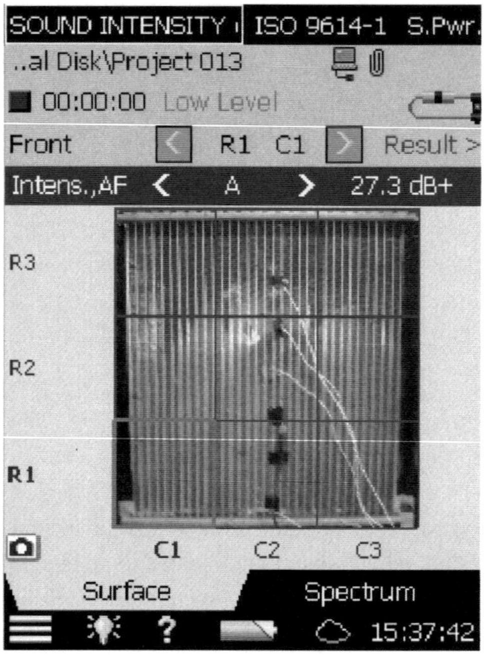

Figure 7. Rectangular grid of car radiator by Hand-held analyzer.

The microphones of sound measurement are placed 1 m away according to standard of noise measurement method as shown in figure 8.

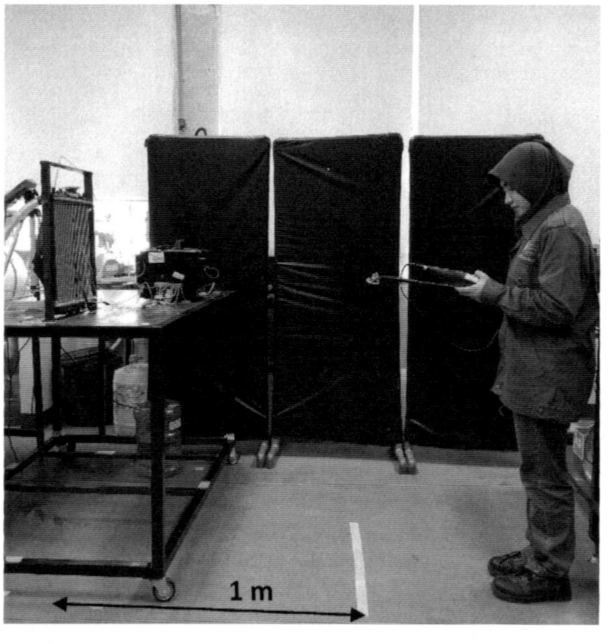

Figure 8. Noise measurement technique during the experiment.

Figure 9 shows sound intensity technique were carried out from the front side of the radiator only. This is according preliminary studies, that location contribute high noise than other sides.

Figure 9. Front view of sound intensity measurement.

During the experiment, the radiator cooling fan as shown in figure 10 plays a key role in this study. It is necessary to know the speed of the cooling fan when the fan is in working condition. A tachometer is used to detect the cooling fan speed reading due to different coolant use will result in different fan speed hence influenced the noise and vibration results.

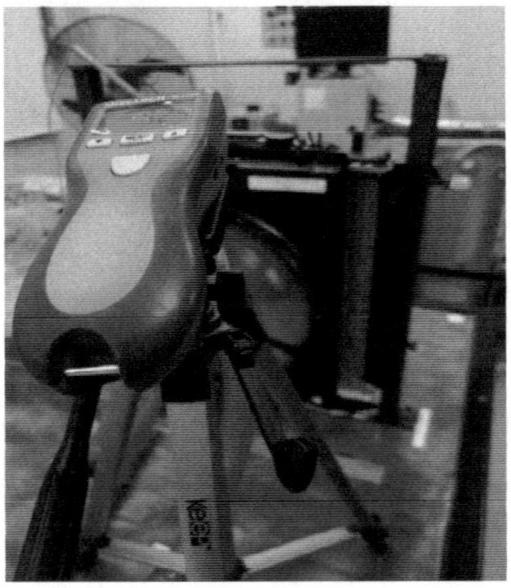

Figure 10. Cooling fan speed measurement using a tachometer.

Finally, the data obtained already displayed on the screen of hand-held analyzer and is interpreted by using post processing software, PULSE LabShop Version 16.1.0. The noisier part of the radiator is determined by observing the colour of its contour mapping. The experiment is repeated for another cooling fan speeds and coolant flowrates.

3. Results and Discussion

Intensity mapping is the contour that explains a more detailed picture of sound field generated by a noise source hence giving a quick overview of the sources in the selected frequency range. The noise measurements are performed in an alternative energy laboratory with less than 20 to 30 dB background noise and already filtered by the hand-held analyser itself. The cooling fan speed is varied from 750 to 1250 rpm (critical condition) and coolant flowrates were 8.0, 11.0, and 14.0 l/min (critical condition). The comparison data between different fan speed also shown in this section.

Each contour mapping presented in this section corresponds to the pressure fluctuation on the suction surface of the rotor blade, which was induced by a different fluid flow through the test section. Figure 11 presents a contour plot of the sound pressure level produced when radiator system operates at constant flowrate 8.0 l/min. It can be seen at 750 rpm of cooling fan speed as shown figure 12(a), maximum sound pressure level located at point 2 with 66.29 dB in the red oval shape area. The lowest

SPL value located at point 1 with 55.97 dB. The SPL value produced ranges from 55 to 66 dB. Next, figure 2(b) shows contour mapping when the cooling fan speed increased to 1250 rpm which is the critical condition in typical operating rate. The maximum SPL value occurred at point 9 with 77.14 dB represented by green colour area and lowest value occurred at point 7 with 66.77 dB. The SPL value produced ranges from 66 to 77 dB.

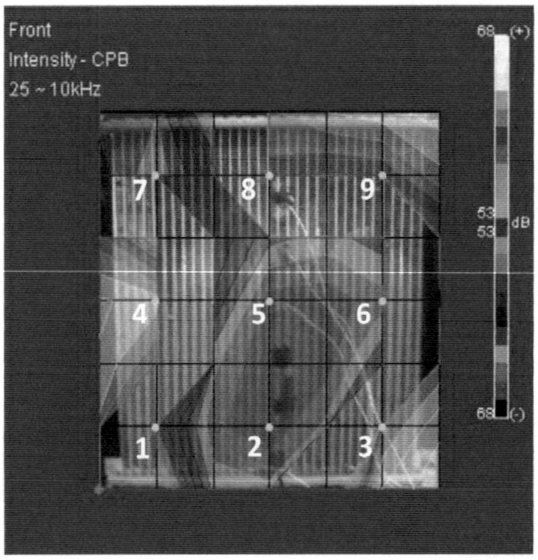

(a)

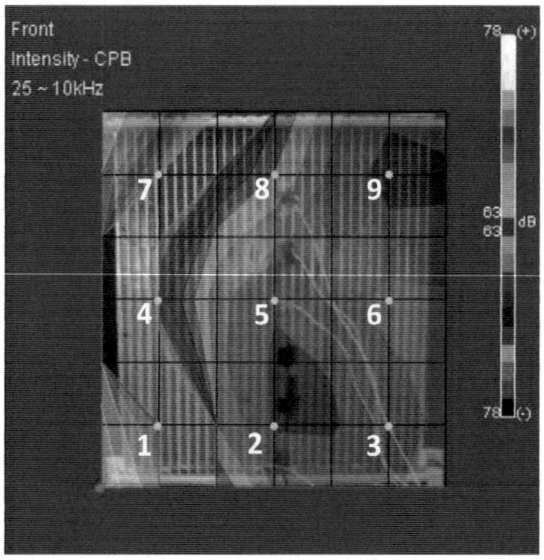

(b)

Figure 11. Contour mapping at 8.0 l/min for (a)750 rpm, (b) 1250 rpm

Table 1 shows the details value of noise level at different location to determine where the highest sound signal radiated from the radiator surface.

Table 1. Sound pressure level at 8.0 l/min.

Location	Sound Pressure Level (dB)	
	750 rpm	1250 rpm
1	55.97	70.01
2	66.29	76.68
3	62.04	75.72
4	64.24	68.03
5	65.74	76.37
6	64.15	74.47
7	63.17	66.77
8	59.90	72.12
9	59.21	77.14

Based on data comparison presented in figure 12, 1250 rpm contribute higher SPL value than 750 rpm at all points of location due to pressure fluctuation at the suction surface of radiator is increasing with fan speed at this constant flowrate 8.0 l/min. This finding is in agreement with the previous studies of sound radiation investigated by Au and Wang [14] which they identified that the higher the speed will generates larger sound pressure.

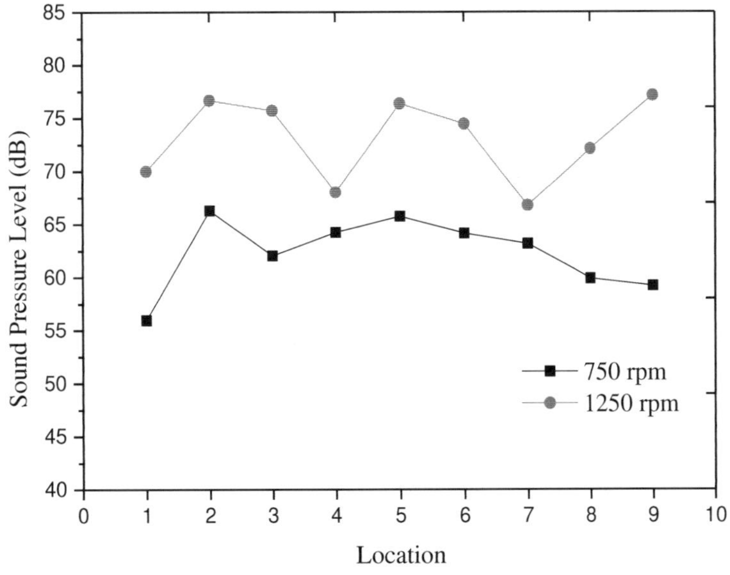

Figure 12. Sound pressure level comparison at 8.0 l/min.

Figure 13 presents a contour plot of the sound pressure level produced when radiator system operates at constant flowrate 11.0 l/min. It can be seen at 750 rpm of cooling fan speed as shown figure 14(a), maximum sound pressure level located at point 1 with 73.6 dB represented by yellow area. The lowest SPL value located at point 2 with 56.25 dB. The SPL value produced ranges from 56 to 73 dB. Next, figure 14(b) shows contour mapping when the cooling fan speed increased to 1250 rpm which is the critical condition in typical operating rate. The maximum SPL value occurred at point 5 with 78.5 dB represented by green colour area and lowest value occurred at point 4 with 66.15 dB. The SPL value produced ranges from 66 to 78 dB.

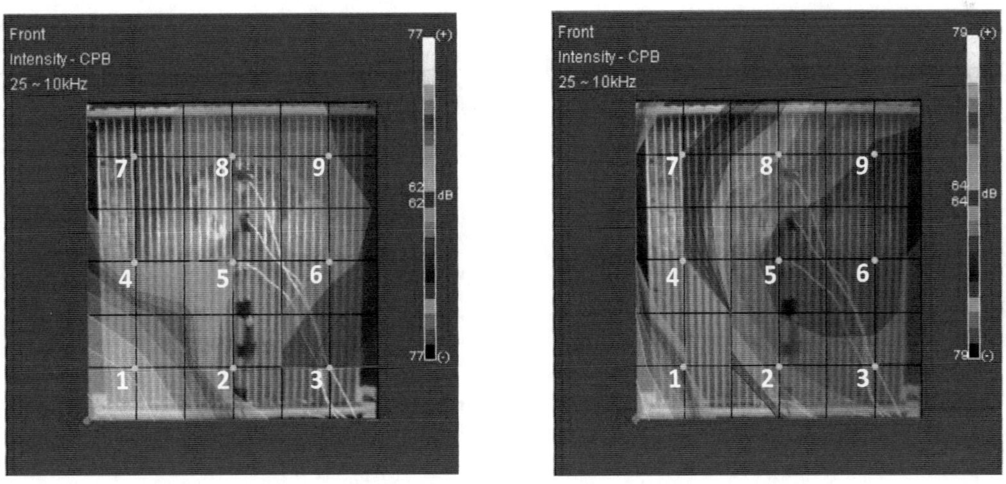

(a) (b)

Figure 13. Contour mapping at 11.0 l/min for (a) 750 rpm, (b) 1250 rpm

Table 2 shows the details value of noise level at different location to determine where the highest sound signal radiated from the radiator surface.

IOP Conf. Series: Journal of Physics: Conf. Series **1262** (2019) 012008 doi:10.1088/1742-6596/1262/1/012008

Table 2. Sound pressure level at 11.0 l/min.

Location	Sound Pressure Level (dB)	
	750 rpm	1250 rpm
1	73.60	70.17
2	56.25	75.69
3	67.17	77.20
4	61.56	66.15
5	64.24	78.50
6	63.96	77.73
7	60.14	69.75
8	63.46	74.18
9	64.88	77.76

Based on data comparison presented in figure 14, 1250 rpm contribute higher SPL value than 750 rpm at most points of location due to pressure fluctuation at the suction surface of radiator is increasing with fan speed at this constant flowrate 11.0 l/min.

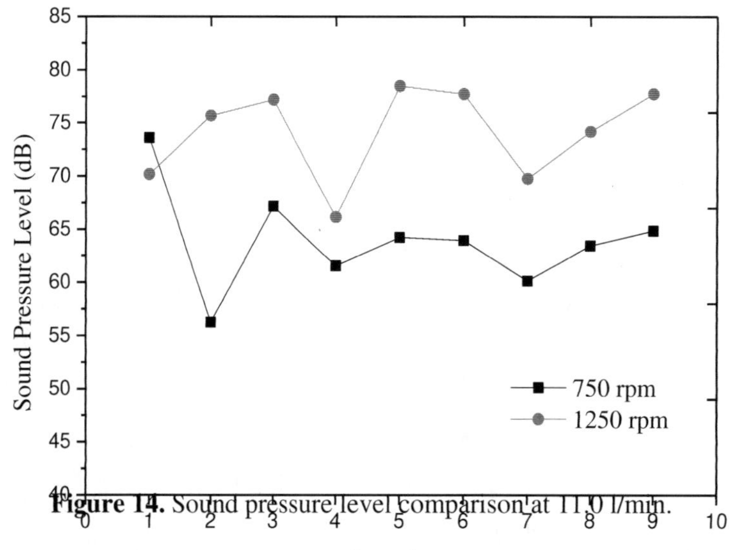

Figure 14. Sound pressure level comparison at 11.0 l/min.

Figure 15 presents a contour plot of the sound pressure level produced when radiator system operates at constant flowrate 14.0 l/min. It can be seen at 750 rpm of cooling fan speed as shown figure 16(a), maximum sound pressure level located at point 5 with 68.62 dB represented by light yellow area. The lowest SPL value located at point 7 with 60.65 dB. The SPL value produced ranges from 60 to 68 dB. Next, figure 16(b) shows contour mapping when the cooling fan speed increased to 1250 rpm which is the critical condition in typical operating rate. The maximum SPL value occurred at point 6 with 78.73 dB represented by green colour area and lowest value occurred at point 7 with 66.91 dB. The SPL value produced ranges from 66 to 78 dB.

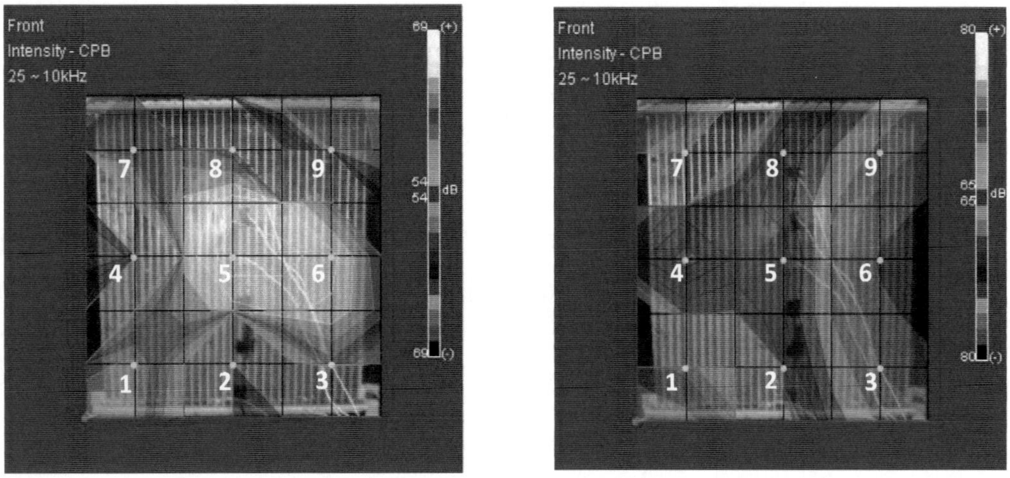

(a) (b)

Figure 15. Contour mapping at 14.0 l/min for (a)750 rpm, (b) 1250 rpm

Table 3 shows the details value of noise level at different location to determine where the highest sound signal radiated from the radiator surface.

Table 3. Sound pressure level at 14.0 l/min.

Location	Sound Pressure Level (dB)	
	750 rpm	1250 rpm
1	63.55	70.50
2	61.43	69.56
3	60.92	77.08
4	61.18	73.23
5	68.62	70.00
6	66.56	78.73
7	60.65	66.91
8	61.12	71.45
9	64.67	76.00

Based on data comparison presented in figure 16, 1250 rpm contribute higher SPL value than 750 rpm at all points of location due to pressure fluctuation at the suction surface of radiator is increasing with fan speed at this constant flowrate 14.0 l/min. This finding is follow the expected trend which is supported by Kim et al. [15] where they studied Noise Source Identification (NSI) of small fan-BLDC motor system for refrigerators. Kim et al. analysed that the overall value of noise will increase as the speed of fan increases.

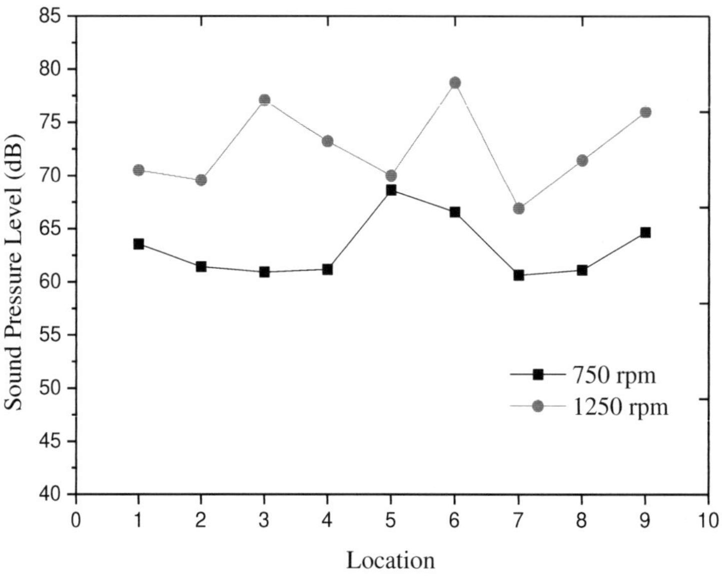

Figure 16. Sound pressure level comparison at 14.0 l/min.

Figure 17 presented data comparison between different flowrate varying from 8.0 to 14.0 l/min at constant speed of 750 rpm and 1250 rpm. Both figures indicate that there is no significant difference of change in coolant flowrate in radiator system to the noise level. Other sources of vibration are hydraulic imbalance, internal recirculation, suction, cavitation, vane passing forces and conditions resulting from pump installation in the system, which significantly affects the noise level due to the structural and air borne noise, and noise radiation behavior with respect to the excitation mechanism.

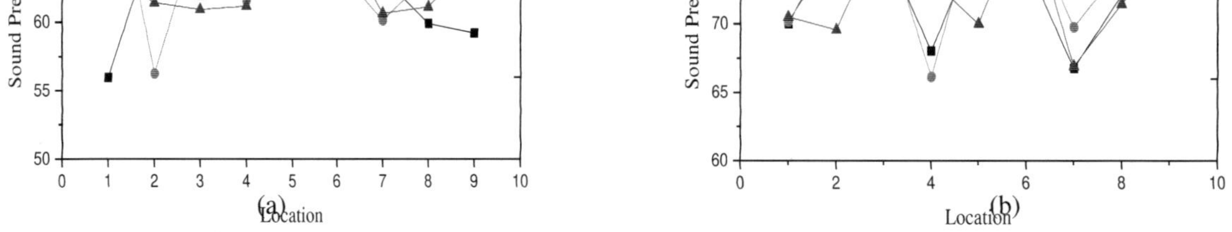

Figure 17. Sound pressure level comparison at (a)750 rpm, (b)1250 rpm

4. Conclusions

Based on measurement result, the location of maximum noise level is identified. It is also analysed that the change of radiator cooling fan speed increases the vibration induced into the system other than fluid flowrate. The results indicate that maximum sound pressure level located on the blade trailing edge of the suction surface. The noise level also is increasing with fan cooling speed where 1250 rpm contribute higher noise level than 750 rpm. The finding also indicates there is no significant difference of change in coolant flowrate to the noise level. On the other hand, sound intensity mapping is an effective method to investigate the maximum noise level that are produced by the automotive radiator. The guidelines that can be used for noise analysis and advancement by process engineers to implement satisfactory operating flow ranges for their existing equipment can be developed through this project. In future, the advancement of vibration in automotive industry for better and safer radiator can be assisted by this study.

Acknowledgments

The author would like to greatly acknowledge the support by focus group of Advanced Structural Integrity of Vibration Research (ASIVR) and Universiti Malaysia Pahang for providing all the equipment used for this project under UMP research grant scheme – RDU 180381.

References

[1] Bruel K C 1984 Measuring Sound. (Naerum, Denmark: DK)
[2] Kang J 2006 *Urban sound environment*: CRC Press)
[3] Beranek L L and Ver I L 1992 Noise and vibration control engineering-principles and applications *Noise and vibration control engineering-Principles and applications John Wiley & Sons, Inc., 814 p.* **1**
[4] ISVR K K and ISVR D T 2007 Noise Source Identification Using Microphone Arrays
[5] Hanif M and Rasid A 2008 Noise source identification of split unit air conditioner system. Universiti Malaysia Pahang)
[6] Ginn K B and Haddad K 2012 Noise source identification techniques: simple to advanced applications. In: *Acoustics 2012,*
[7] Sani M S M, Rahman M M , Baharom M Z and Zaman I 2015 Sound Intensity Mapping of an Engine Dynamometer *International Journal of Automotive and Mechanical Engineering (IJAME)* **11** 2820-9
[8] Davy J L 2010 Building Acoustics and Vibration—Theory and Practice *The Journal of the Acoustical Society of America* **127** 1168-9
[9] Fahy F 2002 *Sound intensity*: CRC Press
[10] Washburn K B, Frazer T and Kunio J 2005 Correlating noise sources identified by beamforming with sound power measurements. SAE Technical Paper
[11] Sani M S M, Rahimi M A F A and Mohamed D 2019 Sound Intensity Mapping of Two Stroke Engine by Using Hemispherical Surface Coordinate Arrangement *IOP Conf. Ser. Mater. Sci. Eng.* 506 012060
[12] Razak N F D, Sani M S M, Azmi W H and Zhang B 2017 Noise and vibration analysis for automotive radiator cooling fan *IOP Conf. Ser. Mater. Sci. Eng.* 257 012083
[13] Sani M S M, Zaman I and Rahman M M 2015 Analysis of Split Air Conditioner Noise using Sound Intensity Mapping *Journal of Advanced Research in Fluid Mechanics and Thermal Sciences* 9(1) 28-33.
[14] Au F T K and Wang M F 2005 Sound radiation from forced vibration of rectangular orthotropic plates under moving loads *Journal of Sound and Vibration* **281** 1057-75
[15] Kim Y-H, Yang B-S and Kim C-J J I J o R M 2006 Noise source identification of small fan-BLDC motor system for refrigerators

IOP Conf. Series: Journal of Physics: Conf. Series **1262** (2019) 012009 doi:10.1088/1742-6596/1262/1/012009

Identification of the nonlinearity in mass-spring system via experimental method

A R Bahari[1], M A Yunus[1*], M N Abdul Rani[1], M A S Aziz Shah[1], W I I Wan Iskandar Mirza[1], N Salleh[2]

[1]Structural Dynamics, Analysis and Validation, Faculty of Mechanical Engineering, Universiti Teknologi MARA (UiTM), 40000 Shah Alam, Selangor
[2]National Space Agency of Malaysia, Pusat Angkasa Negara, 42700 Banting, Selangor

[*]Corresponding author: mayunus@salam.uitm.edu.my

Abstract. Engineering structures operate with a nonlinear dynamics behaviour at certain amplitude of input range which will produce noticeable changes and unpredictable effects. In this paper, the nonlinear property of a spring is investigated. The simplified mass-spring system is developed which consists of two springs attached on a thin plate structure. The translational degree of freedom is considered in the experimental and a preferential direction on the length direction of the spring is investigated. The linear property is first investigated via impact testing for the acceleration measurement. Subsequently, the spectral testing with different amplitude of periodic chirp excitations is carried out to identify the nonlinear property of the springs. The nonlinear dynamics behaviour of the springs are identified from the significant changes of natural frequencies from the frequency response function of the system.

1. Introduction

Structural modelling, computing and testing are very crucial when designing, constructing and maintaining the mechanical systems and engineering structures. However, the approximation approach by assuming the linear model is mostly applied in structural dynamics analysis [1-3]. However, the linear model is consequently invalid and the nonlinear effects are compulsory to take into account when these structures are subjected to dynamics loading with large displacement amplitudes [4-6].

The input energy and the system property are the main factors for the occurrence of nonlinear behaviour of engineering systems and structures [7-8]. Effective investigation is required before the nonlinear dynamic behaviour can be included in the numerical calculations since it is an indicator for the development of structural damage under dynamic excitation.

For example, in the operational of the aerospace flight, the aeronautical fuel consumption reduction can be achieved by increasing the wing aspect-ratio to enhance the flight efficiency while reducing emissions. The improvement in the lift-to-drag ratio however will result in higher deflections which can subsequently cause the nonlinear aero-elastic behaviour [9-11]. Several factors that contribute to the nonlinear behaviour in the mechanical systems are such as large displacement of the system, crack initiation and development, as well as the looseness and presence of friction characteristics of structural joints [12-14].

Nonlinear system identification can be identified numerically and experimentally. The identification in structural dynamics is classified into several categories, namely linearisation, time and frequency domain methods, time-frequency analysis and modal methods [15].

IOP Conf. Series: Journal of Physics: Conf. Series **1262** (2019) 012009 doi:10.1088/1742-6596/1262/1/012009

The structural nonlinear restoring forces can be identified using only partial measurements of the structural responses. The method combines the identification of locations of structural nonlinearities by the introduction of equivalent linear structural system for an original nonlinear system. Then, a power series polynomial is used to approximate structural nonlinear restoring force using the extended Kalman filter to obtain the unknown coefficients of the power series polynomials [16].

A method for detection and characterisation of nonlinear behaviour from a single frequency response function has been presented by [17]. A set of nonlinearity indexes is derived using normalised single mode FRFs together with the Hilbert transform to describe the nonlinear behaviour. This method is capable while being robust in the presence of measured noise.

The backbone curves can be used to identify the linear natural frequencies and the nonlinear parameters. The experimental backbone curve is estimated from the resonant decay data measurement. This method can identify the dynamic behaviour of structures containing discrete nonlinear stiffness in large displacements regimes [18]. On top of that, Londoño et al. applied the approach that is based on estimations of the instantaneous frequency and the envelope amplitude of a decaying response, following a tuned steady-state oscillation of the system, to extract the backbone curves of lightly damped nonlinear systems [19].

Meanwhile, Noël and Kerschen utilized the applied excitations and the corresponding response time series, without knowing the information about the system to identify the nonlinear restoring force which is a direct indicator of the extent of the nonlinearity. Estimation is based on a standard least-square techniques to identify the coefficient of power series polynomial [12].

In modal analysis, the equation of motion for the dynamic response of a linear model to describe the structural motion is a second-order differential equation as in equation (1).

$$M\ddot{x}(t) + C\dot{x}(t) + Kx(t) = F(t) \qquad (1)$$

where M, C and K are the mass, damping and stiffness matrices. These matrices are a square $n \times n$ system. $\ddot{x}$, $\dot{x}$ and x are the $n \times 1$ acceleration, velocity and displacement vectors, respectively. $F(t)$ is an external force vector [20], [21]. For a nonlinear structure, the equation of motion [16], [22] consists of a nonlinear parameter as shown in equation (2).

$$M\ddot{x}(t) + C\dot{x}(t) + Kx(t) + K_{NL} = F(t) \qquad (2)$$

where K_{NL} is a non-linear stiffness term. For a hardening or softening nonlinear restoring force spring with cubic nonlinearities, the equation of motion is known as Duffing's equation [23], [24], [25] as shown in equation (3).

$$M\ddot{x}(t) + C\dot{x}(t) + Kx(t) + K_{NL}x^3(t) = F(t) \qquad (3)$$

The force-deflection characteristic for the hardening spring bends to reduce the displacement with a positive value for the parameter K_{NL} while the softening spring has a negative value for the parameter K_{NL} corresponding to allow for more displacement as the force increases. In this paper, methodology for detecting linear and nonlinear behaviour of a mass-spring system is described. The fundamental property of frequency response function (FRF) which will be exploited for the identification is a frequency-amplitude dependence respectively.

IOP Conf. Series: Journal of Physics: Conf. Series **1262** (2019) 012009 doi:10.1088/1742-6596/1262/1/012009

2. Methodology

2.1. Experimental Set-up

The experimental work consists of measuring the response of the single degree-of-freedom mass-spring system at selected point. The experimental setup is shown in Figure 1. The system consists of a single plate connected by two springs with equal stiffness property, which makes the equilibrium condition of the plate in the horizontal position. The springs were attached to the strings and the fixed base creating the system in free-free boundary conditions.

An electromagnetic shaker provides excitation at the centre of the plate in the vertical direction. The plate was excited symmetrically, therefore this setup is restricted to the study of axisymmetric rigid body modes of the plate that correspond exactly to the stiffness of the spring. An accelerometer was mounted on the plate to measure acceleration response of the mass-spring system.

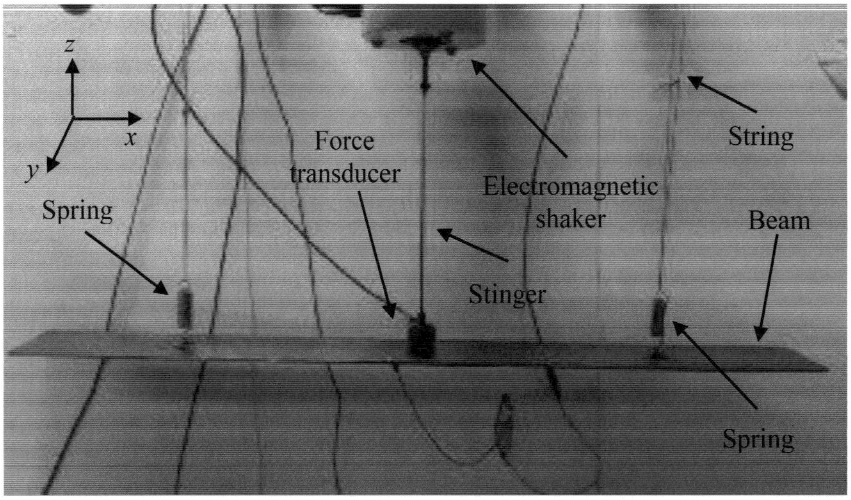

Figure 1. The modal experimental setup.

In this study, the experimental method is based on the comparison between linear and nonlinear identifications of the mass-spring system. The significance of the first part of this methodology is to identify the natural frequency, while the second part will be the nonlinear identification in more details. All the equipment involved in the experimental work are listed in Table 1.

Table 1. Specification of the equipment.

Description	Brand and model
Accelerometer	DYTRAN 3032A
Force transducer	PCB PIEZOTRONICS 208C03
Hammer	DYTRAN 5800
Shaker	DATA PHYSICS V4/T4

2.2. Measurement Techniques

Note that the experiments were carried out to verify the linear and nonlinear experimental identifications, hence the knowledge of the applied force and acceleration response of the system are required. For linear case, modal testing was performed by using the impact hammer excitation to quantify the nonlinear natural frequency as well as the amplitude of the frequency response function, an electromagnetic shaker was used with five different excitation values. While running each test, the acceleration response was measured using a number of accelerometers.

The spring used in this study is a conventional industrial spring. The specification of the spring is listed in Table 2. In order to obtain the stiffness of the spring, manual laboratory measurement was conducted from the elongation of the spring due to the weight's load.

Table 2. Property of the spring.

Parameter	Value
Stiffness (N/mm)	0.105
Length (mm)	19
Diameter (mm)	8

3. Results and Discussion

3.1. *Linear identification of the mass-spring system*

The linear identification was carried out on the basis of experimentally determined frequency response function (FRF) characteristics of free vibration response signals measured from the test system. Conventional modal testing using impact hammer was used in this section to identify the linear modal properties. **Error! Reference source not found.** shows the FRF plot for the mass-spring system. It shows that this system has a single natural frequency of 5.31 Hz. The coherence plot shown in Figure 3 drops in the anti-resonance region and justifies a high confidence output for this measurement.

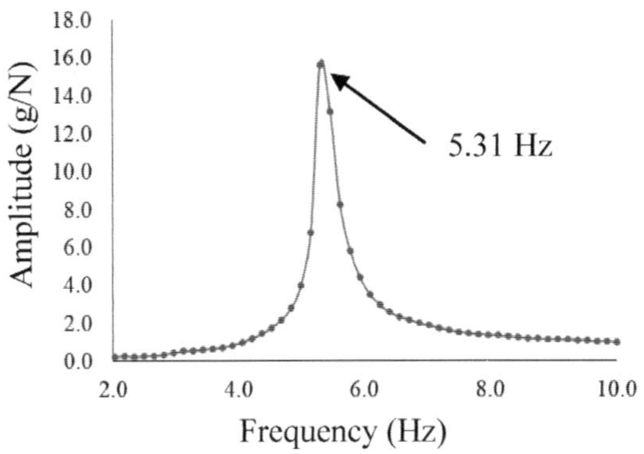

Figure 2. FRF plot by impact testing

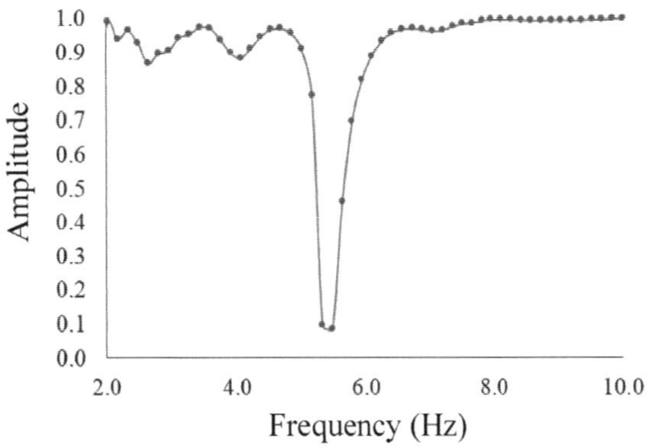

Figure 3. Coherence function

For linear identification in the second method, the FRF plot was obtained by exciting the system using burst random type-excitation. The detail parameters for the burst random experiment are given in Table 3. The bandwidth range is between 0 and 10 Hz. The excitation direction is in negative Z direction.

Table 3. Experimental parameters for the burst random excitation.

Parameter	Value
Excitation type	Burst random
Bandwith (Hz)	10
Spectral line	32
Resolution	0.31
Voltage (V)	0.5
Excitation DOF / direction	1 / -Z

Error! Reference source not found. shows the FRF plot from shaker excitation and associated coherence function is shown in Figure 5. The average recording is ten measurements. It can be seen that between 0 to 10 Hz, this single degree-of-freedom system comprises single natural frequency which is 5.31 Hz. From both results, it is proof that this mass-spring system has a linear natural frequency of 5.31 Hz. In the next section, the nonlinear identification will take place.

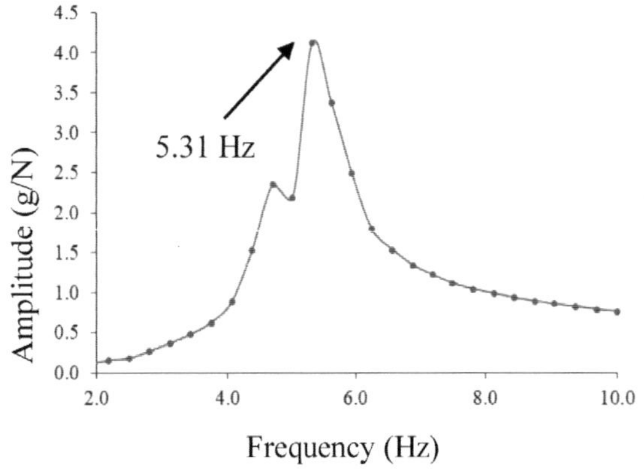

Figure 4. FRF plot by burst random

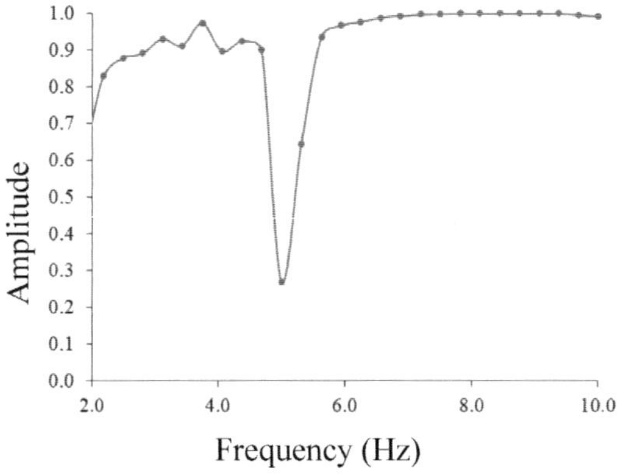

Figure 5. Coherence function

3.2. *Nonlinear Identification*

The sine chirp excitation was used to identify the nonlinear property of this mass-spring system. This selection is chosen because the level of force from low to high frequency can be controlled and the effect of leakage is non-existent providing the steady state response is achieved. The detail of parameters for the nonlinear identification testing is listed in Table 4. The process of underlying the linear behaviour that has been explained in Section 3.1 showed that the linear natural frequency is 5.31 Hz. Therefore, the sweep sine excitation is realized at a low excitation level over a small frequency range between 5 Hz and 6 Hz. The type for the periodic chirp sine testing is linear sweeping up direction. The sweeps were repeated for increasing excitation voltage in increment of 0.2 V.

Table 4. Parameters for the nonlinear identification via sine chirp method.

Parameter	Value
Excitation type	Periodic chirp
Sweep direction	Up
Frequency sweep (Hz)	5 to 6
Bandwith (Hz)	10
Spectral line	32
Resolution	0.31
Voltage (V)	0.2, 0.4, 0.6, 0.8 and 1.0
Excitation DOF	1
Excitation direction	-Z

IOP Conf. Series: Journal of Physics: Conf. Series **1262** (2019) 012009 doi:10.1088/1742-6596/1262/1/012009

The frequency response function (FRF) for each voltage is shown in Figure 6. The variations between FRF for each excitation force are used as an indicator for the detection and characterisation of nonlinear property. The pattern shows that there is a distortion in the FRF plot where there is the natural frequency shifting and amplitude changing when the mass-spring system was excited, with a higher voltage excitation input to the plate.

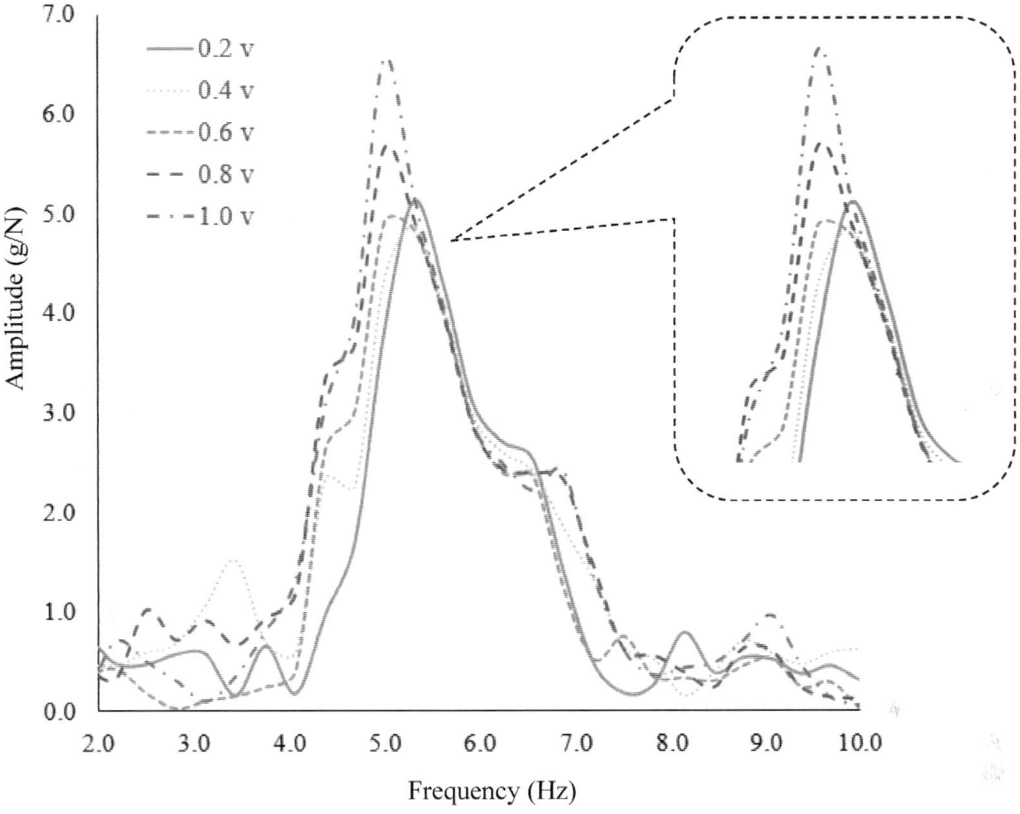

Figure 6. Nonlinear FRF plot by sweep sine excitation

To further evaluate the nonlinear behaviour, the natural frequency and the corresponding amplitude values for each voltage are extracted and listed in Table 5. For lower voltage excitation at 0.2 V and 0.4 V, the natural frequency is constant at 5.31 Hz with decreasing amplitude of the peak from 5.13 g/N to 4.87 g/N. However, the nonlinear behaviour becomes apparent where the natural frequency shifts to lower value at 5.00 Hz when the excitation level is between 0.6 V and 1.0 V. It can be seen that the displacement values, however increase from 4.90 g/N to 6.56 g/N. This result justifies that this mass-spring system has a softening stiffness characteristic, since the natural frequency reduces to lower value when higher voltage excites the system.

Table 5. Magnitude of voltage excitation and the natural frequency of the mass-spring system.

Voltage (V)	Natural Frequency (Hz)	Amplitude (g/N)
0.2	5.31	5.13
0.4	5.31	4.87
0.6	5.00	4.90
0.8	5.00	5.64
1.0	5.00	6.56

4. Conclusion

A linear and nonlinear dynamics behaviour of the mass-spring system has been investigated in this paper. The natural frequency of this single degree-of-freedom system has been obtained by experimental modal analysis. The experimental set up was carefully designed for a high quality investigation of the nonlinear property of a spring. According to the results, the linear identification by impact testing using an impact hammer and burst random excitation testing using an electromagnetic shaker found that the natural frequency is 5.31 Hz. Then, for the nonlinear detection, periodic chirp sine sweep testing with several magnitudes of voltages between 0.2 V and 1.0 V was utilized as an excitation to the system. The results found that the impact testing via the impact hammer cannot capture the nonlinearity effect. This type of excitation has an impulsive with high peak level but low root mean square (r.m.s) level property to detect the nonlinear system. However, the spectral testing via the electromagnetic shaker showed the reduction of natural frequency of the system at higher level of excitation. Moreover, the spring behaves the softening dynamics behaviour. Finally, it can be concluded that the frequency response function measured at a higher excitation level found deviates from the frequency response function of the linear part of the system.

Acknowledgment

The research has been funded by the Malaysian Ministry of Higher Education (MOHE) and Research Management Centre (RMC) of Universiti Teknologi MARA (UiTM) for the Fundamental Research Grant Scheme with the file number 600-IRMI/FRGS 5/3/ (172/2019). The authors would like to express their gratitude to the members of the Centre of Excellence (CoE) Structural Dynamics, Analysis and Validation (SDAV), UiTM Shah Alam for their technical assistance and laboratory facilities.

References

[1] Bahari A R, Yunus M A, Abdul Rani M N, Ayub M A and Nalisa A 2018 *MATEC Web of Conferences* **217** 02008
[2] Claeys M, Sinou J.-J, Lambelin J.-P and Todeschini R 2016 *Mech. Syst. Signal Process.* **70–71** 1097
[3] Gatti G, Brennan M J and Tang B 2019 *Mech. Syst. Signal Process.* **125** 4
[4] Kuether R J and Allen M S 2014 *Mech. Syst. Signal Process.* **46** 1
[5] Blair M, Canfield R and Roberts Jr. R 2005 *J. Aircr.* **42(4)** 832
[6] Frank Pai P, Nguyen B.-A and Sundaresan M J 2013 *Int. J. Non. Linear. Mech.* **54** 85
[7] Delapierre M, Lohaus S H and Pellegrino S 2018 *J. Sound Vib.* **427** 41
[8] Krack M, Scheidt P L and Wallaschek J 2013 *J. Sound Vib.* **332** 6798
[9] Spada C, Afonso F, Lau F and Suleman A 2017 *Aerosp. Sci. Technol.* **63** 363
[10] Changchuan X, Chao A, Yi L and Chao Y 2017*Chinese J. Aeronaut.* **30(2)** 638
[11] Peng C and Han J 2011 *J. Fluids Struct.* **27** 611
[12] Xu B, He J and Masri S F 2012 *Nonlinear Dyn.* **67** 2063
[13] Eder M A, Bitsche R D and Belloni F 2015 *Compos. Struct.* **132** 1075
[14] Gross J, Buhl P, Weber U, Schuler X and Krack M 2018 *Int. J. Non. Linear. Mech.* **99** 240
[15] Noël J P and Kerschen G 2017 *Mech. Syst. Signal Process.* **83** 2
[16] Lei Y, Luo S and He M 2017 *Adv. Struct. Eng.* **20(1)** 69

[17] Ondra V, Sever I A and Schwingshackl C W 2017 *Mech. Syst. Signal Process.* **83** 210
[18] Londoño J M, Cooper J E and Neild S A 2017 *Mech. Syst. Signal Process.* **84** 116
[19] Londoño J M, Neild S A and Cooper J E 2015 *J. Sound Vib.* **348** 224
[20] Bahari A R, Yunus M A, Abdul Rani M N, Wan Iskandar Mirza W I I and Haizuan A R 2019 *AIP Conference Proceedings* **2059**
[21] Su W C, Le T Q, Huang C S and Lin P Y 2017 *Appl. Math. Model* 1
[22] Zhou J, Wang X, Xu D and Bishop S 2015 *J. Sound Vib.* **346** 53
[23] Cveticanin L 2001 *J. Sound Vib.* **247(2)** 279
[24] Hashemi Kachapi S, Dukkipati R V, Hashemi K S G, Hashemi K S M and Hashemi K S 2010 *Nonlinear Anal. Real World Appl.* **11** 1431
[25] Litak G, Seoane J M, Zambrano S and Sanjuan M A F 2012 *Int. J. Bifurc. Chaos* **22(1)** 1

Occupational Noise Exposure Among Airport Workers in Malaysia: An Ergonomic Investigation

Athirah, B.[1] **and Nurul Shahida M.S.**[1,*]
[1] Human Engineering Group, Faculty of Mechanical & Manufacturing Engineering, Universiti Malaysia Pahang, 26600 Pekan, Pahang, Malaysia.

[*]Corresponding author : shahida@ump.edu.my

Abstract: Occupational noise is defined as a disturbance of sound at a workplace received by a worker's auditory system when they are working. A sustained and prolonged exposure to the noise can cause permanent hearing loss or damage to the worker. According to Occupational Safety and Health Act (OSHA), there is a certain tolerable noise level that can be exposed to the workers at a given time. This study aims to investigate the level of noise produced at the landing area in Sultan Ahmad Shah (SAS) Airport, Kuantan, Malaysia. A comprehensive ergonomic assessment divided into two phases is utilized in order to evaluate the noise exposure of the workers. In Phase I, an initial Ergonomics Risk Assessment is applied to gather the information on the ergonomic risk factors of the noise exposure, while for Phase II, an advanced Ergonomics Risk Assessment is used by monitoring the noise using occupational noise level meter. Thirty workers from the landing area which consists of the ground handlers were included as the subjects for this study. The value of noise exposure is measured while the workers perform their job. Results suggested that (a) the workers are aware of the noise exposure as per results shown in initial Ergonomics Risk Assessment in Phase I, and (b) the workers were exposed to high level of noise as suggested by OSHA (> 100 dBA at 2 hours exposure). With the suggested findings, the management of SAS Airport needs to create sufficient awareness to emphasize the usage of hearing protection devices in the workplace.

1. Introduction

Employee's performance and morale can be impacted in both positive and negative ways by the workplace environment. Workplace environment plays a major role in terms of the performance and productivity of an employee. The policy of a workplace that have been established is important to be followed as it can avoid the risk factors that exists at the workplace such as hearing defect, accident, malfunction machine and so on. In this paper, the risk factor that we would like to highlight and discuss is on the hearing defect that might occur during working which is hearing loss and other hearing problems. This significant issue is scarcely discussed especially in the workplace that do not involve high volume of noise or sound. However, occupational noise is one of the major problems that might affect the workers at their workplace with high volume of noise, such as at the airport runaway. Occupational noise and hearing defects are most common effects recognized since the Industrial Revolution. There are many cases reported due to the occupational hearing loss in the workplace. Workers who tend to neglect the hearing problems at their workplace give consequences to their economics and daily life activities. The prolonged exposure of loud noise or high frequency noise will eventually affect their hearing. Thus, we need to find solutions to this problem, specifically through engineering and ergonomics knowledge.

The problem statement of this project is based on five years' recent articles, journals, books, encyclopedia, etc. starting from the year 2014 until 2018. In the findings for more than six decades, the hazards of occupational noise exposure have been recognized and documented. However, it took many years to raise sufficient interest in addressing the problem on a large scale. In 1970, the Occupational Safety and Health Act (OSHA) included a noise standard for all American industries, making it mandatory for them to prevent occupational hearing loss (OSHA, 1970).

Table 1 shows the guideline values of the permissible hours for the workers to be exposed to the noise. NIOSH established the Recommended Exposure Limits (REL) for noise based on the best available science and practice. These guidelines need to be applied to all workers in their workplace including airport which involves aircraft noise. Based on the Guidelines for Control of Occupational Noise 2005 by Department of Occupational Safety and Health (DOSH), Malaysia, these guidelines are intended as a basic practical guidance to the employer in developing noise control measures. With the growth of the aviation industry, Malaysia has to move towards a more systematic and standard aviation management. Malaysia follows the standard of aviation from the International Civil Aviation Organization (ICAO) which is an UN specialized agent that works with the Convention's 192 Member States and industry groups to reach consensus on international civil aviation Standards and Recommended Practices (SARPs) and policies in support of a safe, efficient, secure, economically sustainable and environmentally responsible civil aviation sector [1]. This research, however, utilizes the standard on the ICAO ANNEX 14 which specializes on the Aerodromes - Aerodrome Design and Operations. This manual acts as a guidance for the operation of aerodrome, runway of airports and certain appropriate facilities.

Table 1 Permissible hours for exposure level to the noise in 24 hours

Time to reach 100% noise dose	Exposure level per NIOSH REL	Exposure level per OSHA PEL
8 hours	85 dBA	90 dBA
4 hours	88 dBA	95 dBA
2 hours	91 dBA	100 dBA
1 hour	94 dBA	105 dBA
30 minutes	97 dBA	110 dBA
15 minutes	100 dBA	115 dBA

Although there are many different sources in and around the airport that produces noise, the aircraft itself is the most commonly known source of the noise. This noise is categorized as high decibel noise and affects the hearing performance if being exposed for a long period of time without any protection. Mostly, people will assume that all the noise produced by the aircraft comes from the engine, however, another anatomy of an aircraft that generates noise is the airframe. Many people are not aware of this noise, but it may be of concern as engine efficiency and noise reduction have decreased as much as is technologically possible. Nevertheless, technology has reached a point where there are few things that can be performed to decrease the noise of the engine. Most of the noise generated from the aircraft engines typically occur from the high velocity exhaust gases and the air flow in the fan system.

In conclusion, the occupational noise that occur at the workplace can significantly affect the workers if they are exposed to a high level of noise in a long period of time. It will cause hearing problems and other diseases if the workers neglect the safety procedures and regulations during working hours.

IOP Conf. Series: Journal of Physics: Conf. Series **1262** (2019) 012010 doi:10.1088/1742-6596/1262/1/012010

2. Methodology

This study was conducted in two phases at Sultan Ahmad Shah Airport (Kuantan). The airport is known with its high level of noise exposure by the aircrafts since it operates for two organisations which are Malaysia Airports Holding Berhad and Royal Malaysian Air Force. This research received full support and recognition from the Head Engineer of Malaysia Airport Berhad (Kuantan Branch). Figure 1 shows the details of the research procedure.

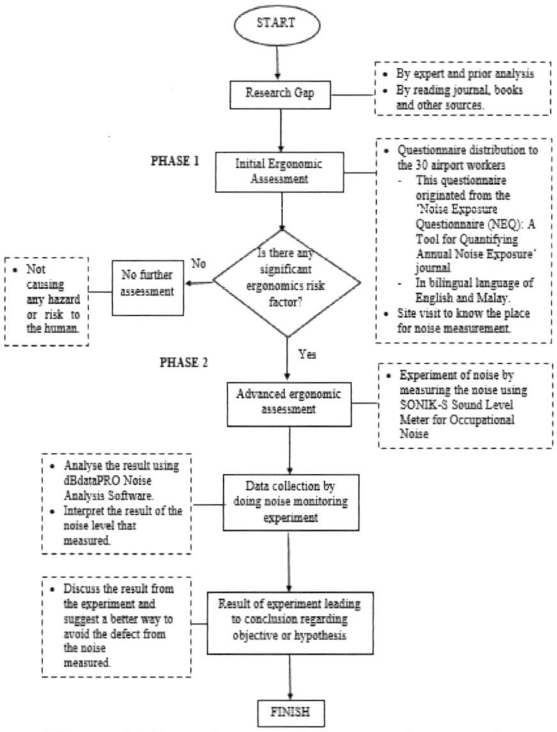

Figure 1 Flow chart of the research procedure

2.1 Phase I

The objective of Phase I is to obtain information on the occupational noise exposure among the workers. In order to achieve the objective, an established 'Noise Exposure Questionnaire' (NEQ) [2] was distributed to ten ground-handlers (n = 10) at the taxing area. The questionnaire has been modified to suit the subjects' background, for instance, the language of the questionnaire was created in both English and Bahasa Melayu. This is important as respondents will response more accurately and rapidly to conditions that placed greater demands on working memory [3].

Each subject completed the self-administered Noise Exposure Questionnaire (NEQ; Appendix A), which required approximately 10 minutes to complete. Subjects were asked to recall participation in specific noisy activities during the past year. The NEQ consists of three sections: 1) basic demographic information (gender and age), 2) six potential screening questions for determining individuals with high-risk noise exposure (Q#1–6), and 3) eleven detailed questions related to participation in loud, noisy activities used to quantify the annual noise exposure (Q#7–17).

2.2 Phase 2

Phase 2 starts as soon as Phase 1 is completed and the information obtained from Phase 1 are further utilized to study the appropriateness in regards with this project's objective. Following from there, Phase 2 is continued with the advanced ergonomic assessment which consists of data collection of noise level and analyses of the result from the experiment. The data collection of noise is obtained by doing the noise monitoring experiment which was conducted by measuring the noise around the aircraft by following the aircraft parking stand.

In the experiment of noise, the measuring was conducted using a device for noise measurement known as SONIK-S Sound Level Meter for Occupational. This device captures the noise produced from the airplane and surrounding. The noise reading is then recorded with the average of 3 consecutive days and it involved the aircraft model of Boeing 737-800. The results were then analysed using an analysis software which sorts out and manages the data file to read the result accurately. The result can be interpreted and the corresponding value of the noise level can be decided.

(a) (b)

Figure 2 The area of noise measurement.

Before the investigation, safety briefing was given at the site and it is crucial to follow the safety requirements strictly as the procedure has stated. Next, the range for measurement of noise are situated near the aircraft, which is approximately 5 meters from the aircraft, as shown in Figure 2 (a) (b), which is in the range of safety lines. Total two readings of noise measurement were taken and the average of the reading is calculated. The reading is taken when the flight is in transit since that is the time when the staff works near the flight. Figure 3 shows the mapping of runway at Sultan Ahmad Shah Airport and the place where the investigation is conducted is in red circle which is the transit area for the airplane. It is called apron and is the area of an airport where aircraft are parked, unloaded or loaded, refuelled, or boarded.

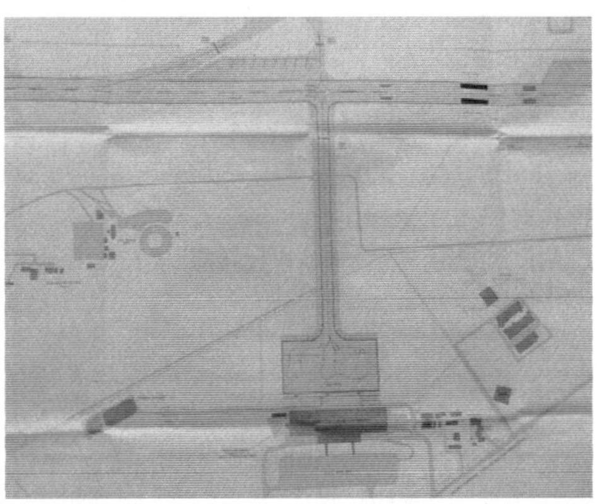

Figure 3 Runway of Sultan Ahmad Shah Airport.

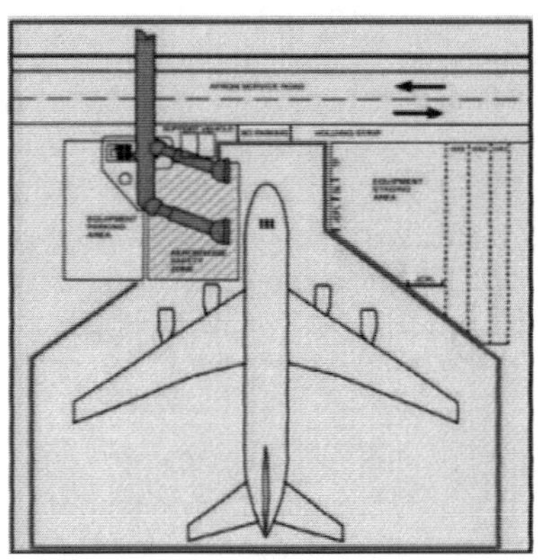

Figure 4 Aircraft parking stand.

The sound level meter is placed near the ears with the human average height of 155 cm to obtain a more accurate reading of noise that enters our ears. On the other hand, Figure 4 shows the aircraft parking stand where the noise reading is taken by following the red line. Furthermore, the ground handler workers work around the aircraft during the transit when the engine of the aircraft in running.

After Phase 2 is completed, the results produced from the experiment are analysed which then leads to the conclusion of this research. The result from the experiment are then discussed and suggests a better way to avoid the defect from the noise measured. Finally, the aim of this project is achieved by determining the degree of noise level that affects hearing especially at the airport and how to resolve this problem that may cause worries to the workers in this field.

2.3 The Experimented Aircraft

Figure 5 shows the aircraft of Boeing 737-800 which is the source of the noise exposed while the ground handler works. Boeing 737-800 is a twin-engine short-to-medium-range narrow body airliner with a capacity of maximum 189 passengers produced by the American manufacturer Boeing Commercial Airplanes. This model of aircraft has 2 turbofan engines with an engine power of 121.4 kN. Three aircrafts of the same model are measured to collect the data for three consecutive days and the average of the noise measurement is calculated. This aircraft produces the noise of more than 85 dBA and through this investigation, the result of the noise level proves the level of noise exposed

Figure 5 Boeing 737-800

3. Results and Discussion

3.1 Phase 1

For this study, total of 11 respondents answered a questionnaire regarding the noise issues at their workplace. The total respondents were from ground handler unit of ramp services, technician, ground equipment operator, aviation security and general workers. From the questionnaire, we determined that three from the total respondents possess hearing problem due to their working condition. The worst scenario that one of the respondent faces is that one of the workers has already received compensation from the Social Security Organization (SOCSO) for his hearing loss due to working environment. He suffered from ruptured eardrum and hearing loss which caused difficulties for him to work. Due to his hearing loss, he was transferred to work in the office where he have worked there for 32 years now. On the other hand, two other respondents experience reduced hearing problem while they work as the ground handler and both of them have work there for more than 10 years. This shows the result of long-time exposure towards high level of noise due to the workplace condition.

3.2 Phase 2

Advanced Ergonomics Assessment stage revealed that the procedure for this investigation is accomplished. This significant study was conducted for three consecutive days by measuring the noise around the aircraft and identify the highest point of noise exposure around the aircraft. In Occupational Safety and Health (Noise Exposure) Regulation 2019 by law of the Occupational Safety and Health Act 1994 (Act 514) [4], the noise exposure limit should not exceed 85dB(A) in daily working time. Following from there, during departure and arrival of aircraft, the reading of noise shows that different impact of noise was emitted from the aircraft. The duration of less than 40 minutes reading was recorded to analyse the impact with the distance of 15 meters away from the aircraft. Day 1 reading of LAeq was 74.8 dBA, day 2 with the reading of 62.4 dBA and day 3 showed LAeq reading of 74.5 dBA, which gives an average of 70.6 dBA. It shows a lower reading of noise than the noise near the aircraft.

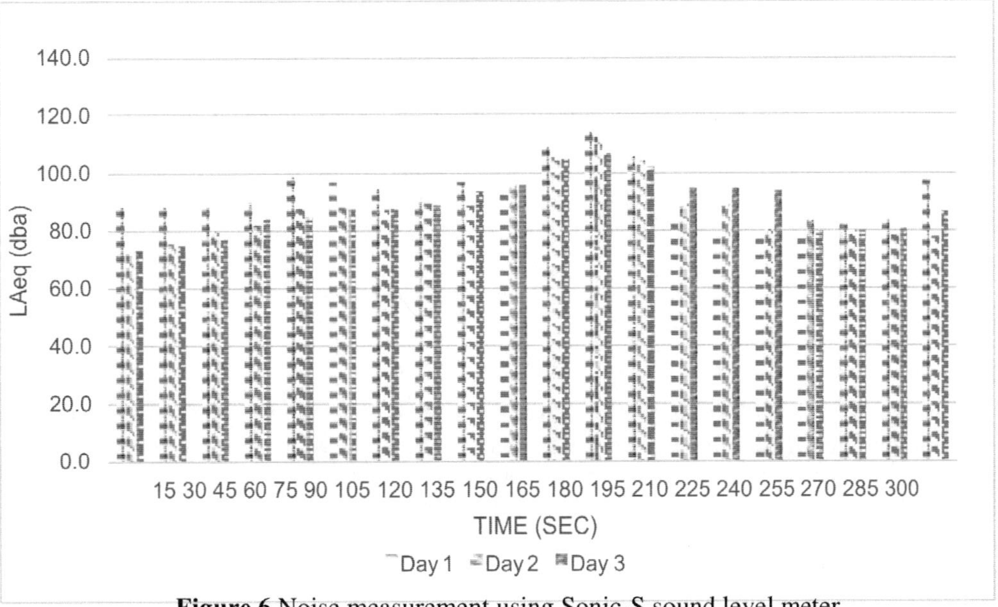

Figure 6 Noise measurement using Sonic-S sound level meter

Figure 6 shows the comparison of bar graph of noise measurement level for three consecutive days at the aircraft parking stand which is less than 5 meters distance from the aircraft. From the graph, it shows that Day 1 mostly have high values of reading while in Day 2 and Day 3, the values of reading are average. The graph was labelled with LAeq value on the *y*-axis and time in second on the *x*-axis. The increasing bar graph in Figure 6 shows the beginning of high level noise detected which was at the undercarriage of the aircraft.

From the bar graph, it shows that the highest source of noise emitted was taken at 165 seconds till 210 seconds. At this time, the highest noise produced an average of 111.4 dBA which exceeds the noise level allowed. Hence, precaution steps or safety awareness need to be taken to avoid the hearing problem among the workers.

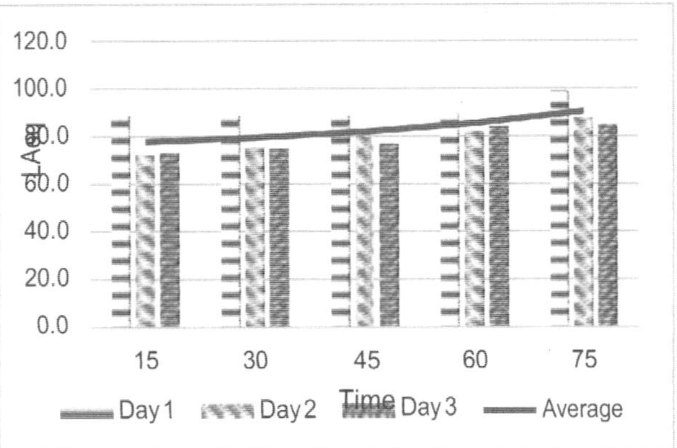

Figure 7 Bar graph of measurement in front of the aircraft

During the early measurement (0 - 75 seconds), the reading of sound level meter were in the range of below than 100 dBA with the highest value recorded on Day 1 at time 75 seconds, which occurred while walking from the front to the centre of the aircraft. The average of noise exposed is between 77.7 - 90.3 dBA and although it did not reach 100 dBA, it still present a small risk of developing hearing disability.

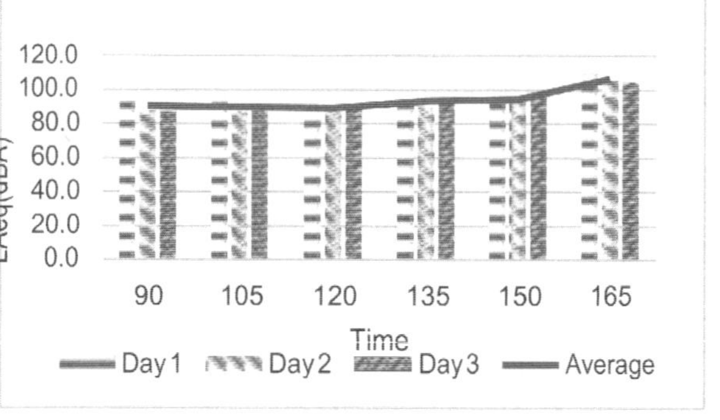

Figure 8 Bar graph at the centre of body aircraft

Next, the bar graph in Figure 8 shows the findings at the centre body of the aircraft which includes the wing and before the tail of the aircraft. Day 1 yield an average of 96.9 dBA during the measurement at the centre of the body which is higher than the reading on Day 2 with 92.7 dBA of reading and Day 3 of 93 dBA. The average value for total 3 days of measurement is 94.2 dBa and this amount may cause hazardous towards the workers. Workers are advised not to be exposed with this high level of noise for more than 8 hours of working.

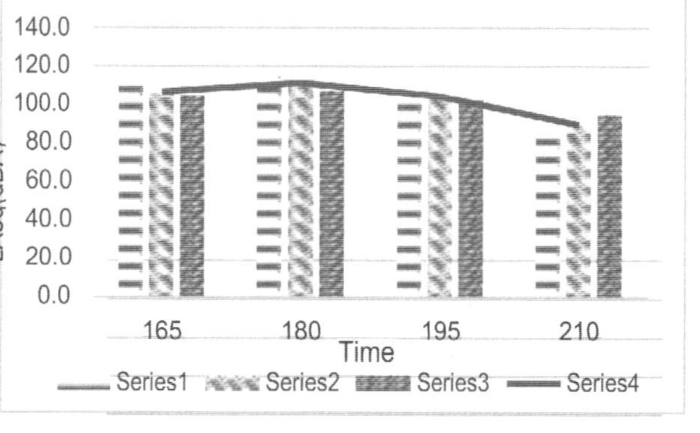

Figure 9 Undercarriage of aircraft.

During the investigation, as shown in Figure 9, we determined that the aircraft recorded the highest value of noise exposure at the undercarriage part of the aircraft with the average value of 111.4 dBA. Undercarriage is a part of aircraft that have landing gear located underneath of the aeroplane. The noise emitted from the aircraft system at undercarriage results with the highest level of noise exposure. Average range at the undercarriage area are between 90 dBA until 120 dBA which is in the hazardous state. During every transit of the airplane, the workers must check the condition of the aircraft including at the undercarriage. Hence, the risk of high level of noise exposure may cause defect to the workers.

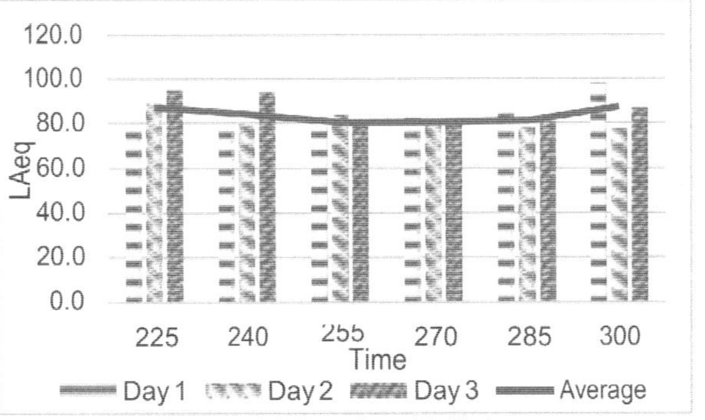

Figure 10 Result at Tail Area

The results show that the average range of noise exposure at tail area is 80 dBA to 90 dBA and this measurement is lower than at the undercarriage. By wearing the complete Personal Protective Equipment (PPE) while working may reduce the risk of hearing problem, thus, can improve the work quality of the ground handler as they are less worried about the damage that might occur. As concerned, excessive noise is a global occupational health hazard with considerable social and physiological impacts. Thus, an act towards this concern needs to be conducted to avoid further mishap towards the society.

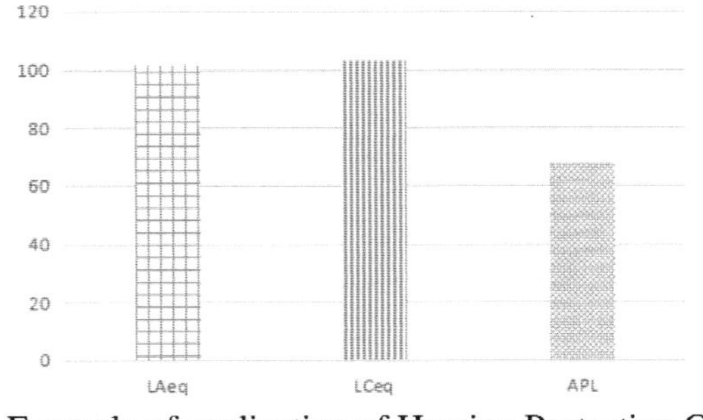

Figure 11 Example of application of Hearing Protection Comparison.

Through this application, a simulation for hearing protection for the noise exposure is performed and with the data from Day 1, a 3M Banded Earplug 1310 was applied to test the efficiency towards the noise. From Figure 11, the LAeq bar graph indicates the data from the log item while the APL bar graph showed the effect of the hearing protector. The noise level at 101.9 dBA can be reduced to the Assumed Protection level (APL) of 67.9 dBA. In other words, it can resist about 30 dBA of noise from the actual reading which is very effective to the user.

Based on the findings, we figured the aircraft's noise based on three categories which are mechanical noise, aerodynamics noise and noise from the aircraft system. This proved that noise pollution produced from the aircraft affected the workers near the aircraft and with today's available technology, manufacturers are trying to overcome this excessive noise problem as it will give good impact to society.

The result of this study shows that the workers were exposed to high level of noise at their workplace. The employer should take such measures to reduce the risk of noise expose to the workers such as providing high quality personal hearing protector to the workers and increasing the awareness and the importance of wearing this protector during working hours. It is also practicable to use the engineering method to reduce this excessive noise, but the employer should use this engineering control together with the administrative control. These controls are to be selected according to a hierarchy that emphasizes engineering solutions (including elimination or substitution) followed by safe work practices, administrative controls, and finally personal protective equipment. Through this precaution step, the impact of noise towards the workers can be reduced. Moreover, the occupational noise is a big concern in our society and many preventions from this problem can be taken by the employers.

4. Conclusions

The proposed objective of this study is in alignment with the noise exposure at the airport which is one of the occupational noise that affects the ground handlers. The predicted and measured noise values near the aircraft shows the same condition of noise exposure which is hazardous and provides high level of noise exposure towards the workers. The highest value of noise measured was at 114.6 dBA on Day 1 and the noise was over the allowable noise exposure of less than 90 dBA for 8 working hours daily. For suggestions, by implementing the control measures of prevention exposed to high level noise, we can ensure and retain an effective work performance of the workers as it can reduce the risk of hearing problem or hearing loss among the ground handlers.

Acknowledgments

The authors gratefully acknowledge Universiti Malaysia Pahang internal grant: RDU180338.

References

[1] Tikka C, Verbeek JH, Kateman E, Morata TC, Dreschler WA, Ferrite S. *Interventions to prevent occupational noise-induced hearing loss*. Cochrane Database Syst Rev [Internet]. 2017 Jul 7 [cited 2019 Feb 20];(7). Available from: http://doi.wiley.com/10.1002/14651858.CD006396.pub4

[2] Filippone, A. (2014). Aircraft noise prediction. *Progress in Aerospace Sciences*, **68,** 27–63. https://doi.org/10.1016/j.asoc.2017.11.004

[3] Franssen, E. A. M., Van Wiechen, C. M. A. G., Nagelkerke, N. J. D., & Lebret, E. (2004). Aircraft noise around a large international airport and its impact on general health and medication use. *Occupational and Environmental Medicine*, **61(5)**, 405–413. https://doi.org/10.1136/oem.2002.005488

[4] Grace Chen. (n.d.). *Sleepless in areas near KLIA - Metro News | The Star Online*. Retrieved December 14, 2018, from https://www.thestar.com.my/metro/metro-news/2018/12/14/sleepless-in-areas-near-klia/#hWH3wSqXmvDxMPAi.99

[5] Ground Handling Services. (n.d.). Retrieved February 26, 2019, from http://www.tibahairports.com/en-EN/Airlines/Pages/ground-handling-services.aspx

[6] International Civil Aviation Organization. (n.d.). Retrieved February 24, 2019, from http://www.mot.gov.my/en/aviation/ICAO-and-international-convention/members-of-ICAO

[7] A. Alkaabi, K. (2017). Studying the Effects of Aircraft Noise Around Abu Dhabi International Airport, UAE on the Surrounding Residential and Workplaces. *Civil Engineering and Urban Planning: An International Journal (CiVEJ),* **4(2),** 59–78. https://doi.org/10.5121/civej.2017.4205

[8] Tikka, C., Verbeek, J. H., Kateman, E., Morata, T. C., Dreschler, W. A., & Ferrite, S. (2017). Interventions to prevent occupational noise-induced hearing loss. *Cochrane Database of Systematic Reviews*, (**7**). https://doi.org/10.1002/14651858.CD006396.pub4

[9] Hearing Protectors: OSH Answers. (n.d.). Retrieved March 22, 2019, from https://www.ccohs.ca/oshanswers/prevention/ppe/ear_prot.html

[10] 3MTM 1310 Banded Earplugs. (n.d.). Retrieved March 22, 2019, from https://www.3m.com.au/3M/en_AU/company-au/all-3m-products/~/3M-1310-Banded-Earplugs/?N=5002385+3294361752&rt=rud

[11] CIAO. (2007). *Review of Noise Abatement Procedure Research & Development and Implementation Results.* Retrieved from http://www.icao.int/environmental-protection/Documents/ReviewNADRD.pdf

[12] EN, I. B. (2001). *Ergonomics of the thermal environment - Medical supervision of individuals exposed to extreme hot or cold environments.*, BS EN ISO(1), 1–51.

[13] International Civil Aviation Organization. (n.d.). Retrieved February 24, 2019, from http://www.mot.gov.my/en/aviation/ICAO-and-international-convention/members-of-ICAO

[14] Maximum, Minimum and Peak Sound Level. (n.d.). Retrieved February 26, 2019, from https://www.noisemeters.com/help/faq/min-max-peak.asp

[15] Revoredo, T., Mora-Camino, F., & Slama, J. (2016). A two-step approach for the prediction of dynamic aircraft noise impact. *Aerospace Science and Technology*, **59**, 122–131. https://doi.org/10.1016/j.ast.2016.10.017

[16] Dan, P. K. (2019). *Federal Government Kesihatan Pekerjaan (Pendedahan Bising) 2019 Occupational Safety And Health (Noise Exposure) Regulations 2019*, (March), 1–26.

[17] Bialystok, E., Craik, F. I. M., Klein, R., & Viswanathan, M. (2004). *Bilingualism, Aging, and Cognitive Control: Evidence From the Simon Task. Psychology and Aging*, **19(2)**, 290–303.

[18] Márquez-Molina, M., Sánchez-Fernández, L. P., Suárez-Guerra, S., & Sánchez-Pérez, L. A. (2014). Aircraft take-off noises classification based on human auditory's matched features extraction. *Applied Acoustics*, **84**, 83–90. https://doi.org/10.1016/j.apacoust.2013.12.003

[19] Johnson, T. A., Cooper, S., Stamper, G. C., & Chertoff, M. (2017*). Noise Exposure Questionnaire: A Tool for Quantifying Annual Noise Exposure.* Journal of the American Academy of Audiology, **28(1)**, 14–35.

[20] Rosso, M., Agius, R., & Calleja, N. (2011). Development and validation of a screening questionnaire for noise-induced hearing loss. *Occupational Medicine*, **61(6),** 416–421. https://doi.org/10.1093/occmed/kqr059

[21] Hiroe, M., Makino, K., Ogata, S., Tamura, A., Suzuki, S., Yamada, I., & Yasuoka, M. (2016). A questionnaire survey on health effects of aircraft noise for residents living in the vicinity of Narita International Airport: Part-2 analysis and result detail. *Proceedings of the INTER-NOISE 2016 - 45th International Congress and Exposition on Noise Control Engineering: Towards a Quieter Future, 2593–2601.* Retrieved from https://www.scopus.com/inward/record.uri?eid=2-s2.0-84994644883&partnerID=40&md5=ccabac8b844d46fe9ca13266aaebe83a

[22] DOSH. (2005). *Guidelines for Control of Occupational Noise.* Retrieved from http://www.dosh.gov.my/index.php/ms/list-of-documents/guidelines/industrial-hygiene-1/1416-01-guidelines-for-control-of-occupational-noise-2005/file

[23] About ICAO. (n.d.). Retrieved February 24, 2019, from https://www.icao.int/about-icao/Pages/default.aspx

[24] Lawson, S. M., & Masterson, E. A. (2019). Hearing Loss among Health Care and Social Assistance Workers: A NIOSH Study. *Hearing Journal*, **72(1)**, 12–13. https://doi.org/10.1097/01.HJ.0000552750.89735.6d

[25] Thepaksorn, P., Siriwong, W., Neitzel, R. L., Somrongthong, R., & Techasrivichien, T. (2018). Relationship Between Noise-Related Risk Perception, Knowledge, and the Use of Hearing Protection Devices Among Para Rubber Wood Sawmill Workers. *Safety and Health at Work*, **9(1)**, 25–29. https://doi.org/10.1016/j.shaw.2017.06.002

IOP Conf. Series: Journal of Physics: Conf. Series **1262** (2019) 012011 doi:10.1088/1742-6596/1262/1/012011

Finite Element Modelling and Model Updating of Riveted Joint Structure

M A S Aziz Shah[1], M A Yunus[1*], M N Abdul Rani[1], M. A. Zulkofli[1]
[1] Structural Dynamics Analysis and Validation, Faculty of Mechanical Engineering, Universiti Teknologi MARA (UiTM), 40000 Shah Alam, Selangor, Malaysia

[*]Corresponding author: mayunus@salam.uitm.edu.my

Abstract. Nowadays, the applications of rivet joints are dramatically increasing because of its capability and efficiency to assemble the engineering structure such as aerospace and mechanical structures. However, every joint such as riveted joint has a significant effect to the overall dynamic behaviour of the assembled structure. This is because, the overall dynamic behaviour of the assembled structure is highly relying on the quality of the riveted joints. Generally, the riveted joints always introduce local effects such as surface contact, clamping force and slip to the surface of the assembled structure. Therefore, this paper is attempted to investigate the dynamic behaviour of the riveted joints structure experimentally and numerically. The experimental modal analysis method is used to measure the dynamic behaviour of the assembled structure. Meanwhile, the predicted data of the dynamic behaviour of the assembled structure is obtained by using finite element method. A few types of element connectors that available in finite element commercial software namely such as RBE, CBAR, CBEAM and CELAS have been investigated in order to represent the rivet joints in the assembled structure. In this study, the results obtained from the finite element analysis (in terms of natural frequencies and mode shapes) are then correlated with experimental counterpart in order to identify the level of the accuracy between element connectors that are used in modelling the rivet joints of the assembled structure. Finally, the reconciliation method using finite element model updating is used to minimise the discrepancies of the initial finite element model based on the most appropriate element connector to experimental data. The results reveal that CBEAM element connector has better capability and most appropriate comparing to other three element connectors to represent rivet joint in the assembled structure.

1. Introduction

In the advancement of the engineering structure the riveted joints have become the most implemented connection method to assemble thousands of sheet metal components in engineering industries such as aerospace and automotive structure. In comparison to the others joining methods such as welded, bolted or adhesive. The riveted joints are much more preferred due to its simplicity in joining process and can be available at very low cost [1-2]. One of the advantages of the riveted joints compared to other jointing methods is that the riveted joint has more flexibility than other type of joints where, the riveted joints can be used to assemble components with dissimilar type of materials. Moreover, the implementation of the riveting process is very fast and easy. Moreover, the riveted joints are having very high connections strength and also very efficient to be used in many industries such as in automotive or aircraft structures [3]. Even though riveted joint having a lot of advantages compared to other type of jointing methods, the presents of a local effects in the jointed structure are not easy to be included or modelled in a large finite element model. This is because the finite element modelling is constructed based on assumptions and simplification on geometry and properties of a structure [4].

For instance, sheet metal substructures and the rivets can easily experience deformation during the riveting process. Furthermore, the quality of the riveted joints structure is depended with axis deviation in which the rivet's axis must be equivalent or parallel with the hole's axis of the structure which it may influence the deformation, stress and strain riveting. The simplification on these problems may lead to the inaccuracy in the prediction of the dynamic behaviour of riveted jointed structure. Previous studies suggested that, it is impractical to model the local effects in finite element modelling precisely because it requires a lot of computational efforts when performing analysis [5-6]. Therefore, the experimental method is used to analyse the dynamic behaviour of riveted joints structure namely experimental modal analysis (EMA). This experimental approach is required as to identify inaccuracies of approximation the initial finite element model. The discrepancies of results between experimental modal analysis and finite element analysis have led to the conciliation procedure through which the finite element model is altered in order to provide good agreement with the experiment result [7-9].

In this study, the dynamic behaviour of the riveted joints structure is investigated numerically and experimentally. Element connectors such as rigid body element (RBE), bar based element (CBAR), beam based element (CBEAM) and spring based element (CELAS) are investigated and their accuracy will be compared with the experiment data in term of natural frequencies and mode shapes. After that, finite element model updating will be applied in order to reduce relative error between the initial finite element model of the assembled structure with experimental data by improving the uncertainties of the input parameters and to ensure the validity of the developed finite element model.

2. Experimental Modal Analysis (EMA)

Experimental modal analysis is a technique to define the dynamic behaviours of the structure from a vibration testing, in terms of natural frequencies, mode shapes and damping ratios [10-12]. The result from the experimental data will be compared with finite element model to identify the correlation between its results. If the results from the test structure is within the acceptable tolerance with predicted model, and then finite element model can be used in further analysis with higher degrees of confidence.

In this study, the riveted structure, which consist of two number of components, namely as 1) hat shape and 2) flat shape plate were tested. The hat shape and flat shape plate substructures were connected by 10 rivets with nominal 5 mm diameter for each rivet and were separated by 70 mm apart. The material used to fabricate the substructures of the structure are mild steel with overall dimension was 300 mm length, 240 mm wide and 5 mm thickness.

The experimental modal analysis was performed to the substructures and assembled structure by using impact hammer and roving accelerometers method as to measure the dynamic behaviour such as natural frequencies and mode shapes. Prior to the measuring process of assembled structure, the structure was setup under free-free boundary conditions whereby four sets of rubber band and strings were used to hang the structure from the specially designed clamps by attaching the strings to the holes on the structure and the other ends of the strings to the clamps. Meanwhile, the frequency bandwidth of interest was set from 1 – 400 Hz. Finally, all the data obtained were processes using LMS SCADAS data acquisition system. The experiment procedures were illustrated in Figure 1, Figure 2 and Figure 3.

Figure 1. Riveted joints structure

Figure 2. Experimental modal analysis on the assembled structure

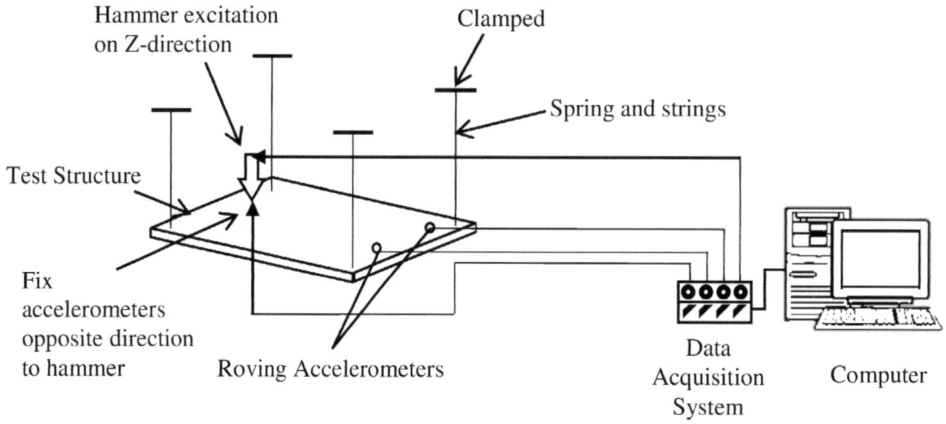

Figure 3. Schematic diagram of experimental modal analysis setup

3. Finite Element Modelling and Analysis

Finite element analysis is a numerical method that involve with the discretisation of the physical domain and the geometry of the structure into elements and in this work, the structure was designed as hat shape and flat shape plate connected by rivets joint.

In the pre-processing stage of the finite element modelling, the MSC PATRAN was used to develop the finite element model of the components and the assembled structure. The components and the assembled structure were modelled using CQUAD4 elements (shell) and the models were meshed into 5 mm meshing size based on the suitability of the meshes size recorded from mesh convergence test. In this study, four types element connectors namely rigid body element (RBE), bar based element (CBAR), beam based element (CBEAM) and spring based element (CELAS) were explored to represent the rivet joints in the modelling of the riveted joined structure. The material properties used in the finite element modelling was based on nominal properties of mild steel is listed in Table 1.

Table 1. Material properties of the laser spot welded structure [13]

Component	Property	Value	Unit
Hat shape plate	Young's Modulus	210	GPa
	Poisson's Ratio	0.30	Unitless
	Mass Density	7700	kg/m^3
Flat shape plate	Young's Modulus	210	GPa
	Poisson's Ratio	0.30	Unitless
	Mass Density	7700	kg/m^3
Rivet	Young's Modulus	70	GPa
	Poisson's Ratio	0.30	Unitless
	Mass Density	2770	kg/m^3

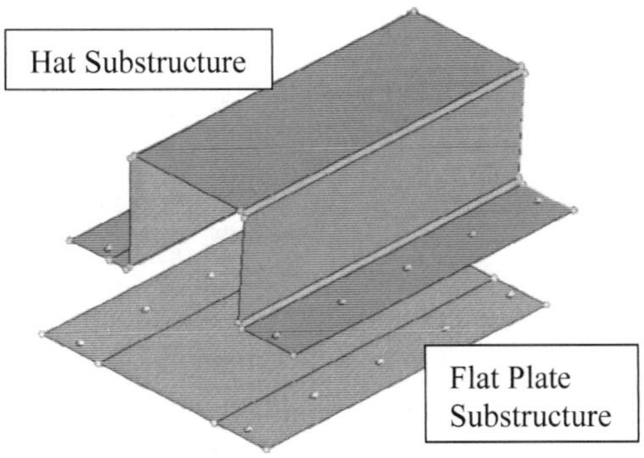

Figure 4. Geometry of the substructures

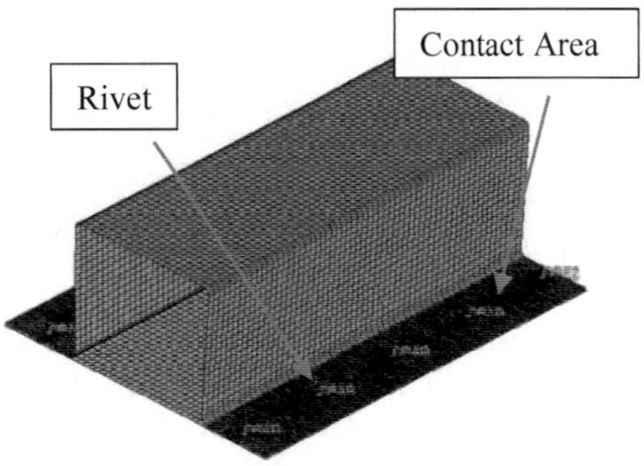

Figure 5. Finite element model of riveted joined structure

In this research, normal mode analysis was calculated by using the normal modes analysis of MSC NASTRAN to identify the natural frequencies and mode shapes of the finite element models of the components and the assembled structure. The equation of motion that are used to discretise the system to a finite element model, also known as 2nd order differential equation is given as

$$\mathbf{M}\ddot{x}(t) + \mathbf{C}\dot{x}(t) + \mathbf{K}x(t) = f(t) \tag{1}$$

where **M**, **C** and **K** are symmetric matrices of mass, damping and stiffness. Meanwhile $\ddot{x}$, $\dot{x}$ and x represent the vector of accelerations, velocities and displacement respectively and f(t) is vector of external forces. The structure that been used in this research were considered having light damping and the effect of damping can be theoretically neglected in finite element modelling. As a result, for the undamped free vibration analysis, the equation (1) can be simplified as

$$\mathbf{M}\ddot{x}(t) + \mathbf{K}x(t) = 0 \tag{2}$$

The equation (2) can be solved by assuming the harmonic solution in the form of

$$x = \phi sin\omega t \tag{3}$$

where ω and ϕ are the mode shape and natural frequency of the system. If the differentiation of the assumed harmonic solution is performed and substituted in equation (2), the equation of motion yields and simplified to the following

$$(\mathbf{K}-\omega^2\mathbf{M})\phi = 0 \tag{4}$$

The natural frequencies and mode shapes of the laser spot welded structure can be predicted by solving the equation (4) using finite element commercial software such as MSC NASTRAN.

.

4. Finite element model updating

Model updating is one of the methods used to improve the correlation of finite element models and the test structure by correcting the invalid assumptions of the model to an acceptable level of accuracy and the process is carried out by tuning some uncertain model parameters systematically. However, it is difficult to select the right parameters that sensitive to the change of modal properties. This is because the selection of the parameters should be justified by engineering understanding of the structure and the number of selected parameters should be kept to a minimum in order to avoid ill-condition problems.

The selection of the most influential parameters was performed using sensitivity analysis and can be expressed as,

$$\mathbf{S} = \boldsymbol{\phi}_i^{\mathbf{T}}\left[\frac{\partial\mathbf{K}}{\partial\theta_j} - \lambda_i\frac{\partial\mathbf{M}}{\partial\theta_j}\right]\boldsymbol{\phi}_\mathbf{i} \tag{5}$$

where, S shows the sensitivity matrix, K and M are the stiffness and mass matrices respectively, while ϕ, λ and θ represent eigenvector, eigenvalue and parameter respectively. Furthermore, i indicates the i-th eigenvalue and j is the j-th parameter.

In order to perform model updating, the NASTRAN optimisation code (SOL 200) is used in which it allows the objective function (J) to be minimised by adjusting the eigenvalues of the initial finite element model until the objective function is converged. The objective function can be expressed as,

$$J = \sum_{i=1}^{n} W_i\left(\frac{\lambda_i^{\mathrm{fe}}}{\lambda_i^{\exp}} - 1\right)^2 \tag{6}$$

where, $\lambda_i^{\exp}$ is the i-th experimental eigenvalue and λ_i^{fe} is the i-th predicted eigenvalue from the finite element model and n is the number of eigenvalues involved in the updating procedure.

5. Results and Discussion

In this study, the first five natural frequencies and mode shapes of the riveted joints structure were calculated experimentally and numerically. The accuracy of the finite element models of the structure are using four different types element connectors which to represent the rivet joint namely RBE2, CBAR, CBEAM and CELAS was investigated. The results shown in Table 2 consists the comparisons between experiment data and initial finite element data. From the table, it shows that CELAS based FE model obtained the most relative total error which is 320.72%, meanwhile total relative error of RBE based FE model is 21.86%, CBAR based FE model is 31.3% and the less total relative error is CBEAM based FE with 31.22 %. The results in Table 1 clearly shows that the CELAS based FE model is the highest relative total error. The results suggest that the CELAS element connector is not suitable to be represent as rivet joints. This is because CELAS element cannot produce adequate rigidity to the assembled structure. Furthermore, the inaccuracy in the finite element results is happen due to the influenced of the local effects in the rivet joints that correspond badly to the sheet metal structure [14-16]. Moreover, the invalid assumptions of the input parameters on the structure such as Young's

IOP Conf. Series: Journal of Physics: Conf. Series **1262** (2019) 012011 doi:10.1088/1742-6596/1262/1/012011

modulus, Poisson's ratio, sheet metal thickness and diameter of rivets also effect the dynamic behaviour of the riveted joints structure.

Based on the result in Table 2, RBE based FE model has shown the lowest error in comparison with the other element connectors. However, CBEAM based FE model was selected as the most appropriate element connector to represent riveted joint in the assembled structure. This is because RBE element connector do not have any material properties and on top of that, the RBE properties is based on infinite stiffness configuration [17-19]. This shortcoming makes RBE based FE model not suitable to be used in model updating. Therefore, CBEAM based FE model is more preferable. In order to reduce discrepancies of the CBEAM based FE model with experimental data counterpart, finite element model updating was performed by minimising the objective function with the experimental data was used as benchmark. The result of the dynamic behaviour of the updated finite element model of the structure is shown in Table 3. The result shown that model updating method managed to reduce the total relative error of the CBEAM based FE model from 31.22% to 16.44%. The modal assurance Criterion (MAC) values of the updated finite element model also shown a good improvement.

Table 2. Comparison different types of element connectors.

Mode	Test	RBE		CBAR		CELAS		CBEAM	
		FE	Error	FE	Error	FE	Error	FE	Error
	(Hz)	(Hz)	(%)	(Hz)	(%)	(Hz)	(%)	(Hz)	(%)
1	153.72	149.95	2.45	146.68	4.58	51.34	66.60	146.73	4.55
2	159.78	157.38	1.50	156.24	2.22	62.96	60.60	156.24	2.22
3	228.74	224.13	2.02	220.65	3.54	93.10	59.30	220.86	3.45
4	321.47	291.09	9.45	284.40	11.53	97.25	69.75	284.65	11.45
5	331.34	310.01	6.44	300.10	9.43	117.70	64.48	299.67	9.56
Total Error			21.86		31.30		320.72		31.22

Table 3. Comparison between the measured, predicted and updated results of the structure.

Mode	Test	Initial CBEAM			Updated CBEAM		
		FE	Error	MAC	FE	Error	MAC
	(Hz)	(Hz)	(%)		(Hz)	(%)	
1	153.72	146.73	4.55	0.85	149.85	2.52	0.92
2	159.78	156.24	2.22	0.81	158.14	1.03	0.90
3	228.74	220.86	3.45	0.84	224.17	2.00	0.89
4	321.47	284.65	11.45	0.74	301.91	6.09	0.88
5	331.34	299.67	9.56	0.79	315.42	4.80	0.88
Total Error			31.22			16.44	

6. Conclusions

In this paper, the dynamic behaviour of a riveted joints structure was analysed experimentally and numerically. The main objective was to identify the most appropriate element connector to represent the rivet joint accurately in the assembled structure. The discrepancies of the initial finite element model and the test structure was determined by comparing natural frequencies and mode shapes of the structure. The result shows that CBEAM element was the best element connector to represent as rivet joints in the assembled structure. Therefore, the initial CBEAM based FE model was used in finite element model updating method to reduce the discrepancies with experimental data. Based on the result obtained, it is recommended to use CBEAM element connector to represent as rivet joints. However, there is still an improvement that can be made in the future such as the inclusion of local effects in the finite element modelling that may improve the prediction result of the assembled structure.

Acknowledgments
The authors wish to acknowledge the Malaysia Ministry of Higher Education (MOHE), Research Management Institute (RMI) and Institute of Graduate Studies of Universiti Teknologi MARA (UiTM) for providing financial support for this research through the fundamental research grant scheme (FRGS) 600-IRMI/FRGS 5/3 (172/2019). They would also like to extend their sincere gratitude for the support and help given by, Prof. Hadariah Bahron and all SDAV members.

References
[1] A. L. L. da Silva *et al.*, "Fatigue characterization of a beam-to-column riveted joint," *Eng. Fail. Anal.*, vol. 103, pp. 95–123, Sep. 2019.
[2] L. Huang, H. Guo, Y. Shi, S. Huang, and X. Su, "Fatigue behavior and modeling of self-piercing riveted joints in aluminum alloy 6111," *Int. J. Fatigue*, vol. 100, pp. 274–284, Jul. 2017.
[3] M. A. Yunus, M. N. A. Rani, M. S. M. Sani, and M. A. S. A. Shah, "Finite element model updating of riveted joints of simplified model aircraft structure," *AIP Conf. Proc.*, vol. 1952, no. 1, p. 20013, 2018.
[4] M. S. A. Sulaiman, M. A. Yunus, M. N. A. Rani, A. M. Saman, Z. Mohamed, and M. A. S. A. Shah, "Characterisation of the dynamic characteristics of a complex jointed structure," in *MATEC Web of Conferences*, 2017.
[5] M. A. S. A. Shah, M. A. Yunus, A. M. Saman, M. N. A. Rani, A. R. Bahari, and C. Peter, "Finite element model updating of laser stitch welded structure," *IOP Conf. Ser. Mater. Sci. Eng.*, vol. 506, p. 012010, 2019.
[6] H. H. Khodaparast *et al.*, "Fuzzy finite element model updating of the DLR AIRMOD test structure," *Appl. Math. Model.*, vol. 52, pp. 512–526, Dec. 2017.
[7] Ø. W. Petersen and O. Øiseth, "Sensitivity-based finite element model updating of a pontoon bridge," *Eng. Struct.*, vol. 150, pp. 573–584, Nov. 2017.
[8] V. Panwar, P. Gupta, A. K. Bagha, and N. Chauhan, "A review on studies of finite element model updating and updating of composite materials," *Mater. Today Proc.*, vol. 5, no. 14, pp. 27912–27918, Jan. 2018.
[9] M. A. S. Aziz Shah, M. A. Yunus, M. N. Abdul Rani, N. A. Tormidi, A. A. Mat Isa, and W. I. I. Wan Iskandar Mirza, "Closed mode separation of the symmetrical structure," *J. Mech. Eng.*, vol. 5, no. 3, pp. 88–97, 2018.
[10] M. A. Yunus, M. A. S. Aziz Shah, M. N. Abdul Rani, W. I. I. Wan Iskandar Mirza, and S. Kushairi, "Experimental and finite element analysis of laser stitch welded structure," *J. Mech. Eng.*, vol. 4, no. 3, pp. 155–162, 2017.
[11] F. Adel, S. Shokrollahi, M. Jamal-Omidi, and H. Ahmadian, "A model updating method for hybrid composite/aluminum bolted joints using modal test data," *J. Sound Vib.*, vol. 396, pp. 172–185, May 2017.
[12] J. M. W. Brownjohn, A. Raby, J. Bassitt, A. Antonini, E. Hudson, and P. Dobson, "Experimental modal analysis of British rock lighthouses," *Mar. Struct.*, vol. 62, pp. 1–22, Nov. 2018.
[13] M. A. Yunus, "Finite element modelling and updating of structure of sheet metal with bolted and welded joints," University of Liverpool, 2011.
[14] J. Q. Yin, Z. Q. Wang, Y. G. Kang, and N. Li, "Finite element analysis of axis deviation between rivet and plate hole," *Appl. Mech. Mater.*, vol. 321–324, pp. 188–191, 2013.
[15] X. C. He, "Sheet material property effects upon dynamic behavior in self-pierce riveted joints," *Mater. Sci. Forum*, vol. 675–677, pp. 999–1002, 2011.
[16] S. M. S. . Hassin.N.M, Yunus.M.A, Abdul Rani.M.N, "Experimental and finite element analysis of riveted joints structure of a simplified model of aluminium crash box," *J. Eng. Appl. Sci.*, vol. 11, pp. 1451–1455, 2016.

IOP Conf. Series: Journal of Physics: Conf. Series **1262** (2019) 012011 doi:10.1088/1742-6596/1262/1/012011

[17] N. M. Hassin, M. A. Yunus, A. Radzi, A. Ghani, N. A. Rani, and S. Kasolang, "Using updated model for the crash analysis," *Int. Congr. Sound Vib.*, pp. 1–8, 2016.

[18] Abdullah N A Z, Sani M S M, Hussain N A, Rahman M M and Zaman I 2017 Dynamics properties of a Go-kart chassis structure and its prediction improvement using model updating approach *International Journal of Automotive and Mechanical Engineering (IJAME)* 14 3887-97

[19] Fouzi M S M, Jelani K M, Nazri N A and Sani M S M 2018 Finite Element Modelling and Updating of Welded Thin-Walled Beam *International Journal of Automotive and Mechanical Engineering (IJAME)* 15(4) 5874-89

IOP Conf. Series: Journal of Physics: Conf. Series **1262** (2019) 012012 doi:10.1088/1742-6596/1262/1/012012

Vibration Analysis on Palm Oil Methyl Ester Biodiesel as a Fuel with The Additional of Butanol

J.M. Zikri[1], M.S.M. Sani[1,3*], Erdiwansyah[1], A.F. Yusop[2],I. Izzudin[2], S. Sapee[2]
[1]Advanced Structural Integrity & Vibration Research, Faculty of Mechanical and Manufacturing Engineering, Universiti Malayisa Pahang, 26600 Pekan, Pahang, Malaysia
[2]Advanced Fluid Focus Group, Faculty of Mechanical and Manufacturing Engineering, Universiti Malaysia Pahang, 26600 Pekan, Pahang, Malaysia
[3]Automotive Engineering Centre, Universiti Malaysia Pahang, 26600 Pekan, Pahang, Malaysia

*Corresponding author: mshahrir@ump.edu.my

Abstract. Along with the development of alternative fuel to replace the usage of petroleum-derivatives, the vibration in the diesel engine has become one of the main focus by researchers recently. The biodiesel has been investigated numerously since the application does not need any modification of the engine and with the different fuel properties, the biodiesel can enhance or deteriorates the engine. Hence, this study will be carried out to investigate the effect adding butanol in two different biodiesel blends by using vibration analysis. The vibration level will be determined in velocity term in RMS by using uniaxial accelerometer. The results indicated that the B10Bu10 blend shows a better reduction in vibration level with the decrement more than 30% for both speed in 25% load while in the 50% load, the decrement was observed to be 13.46% for 1800 RPM and 22.9% for the 2100 RPM. Finally, it can be conclude that the usage of biodiesel with the additional of butanol can improve the vibration level in the diesel engine, however, with an additional volume of biodiesel in the blend tends to increase the vibration level in certain condition.

1. Introduction

With the development of the technologies nowadays has made the engine used widely in many sectors such as automotive, construction, agriculture, etc [1-3]. The application of the engine was chosen based on their characteristics such power produced, fuel consumption and the efficiency. As an option that covered all the criteria, the diesel engine has been used due to its high power produced, good fuel consumption and high efficiency [1-6].

Despite of all the advantages of the diesel engine, there are several major drawbacks from the usage of the engine and of them is the depletion sources of fossil fuel. Since the diesel were produced from a non-renewable source, it is a must to consider the availability to preserve it for the future use. It has been stated that based on current fossil fuel reserves, the source will be completely exhausted by 2055 [7-9]. This phenomenon has led to the many invention of alternative fuel in order to replace the hundred percent use of diesel as a fuel. By reducing the dependency on the fossil fuel, the alternative fuel can prolong the estimated period for the fossil fuel to be exhausted. As a promising solution, one of the inventions that seems to have a possibility to be a substitution for the diesel is the biodiesel [10-13].

Biodiesel is a biodegradable fuel that were produced from a renewable sources such as vegetable oils, animal fats and waste products that has gone under some specific processes to produce almost likely diesel fuel with some improvements in terms of emission and eco-friendly [14-15, 43-44]. There are several ways to produce the biodiesel but the most common is the transesterification which is through the chemical process and reaction [16-21]. Can be utilized up to 20% percent of blending ratio with the diesel without any major modification on the engine has lead the biodiesel to be a way better than other alternative fuels [17,22-27].

However, with the utilization of biodiesel as an alternative fuel also had a few detonations and one of the most significant is the vibration level in the engine[28-31]. This matter has become a serious issue among the researcher since the high blend ratio of biodiesel will produces a higher vibration level which are the consequences effect from the high combustion pressure[32-35]. Kadam et al.[36] has performed an optimization of compression ignition (CI) engine operated with Simarouba Biodiesel with four controlled factors namely as fuel fraction, compression ratio, injection pressure and injection timing with four different level. It was observed that among those input parameters, the compression ratio is the most significant parameter for the vibration response and followed by fuel fraction, injection pressure and injection timing. They also use a multiple regression analysis by least square method to identify the correlation between the parameters and the output which is vibration level. It was revealed that the 20% of fuel fraction has given the lowest vibration level and it can be concluded that by increasing the fuel fraction, the vibration levels were found to be decrease. For the other parameters, the vibration level at the lowest value when the compression ratio was reduced, also when the injection pressure at the highest value with 270 bar and the injection timing at 25° before Top Dead Center (bTDC) which means that 1° more than diesel engine manufacturer's recommendation.

While in different work, Ahmet Çalık has investigates the vibration characteristics of CI engine operated with diesel and Waste Cooking Oil (WCO) biodiesel enriched with hydrogen gas. The investigation has been done by using different types of fuel and engine speed. The results indicated that the highest vibration levels were measured in z-axis which is vertical due to the upward and downward motion of the piston. It has been stated that in all engine speeds, the vibration levels were found to be highest with the use of conventional diesel fuel and the decrement was observed about 3.6% when using WCO. Further decrement was recorded when the hydrogen was introduced, by diminishing about 1.5% and 4% of vibration level respectively when using conventional diesel and WCO biodiesel [31].

With the same type of biodiesel which is WCO, Chiatti et al.[37] has carried out a work to identify the vibration characteristics in the city car engine without the additional gas introduced to the cylinder. They have concluded that the RMS of acceleration is significantly affected by the engine speed. The RMS values was observed to be increased with the increment of the engine speed due to the working principle of mechanical components in the engine and this was confirmed by [32]. Out of three types of blending ratio, B20 was found to have the lowest vibration level compared to B10 and B40.

By considering the presence of bio-alcohol in the biodiesel, a work has been done by Madhava et al. to investigate the effects of blending the Neem Methyl Ester (NME) with the bio-alcohol namely as methanol in term of vibration. It was found that at full load, the vibration levels were reduced by 36.4%, 50.7% and 40.4% for NME compared to diesel at first, second and third highest peak amplitude. They also found that with further increment of methanol in the blending ratio will increase the vibration level in the engine. Hence, it can be said that the presence of methanol will help to improve the vibration level and the blending ratio must be suitable in order to reduce the it [38].

After doing a thorough literature survey, it was found that there is still a research gap using a Palm Oil Methyl Ester (POME) blends with the butanol. Hence, this work is proposed to investigate the vibration level in the diesel engine fuelled with the POME + Butanol by carry out the investigation with the different engine speed and load.

2. Experimental Setup

2.1 Compression Ignition Engine

The diesel engine (YANMAR TF 120M) that will be used in this experiment is a single cylinder, direct injection and coupled with a hydraulic dynamometer to give load to the engine. The load will be given by 25% and 50% that will be clarified as low and medium load to give a real-application of the engine. The engine was shown in figure 1 and the specifications as listed in table 1.

Figure. 1 YANMAR TF 120M CI Engine

Table 1. Engine Specifications

Specifications	YANMAR TF 120M
No of Cylinder	1
Cylinder Bore x Stroke, mm	92 x 96
Displacement, cc	0.638
Cooling System	Hopper/Radiator
Dimensions: Length x Width x Height, mm	695 x 348.5 x 530
Compression Ratio	17.7

2.2 Preparation of the Fuel

In order to run the experiments, the fuel must be prepared just right before the experiment to ensure that the mixture was completely dissolve together. There are three types of fuel that will be used where the diesel as a baseline fuel, the second one is the POME10+Bu10 and the last one is the POME15+Bu10. The mixtures were prepared using the mechanical stirrer (IKA RW 20 digital) with the rpm range of 700-800rpm for 15 minutes. The preparation of the fuel can be seen in the figure 2 below.

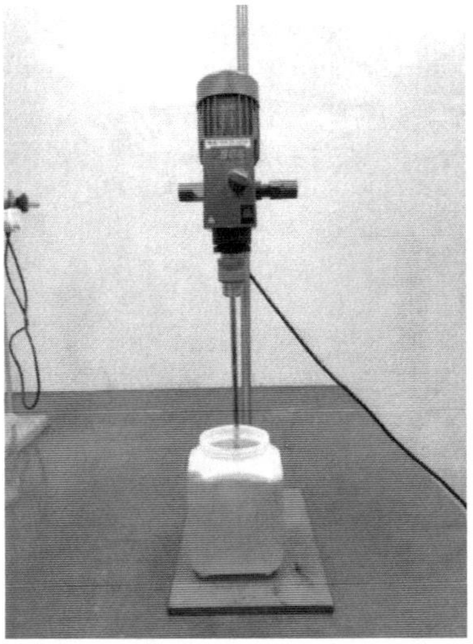

Figure.2 Blending process of the fuel

2.3 Vibration Measurement

For collecting the vibration data, NI-DAQ max instrument and uni-axial accelerometer were used in this work. The accelerometer sensor was placed right on the top of cylinder head to measure the vibration level from the piston movement since among the three axis which consists of vertical, longitudinal and lateral, the vertical and the longitudinal axis always shows a higher peak compared to the lateral. This condition will depend on the type of the engine used due to different types of working principle. For every experiment, the engine will be warmed up and let be stabled before recording the vibration measurement. The connection of the instruments and the location of the sensor were depicted in figure 3 and 4 below.

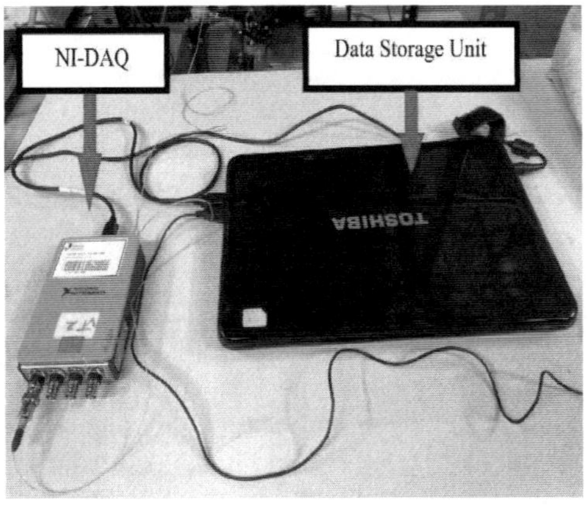

Figure 3. Connection of Instrument

Figure 4. Sensor Location

3. Results and Discussion

All the data that has been taken from the measurements has been plotted and analysed using MATLAB according to the engine speed and load. Figure 6 and figure 7 below will indicates the results on 25% load applied to the engine in 1800 RPM and 2100 RPM while figure 8 and figure 9 will show the results for the 1800 RPM and 2100 RPM with 50% of load. From all the graph shown below, it can be said that for every parameter the peak will happened at double frequency from the first peak frequency. This findings were consistent with findings of past studies by Patel et al. [39] which found that the peak of in-cylinder pressure occur in multiples of 12.5Hz. Since the vibration level were significant affected from the fluctuations of the in-cylinder pressure, then this theory will be related.

In order to determine the dynamic properties of the engine, a bump test was carried out. This test was intentionally to obtain the natural frequency to avoid the effect of resonant frequency in the combustion frequency. By giving force through impact hammer as the input and the reading from accelerometer as the output, the frequency response function of the engine will be determined. From the figure 5 below, it can be seen that two natural frequencies happened in the engine at 13Hz with the amplitude of 0.00927 g/N and 24Hz with 0.00254 g/N. Hence, these two frequencies will be avoided during the experiment in order to know the exact frequency that happened by combustion excitation forces.

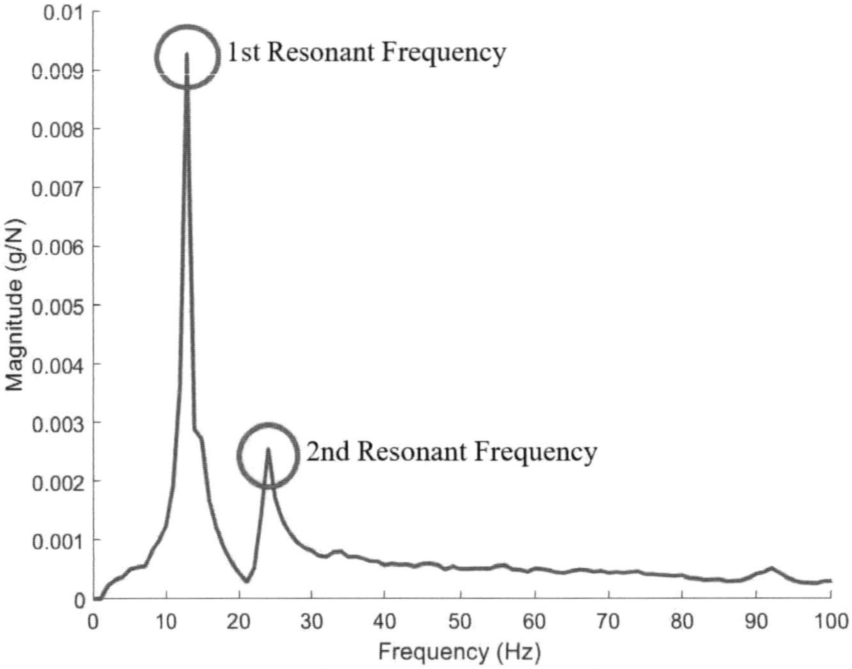

Figure 5. Bump Test of The Engine

By focussing at the operational frequency for each parameter, it was found out that there is a significance changes between each fuel. From figure 6, it is notable that the diesel was the highest vibration level compared to B10Bu10 and B15Bu10 with the value of 0.002123, 0.001343 and 0.001605 m/s. The decrement observed when using B10Bu10 was 36.74% while the B15Bu10 reduced about 24.4%. For the speed of 2100 RPM that were depicted in figure 7, the results still show the same trend where the highest vibration level was diesel. However, the overall vibration level for this speed was decreased compared to previous speed. When increasing the speed, with the value of 0.000891, 0.000455 and 0.000765 m/s for diesel, B10Bu10 and B15Bu10, the B10Bu10 tends to show a better reduction with 34.34% while the B15Bu10 still reduce the vibration but only with small percentage which is 14.14%. This condition may happen due to the high engine speed that lead to short premixing combustion phase hence will lead to incomplete combustion.

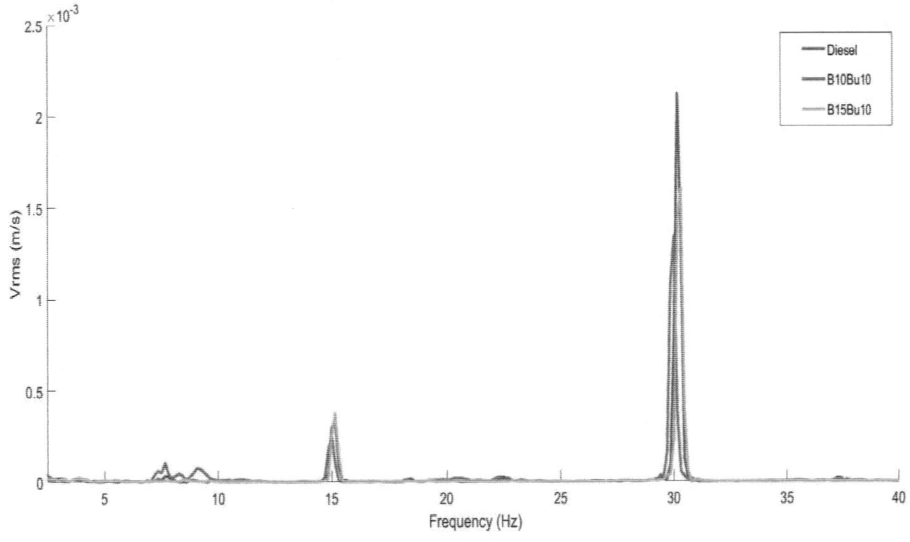

Figure 6. 1800 RPM with 25% of Load

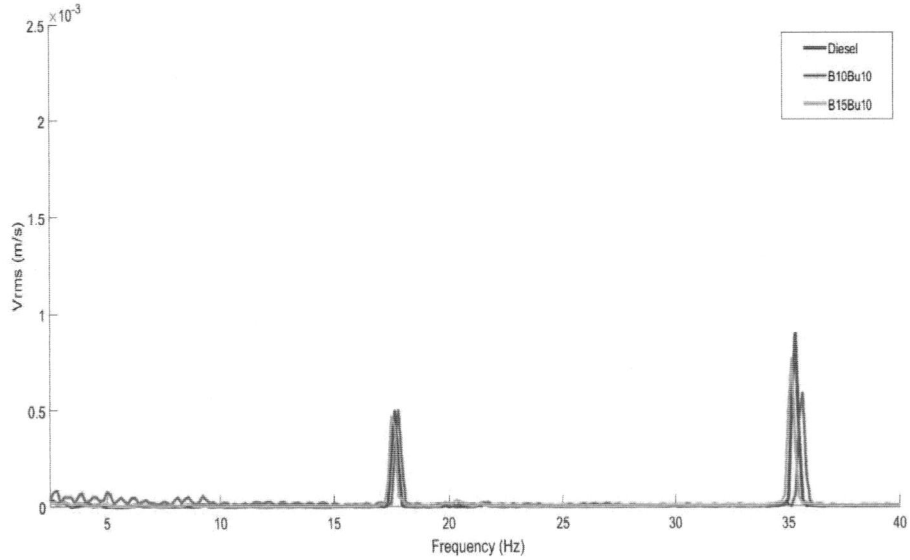

Figure 7. 2100 RPM with 25% of Load

Despites the peak frequencies in 25% load were dominated by diesel fuel, in 50% load, the B15Bu10 tends to shows a significance differences where it contributes a highest peak frequency. As can be observed from figure 8 below, the B15Bu10 contributes to higher vibration level as the load increases. It can be said that by increasing the biodiesel amount in the blends will increase the viscosity of the fuel that will affect the spray penetration and atomization of the fuel [37]. This theory can be confirmed when the B10Bu10 shows a lower peak when compared to diesel and B15Bu10. Being added that at higher loads, the air-fuel mixture will be uneven hence will leads to poor combustion [40-42]. However, it was noticed that when the speed was increased, the overall vibration level will decrease. The same phenomenon also happened in 25% load. With the value of 0.001931, 0.001671 and 0.002198 m/s for diesel, B10Bu10 and B15Bu10, the decrement when using B10Bu10 was 13.46% and when using B15Bu10, a sudden increment with 12.15% was observed. It is notable that when using biodiesel+butanol as a fuel owing to slightly early combustion frequency. This condition may happen due to the high cetane number of the biodiesel+butanol blends which lead to shorter ignition delay. From figure 9, it can be noticed that the diesel still contributes higher vibration level with the value of 0.000987 m/s while the B10Bu10 and B15Bu10 with 0.000761 and 0.000913 m/s. The decrement was observed to be 22.9% for the B10Bu10 and 7.5% for the B15Bu10. It can be concluded that the B10Bu10 does show a positive result by reducing the vibration level in both load condition compared to those two fuels. These conditions had proven that the B10Bu10 has a better fuel property with its blend ratios between diesel, biodiesel and butanol.

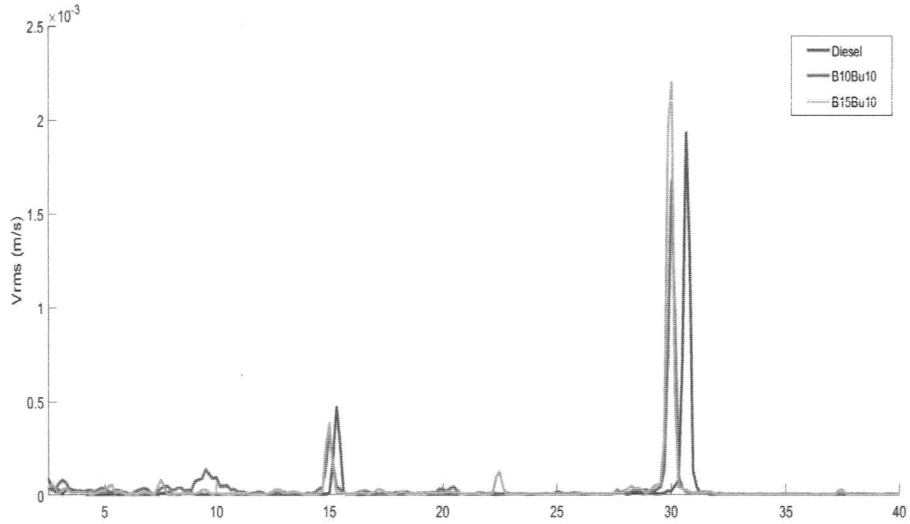

Figure 8. 1800 RPM with 50% of Load

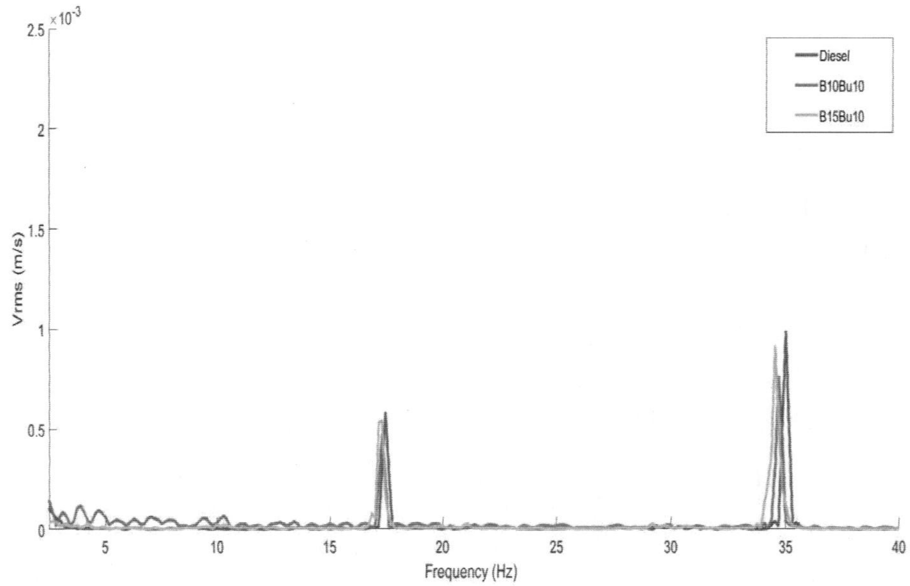

Figure 9. 2100 RPM with 50% of Load

4. Conclusion

After all the experiments and the analyzation of data were done, it was found that out of the three fuel that has being used, the B10Bu10 shows the best results by reducing the vibration level in both load conditions as high as 49%. With a further increment in biodiesel volume in the blends will increase the vibration level at high load which is related with the kinematic viscosity of the fuel that can affect the fuel atomization hence will lead to incomplete combustion. However, the best fuel still needed to be investigated by other analysis in order to ensure that it can replace the fully usage of diesel in current engine.

Acknowledgement

The author would like to greatly acknowledge the support by focus group of Advanced Structural Integrity of Vibration Research (ASIVR) and Universiti Malaysia Pahang for providing all the equipment used for this project under UMP research grant scheme – RDU 180381.

References

[1] Hoseini S S *et al.* 2017 The effect of combustion management on diesel engine emissions fueled with biodiesel-diesel blends *Renewable and Sustainable Energy Reviews* 73 307–331
[2] Çelebi Y and Aydın H 2019 An overview on the light alcohol fuels in diesel engines *Fuel* 236 890–911
[3] Prasad R and Bella V R 2011 A Review on Diesel Soot Emission, its Effect and Control *Bull. Chem. React. Eng. Catal.* 5(2)
[4] Reif K 2014 Fundamentals of Automotive and Engine Technology *Springer* 130–141
[5] Naik RT K S. and Maheswara Rao S U 2016 Performance and Emissions Characteristics of Variable Compression Ignition Engine *Adv. Automob. Eng.* 05(2)
[6] Li Y, Li H, Guo H, Wang H, and Yao M 2018 A numerical study on the chemical kinetics process during auto-ignition of n-heptane in a direct injection compression ignition engine *Appl. Energy* 212 909–918
[7] Silitonga A S *et al.* 2019 Intensification of Reutealis trisperma biodiesel production using infrared radiation: Simulation, optimisation and validation *Renew. Energy* 133 520–527
[8] Ullah K *et al.* 2015 Synthesis and characterization of biodiesel from Aamla oil: A promoting non-edible oil source for bioenergy industry *Fuel Process. Technol.* 133 173–182
[9] Damanik N, Ong H C, Tong C W, Mahlia T M I and Silitonga A S. 2018 A review on the engine performance and exhaust emission characteristics of diesel engines fueled with biodiesel blends *Environ. Sci. Pollut. Res.* 25(16) 15307–15325
[10] Shan R, Lu L, Shi Y, Yuan H, and Shi J 2018 Catalysts from renewable resources for biodiesel production *Energy Convers. Manag.*178 277–289
[11] Ali B, Yusup S, Quitain A T, Alnarabiji M S, Kamil R N M and Kida T 2018 Synthesis of novel graphene oxide/bentonite bi-functional heterogeneous catalyst for one-pot esterification and transesterification reactions *Energy Convers. Manag.* 171 1801–1812
[12] Gebremariam S N and Marchetti J M 2018 Economics of biodiesel production: Review *Energy Convers. Manag.* 168 74–84
[13] Arumugam A and Ponnusami V 2019 Biodiesel production from Calophyllum inophyllum oil a potential non-edible feedstock: An overview *Renew. Energy* 131 459–471
[14] Faried M, Samer M, Abdelsalam E, Yousef R S, Attia Y A and Ali A S 2017 Biodiesel production from microalgae: Processes, technologies and recent advancements *Renew. Sustain. Energy Rev.*, 79 893–913.
[15] Cancela A, Pérez L, Febrero A, Sánchez A, Salgueiro J L and Ortiz L 2018 Exploitation of Nannochloropsis gaditana biomass for biodiesel and pellet production *Renew. Energy* 133 725-730
[16] Rattanaphra D, Soodjit P, Thanapimmetha A, Saisriyoot M and Srinophakun P 2019 Synthesis, characterization and catalytic activity studies of lanthanum oxide from Thai monazite ore for biodiesel production *Renew. Energy* 131 1128–1137

[17] Vardast N, Haghighi M, and Dehghani S 2019 Sono-dispersion of calcium over Al-MCM-41used as a nanocatalyst for biodiesel production from sunflower oil: Influence of ultrasound irradiation and calcium content on catalytic properties and performance *Renew. Energy* 132 979–988.
[18] Nayebzadeh H, Saghatoleslami N, Haghighi M and Tabasizadeh M 2017 Influence of fuel type on microwave-enhanced fabrication of KOH/Ca12Al14O33nanocatalyst for biodiesel production via microwave heating *J. Taiwan Inst. Chem. Eng.* 75 148–155
[19] Rahmani V B and Haghighi M 2017 Biodiesel production from sunflower oil over MgO/MgAl2O4nanocatalyst: Effect of fuel type on catalyst nanostructure and performance *Energy Convers. Manag.*134 290–300
[20] Simasatitkul L, Gani R, and Arpornwichanop A 2012 Optimal design of biodiesel production process from waste cooking palm oil *Procedia Eng.* 42 1292–1301
[21] Endalew A K, Kiros Y, and Zanzi R 2011 Heterogeneous catalysis for biodiesel production from Jatropha curcas oil (JCO) *Energy* 36(5) 2693–2700
[22] Moradi G R, Dehghani S and Ghanei R 2012 Measurements of physical properties during transesterification of soybean oil to biodiesel for prediction of reaction progress *Energy Convers. Manag.* 61 67–70.
[23] Malhotra R and Ali A 2019 5-Na/ZnO doped mesoporous silica as reusable solid catalyst for biodiesel production via transesteri fi cation of virgin cottonseed oil *Renew. Energy* 133 606–619
[24] Somnuk K, Soysuwan N, and Prateepchaikul G 2018 Continuous process for biodiesel production from palm fatty acid distillate (PFAD) using helical static mixers as reactors *Renew. Energy* 131 100–110
[25] Tongroon M, Suebwong A, Kananont M, Aunchaisri J and Chollacoop N 2017 High quality jatropha biodiesel (H-FAME) and its application in a common rail diesel engine *Renew. Energy* 113 660–668
[26] Ouanji F, Khachani M, Boualag M, Kacimi M and Ziyad M 2016 Large-scale biodiesel production from Moroccan used frying oil *Int. J. Hydrogen Energy* 41(45) 21022–21029.
[27] Murugesan A, Umarani C, Subramanian R and Nedunchezhian N 2009 Bio-diesel as an alternative fuel for diesel engines-A review *Renew. Sustain. Energy Rev.* 13(3) 653–662.
[28] Uludamar E *et al.* 2017 Evaluation of vibration characteristics of a hydroxyl (HHO) gas generator installed diesel engine fuelled with different diesel–biodiesel blends *Int. J. Hydrogen Energy* 42 (36) 23352–23360.
[29] Uludamar E, Yildizhan Ş, Aydin K and Özcanli M 2016 Vibration, noise and exhaust emissions analyses of an unmodified compression ignition engine fuelled with low sulphur diesel and biodiesel blends with hydrogen addition *Int. J. Hydrogen Energy* 4(26) 11481–11490
[30] Uludamar E, Tosun E and AydIn K 2016 Experimental and regression analysis of noise and vibration of a compression ignition engine fuelled with various biodiesels *Fuel* 177 326–333.
[31] Çalık A 2018 Determination of vibration characteristics of a compression ignition engine operated by hydrogen enriched diesel and biodiesel fuels *Fuel* 230 355–358.
[32] Heidary B, Hassan-beygi S R, Ghobadian B and Taghizadeh A 2013 Vibration analysis of a small diesel engine using diesel-biodiesel fuel blends *Agric. Eng. Int. CIGR J.* 15(3) 117–126.
[33] Selim M Y E 2001 Pressure-time characteristics in diesel engine fueled with natural gas *Renew. energy* 22(4) 473–489
[34] Taghizadeh A and Mahdavian A 2019 Fault detection of injectors in diesel engines using vibration time-frequency analysis *Appl. Acoust.* 143 48–58.
[35] Manieniyan V and Sivaprakasam S 2013 Vibration Analysis in DI Diesel Engine Using Diesel and Biodiesel 4(8) 3586–3589.
[36] Kadam D V and Jadhav S D 2017 Optimization of Vibration , Performance and Emission of C . I . Engine Operated on Simarouba Biodiesel Using Taguchi and Multiple Regression Analysis *AeroTech Congress & Exhibition*
[37] Chiatti G, Chiavola O and Palmieri F 2017 Vibration and acoustic characteristics of a city-car

engine fueled with biodiesel blends *Appl. Energy* 185 664–670.

[38] Budharaju M V, Naradasu R K and Guvvala P 2017 Effect of oxygenated fuels on performance , combustion, emission and vibration characteristics of a compression ignition engine *Journal Biofuels* 7269 1–9.

[39] Patel C, Agarwal A K, Tiwari N, Lee S, Lee C S and Park S 2016 Combustion, noise, vibrations and spray characterization for Karanja biodiesel fuelled engine *Appl. Therm. Eng.* 106 506–517.

[40] Ravi M, Vijaya Kumar K C K and Murugesan A 2015 Certain investigations on the performance of emission, vibration and noise characteristics of C.I engine using bio gas and bio diesel as alternate fuel *Int. J. PharmTech Res.* 8(1) 11–19.

[41] Erdiwansyah, Mamat R, Sani M S M, Sudhakar K, Kadarohman A and Sardjono R E 2019 An overview of Higher alcohol and biodiesel as alternative fuels in engines *Energy Reports* 5 467-479.

[42] Erdiwansyah, Sani M S M, Mamat R, Khoerunnisa F, Rajkumar A R, Razak N F D and Sardjono R E 2018 Vibration Analysis of the Engine Using Biofuel Blends: A Review *MATEC Web of Conferences* 225 (01010)

[43] Erdiwansyah, Mamat R, Sani M S M and Sudhakar K 2019 Renewable energy in Southeast Asia: Policies and recommendations *Science of The Total Environment* 670 1095-1102.

[44] Erdiwansyah, Sani M S M, Mamat R, Khoerunnisa F, Sardjono R E, Ali O M and Ibrahim T K 2018 Effects of Diesel-Biodiesel Blends in Diesel Engine Single Cylinder on the Emission Characteristic *MATEC Web of Conferences* 225 (01013)

IOP Conf. Series: Journal of Physics: Conf. Series **1262** (2019) 012013 doi:10.1088/1742-6596/1262/1/012013

Finite element modeling and updating of the composite plate structure

R M Yaacob[1,2], M A H Hashim[1], and M S M Sani[1,3*]
[1]Advanced Structural Integrity & Vibration Research (ASIVR), Faculty of Mechanical and Manufacturing Engineering, Universiti Malaysia Pahang, 26600 Pekan, Pahang, Malaysia
[2]Faculty of Engineering and Technology, DRB-HICOM University of Automotive Malaysia (DHU), 26607 Pekan, Pahang, Malaysia.
[3]Automotive Engineering Centre, Universiti Malaysia Pahang, 26600 Pekan, Pahang, Malaysia

*Corresponding author: mshahrir@ump.edu.my

Abstract. Composite plate structure is broadly used in engineering structures for automotive and aerospace applications. Furthermore, the problem in vibrational is one of the industries main challenges. Composite materials are frequently used in industry for weight and cost reduction to enhance mechanical properties in structural components. This research aims to execute a composite plate structure model updating procedure. Through finite element analysis (FEA) and experimental modal analysis (EMA), modal parameters such as natural frequencies, mode shapes, and damping ratios are collected. Composite plate structure is set to be free-free boundary condition and node equivalence is done in the surface plate area. Correlation is executed between these two data sets. The discrepancies in natural frequencies between FEA and EMA were reduced with the selected parameters identified to perform structure updating model using sensitivity analysis.

1. Introduction

Natural fiber composites have been one of the major research topics in recent decades for a number of reasons, including their potential to improve sustainability at a lower cost to replace synthetic fiber reinforced plastics [1]. Several literature studies have been conducted to report natural plant-based cellulose fibers such as ramie, jute, flax, sisal, henequen, and pineapple as reinforcement of composites due to low cost, low density, high specific strength, and modulus, as well as good insulation and acoustic properties [2-7]. Natural fiber-reinforced bio-composites was a good alternative to conventional materials for applications in aerospace, underwater, and transport [8]. Composite materials, therefore, came into play, which serves as a confining material for all applications requiring weightless and high resistance applications in structural and non-structural fields. Combining two or more constituents that should have different chemical or physical properties forms the composite material. One constituent is referred to as reinforced material or fiber, and the other is referred to as the matrix embedding fiber. This newly formed material has different characteristics and is known as a composite material. These individual components remain separate and within the finished structure can be in a different orientation with each other [9]. Composites have good vibrational properties, especially when added with a natural fiber like reinforcements. India has significant natural fiber availability like a fiber, jute fiber, coir fiber, sisal, pineapple fiber, banana fiber, and ramie fiber [10].

Based on previous literature studies, composite materials have some unique properties at the microscopic level, such as homogeneous and heterogeneous, it has high strength, rigidity, toughness and results in weight and cost reduction [10]. Fiber's function is to enhance composite properties while the matrix of composites used to support and protect the fibers. It can be increased to enhance the mechanical strength of these composites by improving molecular bonding in the matrix by adding materials such as acrylic resin, etc. that have good molecular structured material properties [9].

The method of finite element (FE) is one of the practical methods of engineering analytical modeling. The construction of large and sophisticated analytical models has been made possible by modern computers capable of processing large matrix problems at high speed. Studies have found that the finite element analysis is broadly used to model structural dynamics to simulate the behavior of real systems because accurate results are considered to be reliable. According to Zienkiewics et al, FE was the most suitable tool in structural engineering for numerical modeling due to the ability to handle complex geometry structures, large complex assemblies of structural components and the ability to perform many different types of analysis [11]. In 2007, Zivanovic, S. et al. [12] pointed that the simulation of near-resonant dynamics is very sensitive to even small variations in modal parameters (natural frequencies, damping ratio, and mode shape) when performing design response calculations. Structural analysts are constantly challenged to produce better designs to meet the authorities ' requirements for safety, economic and environmental regulations. A current study attempts to cover a wide variety of modal analysis that focuses on determining the dynamic behavior of a test structure using experimental approaches via experimental modal analysis (EMA) and a numerical prediction technique using finite element analysis (FEA).

Updating the finite element model is an updating technique that provides more accurate data-related results than the other initial models [13]. The main purpose of model updating is to reduce the discrepancies for both numerical and experimental results by updating selected parameters from resonance frequencies and mode shapes [14-15]. The updating technique has two methods that are sensitivity method and the direct method. The most important task in model updating is selecting the updating parameter. If numerical predictions are insensitive to a parameter chosen, then updating will lead to a change in the uncertain value parameter. The difference between results and predictions was reconciled with changes to other (more sensitive) parameters that may require fewer updates. This is due to selecting only sensitive parameters to minimize percentage error discrepancies. Before the updating process, a sensitivity analysis had to be performed to identify which parameters were the most influential to choose from. As mentioned by [16], the selected parameters should be justified by an engineering understanding of the structure and the number of parameters should be kept to a minimum in order to avoid problems with un-conditioning. A few published papers managed to reduce the discrepancies between predicted results of FEA and measured data from EMA using FE model updating with MSC Nastran optimization algorithm, SOL200 [17-21].

The aim of this study is to examine the dynamic properties of the composite plate structure to be performed in numerical and experimental analysis for the determination of modal properties. Sensitivity method is one of the techniques of updating that focuses on model updating problem approach. Fabrication is carried out using the hand lay-up method and compression molding technique at a pressure of 100 – 110 bar and temperature of 140 – 150°C.

2. Experimental set-up

2.1 Materials

Ramie woven fibers are used as reinforcement material and this composite uses epoxy resin as a polymer matrix material.

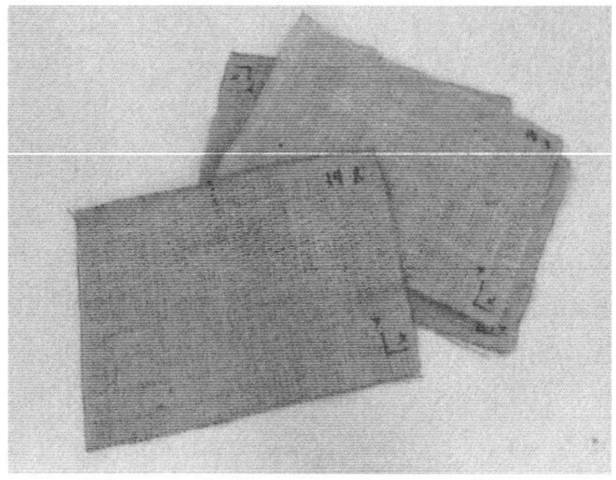

Figure 1. Ramie woven fiber

2.2 Hardener

The curing action of Epoxy is performed by mixing a substance or mixture called a hardener. It is added in an epoxy ratio of 1:2.

2.3 Fabrication of material

Fabrication of the composite ramie-enhanced polymer matrix is prepared and cured at room temperature [22 – 23, 28]. Resin and Hardener needed quantities were mixed in a beaker. The mixture of the Ramie Epoxy was stirring by using the chopstick for a few minutes until the mixture is fully penetrated. The mixture needs to be stirred in slow and constant speed to avoid the increasing number of air bubbles. As the speed increase, the reaction of the resin and hardener will increase, then it will produce more air bubbles. The reason for minimizing the mixture's air bubbles is to reduce the number of voids on the composite plate surface.

2.4 Mold preparation

The mold is made of 350 x 250 mm aluminum for the base of the mold and 263 x 193 mm for the composite plate mold frame. The aluminum plate for the frame structure was marked with the marker pen and cut using the hand grinder. Figure 2 (a-b).

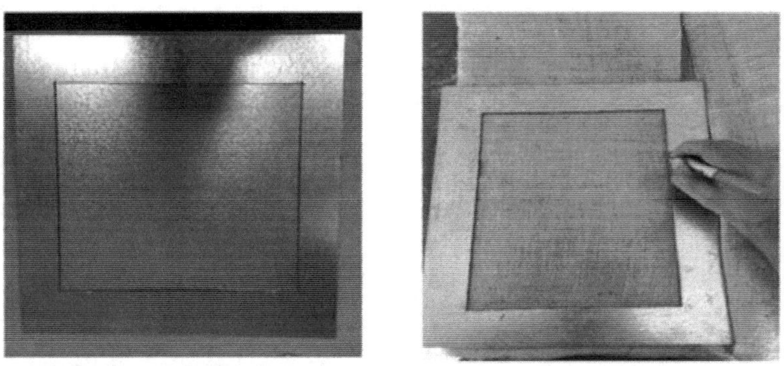

(a) (b)

Figure 2. (a) Aluminum Mold; (b) Marking Process on Ramie Woven Fibre

2.5 Curing of castings

Curing takes place at room temperature for approximately 24 hours at 25 degrees Celsius. Once the healing has been completed, the mold has been opened and the specimen carefully removed. The polystyrene barriers were created to block the excessive epoxy from the composite plate to prevent damaged the surface of the machine (Figure 3).

Figure 3. Compression Machine

3. Experimental modal analysis

Recent developments in the field of capabilities in data acquisition and processing have led to significant advances in experimental modal analysis or modal testing. Modal testing defines a study of a mechanical structure's dynamic characteristics to extract modal parameters (mode shape, damping ratio, and natural frequency The instrumentation and setup used in modal testing to influence the experimental result [24-27]; therefore, modal testing should be performed in free-free boundary condition in order to obtain the accurate result. A certain input (hammer, shaker) will excite the structure and the sensing mechanism will be used to measure the input force to produce a frequency response (FRF's) set. EMA plays an important role in structure design and analysis and validates the results of simulation models before they can be used for further detailed analysis. To achieve an adequate spatial resolution of global structural mode shapes, the composite plate structure was divided into 24 grid points. In order to avoid nodal points, the location and measuring points were carefully selected. Figure 4 shows the placement of grid points on the structure of the composite plate.

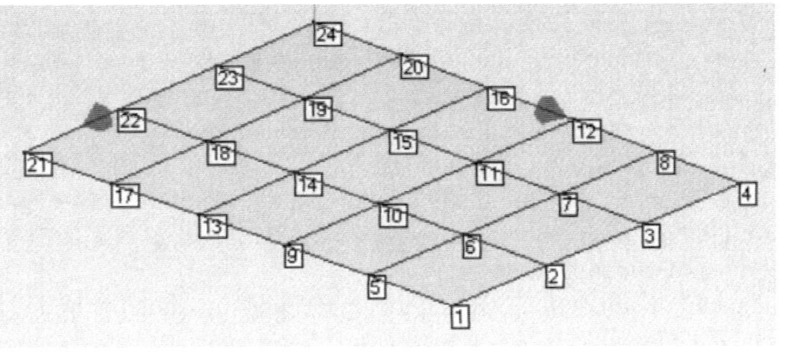

Figure 4. Composite plate structure with labeled grid points

In this study, the excitation method used impacts hammer with rowing accelerometer. Then, the impact hammer is attached to exciting the structure with a force transducer on its head. While a uni-axial accelerometer attached to the structure detected the response and the data analyzer was used to convert the signal response to the frequency domain in the time domain. The DAQ (Data Acquisition System) records the dynamic force signal and transfers the data to the computer to obtain FRF's by comparing the excitation of force and the acceleration of response signals. Figure 5 shows an experimental configuration for measuring systems.

IOP Conf. Series: Journal of Physics: Conf. Series **1262** (2019) 012013 doi:10.1088/1742-6596/1262/1/012013

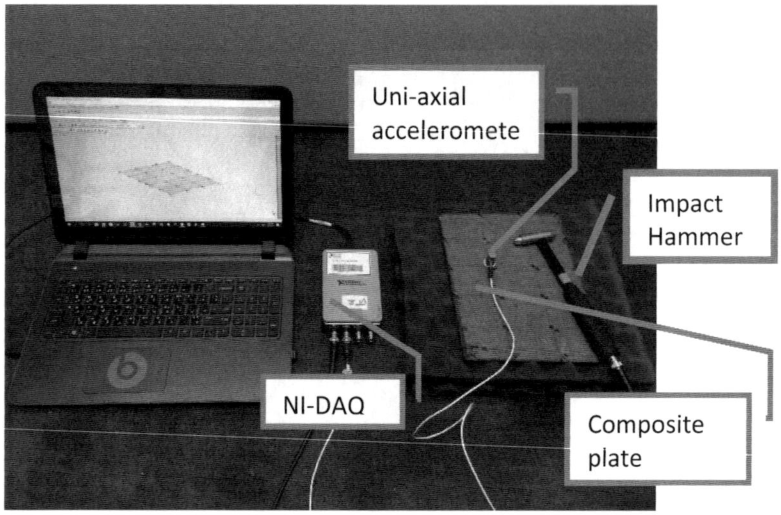

Figure 5. Experimental modal analysis setup.

Natural frequencies and mode shapes have been extracted from ME'scopeVES software using a curve fitting method and are summarized in Table 1.

Table 1: Resuts of experimental natural frequencies

Mode	Natural Frequency (Hz)
1	114
2	158
3	272

4. Finite element analysis

In this study, using MSC Nastran/Patran Software, FEA was performed. Modeling of finite elements took place in the free-free boundary condition. The interface FEA model consisting of 60 nodes and 40 shell elements (CQUAD4). Furthermore, the model properties are shown in Table 2. Node equivalence was performed in the model's entire area to represent the surface rigidity in the structure. In order to achieve better results, a higher percentage of mesh size is set, but a supercomputer is required to carry out the analysis [16]. For this study, 20-30 mode shapes were analyzed and the test specimen was not subject to any loading or boundary condition. By supporting the structure with a soft material such as a sponge, the free-free boundary condition was simulated. The simulation was performed part by part to compare it later with the experimental analysis.

Modal properties calculation in MSC.Natran/Patran was performed using SOL 103, the solution for normal mode analysis [26]. The mode shapes and natural frequencies of the structure can be calculated using normal mode analysis. The equation of motion in normal modes analysis is indicated as Eq. (1).

$$\boldsymbol{M}u + \boldsymbol{K}u = 0 \tag{1}$$

where K and M are respectively rigidity and mass matrices. MSC Nastran/Patran calculates these system matrices automatically based on the FEA model's geometry and properties. From Eq. 1, it can be reduced by assuming a harmonic solution to an eigenvalue problem which is indicated as Eq. (2).

$$[\boldsymbol{K} - \lambda_i \boldsymbol{M}]\{\emptyset_i\} = 0 \tag{2}$$

Where $\{\phi\}$is the eigenvector (mode shape) equivalent to the eigenvalue λ (natural frequency). Generally, each eigenvalue is proportional to a natural frequency and corresponds to eigenvector. The eigenvalues are correlated to the natural frequencies as Eq. (3).

$$f = \frac{\sqrt{\lambda_i}}{2\pi} \tag{3}$$

Table 2. Material properties of the plate structure

Properties		Nominal value
Material properties (Composite)	Young's Modulus, E	7.3 GPa
	Poisson ratio	0.3
	Density	1243 kg/m^3
Shell properties	Thickness	0.004 m

5. Correlation between EMA and FEA

The objective of the FEA-EMA correlation is to clarify the results obtained from this research, where discrepancies will be compared between both results and the percentage error will be captured. Table 3 indicates the natural frequencies attained from the finite element model and the structure experimental modal testing and the percentage of error in the various cases. Mode 1 had the lowest error percentage while Mode 2 had the highest error percentage. Table 4 illustrates the mode shape between FEA and EMA.

Table 3. Correlation between natural frequency, FEA, and EMA

Mode	FEA	EMA	Error (%)
1	114.14	114	0.12
2	131.98	158	16.47
3	254.21	272	6.54

This study indicates that the inclusion of residual stresses for both numerical and experimental results occurs when the plate tends to increase the natural frequency, as these stresses are primarily tensile in nature, thus increasing the lateral stiffness of the plate, resulting in an increase in natural frequency. This is due to residual stress is small as these stresses act on small areas relative to the plate area.

Table 4. Correlation between mode shape, FEA and EMA

Mode	Mode Shape	
	FEA	EMA
1	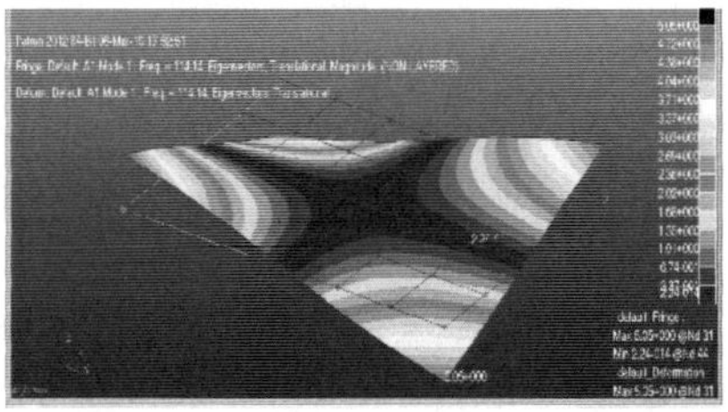	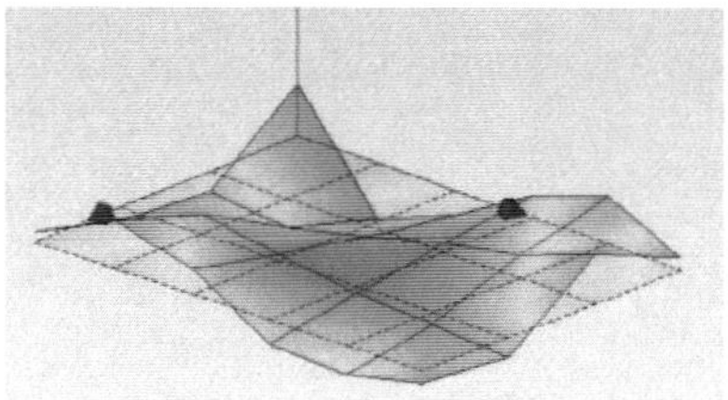

2	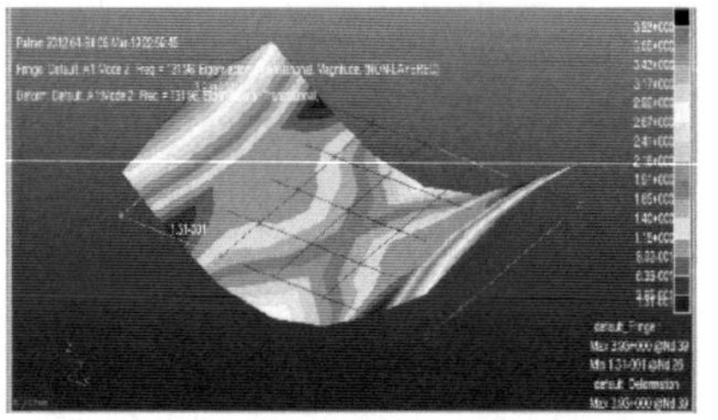	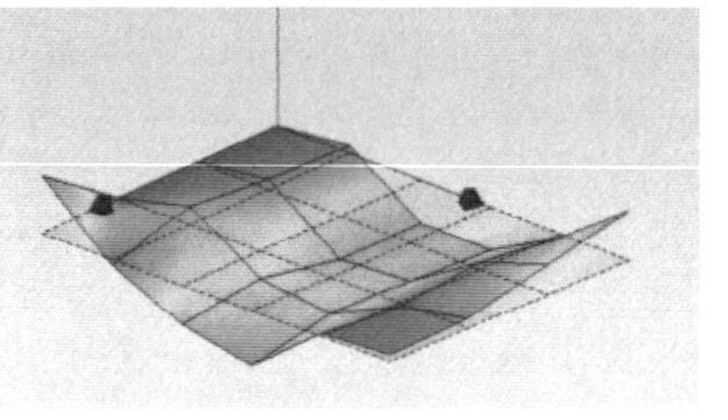
3	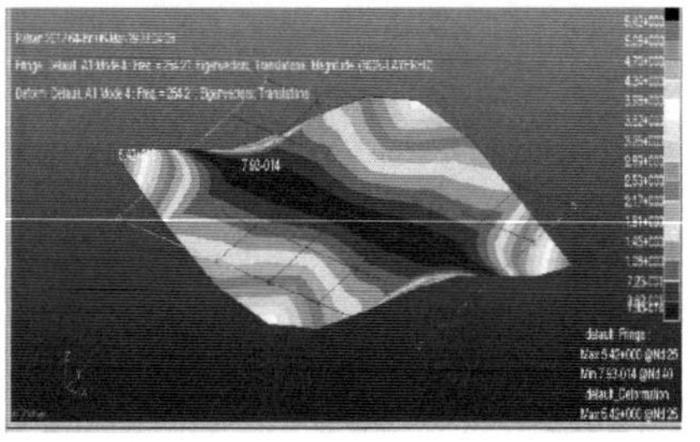	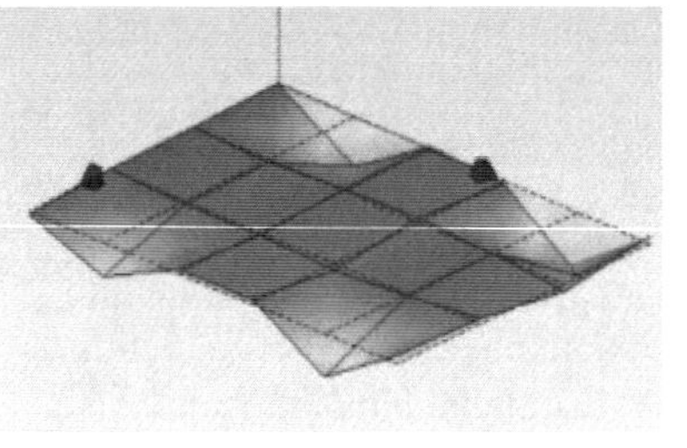

6. FE model updating

In finite element analysis, model updating is a process modifying selective parameters to decrease the discrepancies of the results for both experimental and numerical. The correlation objective is to minimize the error percentage between FEA and EMA. This can be accomplished by model updating. The iterative methods using modal data has been used in this study with the adoption of SOL200 optimization algorithm supported by MSC Nastran.

It is very important to perform model updating with a selection of sensitive parameters to minimize the discrepancies. In this study, which resulted in three parameters were picked, namely Modulus Young, plate density and thickness. Table 5 represents the results of the plate structure before updating, comparison of natural frequencies in FEA between before and after updating. Significant errors were observed in Mode 2 when the model updated reduced the error rate from 16.47% to 10.13%. Mode 1 shows an increasing percentage of error, but other modes show the percentage of error significantly reduced.

Table 5. Comparison values of natural frequencies for initial and updated results.

Mode	EMA natural frequency (Hz)	Initial FEA natural frequency (Hz)	Initial percentage of error (%)	Updated numerical natural frequency (Hz)	Percentage of error after Model Updating (%)
1	114	114.14	0.12	123	7.89
2	158	131.98	16.47	142	10.13
3	272	254.21	6.54	274	0.74
		Total Error	**4.63**	Total Error	**3.75**

Three parameters were identified to reduce natural frequency discrepancies between EMA and FEA through model updating. Table 6 indicates the update parameter deviation and explains that Young's Modulus, density, and thickness play a major role in reducing discrepancies. The thickness of the plate shows a higher sensitivity value when analyzing using sensitivity analysis.

This study indicated that the error percentage of mode 1 after updating was increased due to numerical analysis. This is because in this studies it using homogenous surface in FEA for the composite plate. In this case, approached of modeling plate perhaps not appropriate for composite because it has other methods such as solid surface. Besides that, it also can try and error for materials properties in FEA using composite structure based on a layer by layer for each fiber and matrix. As overall, the total percentage of error has decreased which is it achieve the main objective in updating the error from the initial result. In the other hand, the results also effect from updating method because it used the same weighted for each parameter. In future research, it can improve using different approaching of updating method.

Table 6. Changes in updating parameters from the initial values.

Parameter	A Initial Value	B Updated Value	Deviation (%) $= \left\|\frac{B-A}{A}\right\| \times 100$
Young's Modulus (GPa)	7.3	7.49	2.54
Thickness	.004	.0042	4.76
Density	1246	1208.62	3.09

7. Conclusion

This paper presented results that reduced the discrepancies on the composite plate between EMA and FEA. The equivalence of nodes is performed at the surface area in FEA with the free-free boundary condition applied to the composite plate. The findings of this study indicate that by selecting appropriate fabrication techniques, the bonding between fibers and matrix can be improved. In this study, three parameters were selected to update parameters. In an earlier phase, parameters were assumed when performing FEA, resulting in natural frequency discrepancies, this gap is narrowed down with the updating procedure. By applying the model updating technique using optimization of sensitivity analysis, the discrepancies between EMA and FEA have been reduced. This research has raised many questions that need to be investigated further. Further work needs to be done by focusing on updating the composite plate's different material properties and thickness, defining the element can likely reduce the discrepancies between EMA and FEA.

Acknowledgement

The authors of this paper would like to acknowledge and encourage the focus group of Advanced Structural Integrity of Vibration Research (ASIVR), Universiti Malaysia Pahang (UMP) to supply all the equipment used for this work.

References

[1] Pickering K L, Efendy M G A and Le T M 2016 A review of recent developments in natural fibre composites and their mechanical performance *Composites: Part A* 83 98-112

[2] Cazaurang-Martinez M N, Herrera-Franco P J, Gonzalez-Chi P I and Aguilar-Vega M J 1991 Physical and Mechanical Properties of Henequen fibers *Journal of Applied Polymer Science* 43(4) 749-756

[3] Joseph K, Thomas S, Pavithran C and Brahmakumar M 1993 Tensile Properties of Short Sisal Fiber-reinforced Polyetylene Composites *Journal of Applied Polymer Science* 47(10) 1731-1739

[4] Luo S and Netravali A N 1999 Interfacial and mechanical properties of environment-friendly "green" composites made from pineapple fibers and poly(hydroxybutyrate-co-valerate) resin *Journal of Materials Science* 34(15) 3709–3719.

[5] Luo S and Netravali A N 1999. Mechanical and Thermal Properties of Environment-Friendly "Green" Composites Made from Pineapple Leaf Fibers and Poly(Hydroxybutyrate-Co-Valerate) Resin *Journal of Polymer Composites* 20(3) 367-378.

[6] Luo S and Netravali A N 2001 Characterization of Henequen Fiber-poly(hydroxybutyrateco-hydroxyvalerate) Interface *Journal Adhesive Science Technology* 15(4) 423 – 437

[7] Lodha P and Netravali A N 2002 Characterization of Interfacial and Mechanical Properties of "green" Composites with Soy Protein Isolate and Ramie Fiber *Journal of Materials Science* 37(17) 3657–3665.

[8] Chen Y,Sun L, Chiparus O, Negulescu I, Yachmenev V and Warnock M 2005 Kenaf/Ramie Composite for Automotive Headliner *Journal of Polymers and the Environment* 13(2) 107-114.

[9] Chentan M R, Venkat P S N, Krishna G S G, Chennakesava R and Vijay P 2018 Dynamic Vibrational Analysis on Areca Sheath fibre reinforced bio composites by Fast Fourier Analysis *ICMPC. Materials Today: Proceedings* 5 19330 – 19339

[10] Panwar V, Gupta P, Bagha A K and Chauhan N 2018 A review on Studies of Finite Element Model Updating and Updating of Composite Materials *ICCMMEMS. Materials Today: Proceedings 5* 27912 – 27918.

[11] Wan Iskandar Mirza W I I, Rani M N A, Othman M H, Kasolang S and Yunus M A 2016 Reduced order model for model updating of a jointed structure *Jornal of Engineering and Applied Sciences* 11 2383–6

[12] Živanović S, Pavic A and Reynolds P 2007 Finite element modelling and updating of a lively footbridge: The complete process *Journal of Sound and Vibration* 301 126-45.

[13] Friswell M I and Mottershead J E 1995 Finite Element Model Updating in Structural Dynamics *Kluwer Academic Publishers Group* Norwell, MA

[14] Abdullah N A Z, Sani M S M, Rahman M M and Zaman I 2015 A review on model updating in structural dynamics *IOP Conf. Ser. Mater. Sci. Eng.* 100 012015

[15] Zahari S N, Zakaria A A R, Sani M S M and Bujang I 2016 A review on Model updating of joint structure for dynamic analysis purpose *MATEC Web of Conferences* 74 (00023)

[16] Husain N A, Khodaparast H H and Ouyang H 2010. FE model updating of welded structures for identification of defects *International Journal of Vehicle Noise and Vibration* 6 163-75.

[17] Zahari S N, Sani M S M and Ishak M 2017 Finite element modelling and updating of friction stir welding (FSW) joint for vibration analysis *MATEC Web of Conferences* 90 (01021)

[18] Omar R, Rani M N A, Wan Iskandar Mirza W I I, Yunus M A and Othman M H 2017 Finite Element Modelling and Updating for Bolted Lap Joints *Journal of Mechanical Engineering* 4 202–22

[19] Izham M H M, Abdullah N A Z, Zahari S N and Sani M S M 2017 Structural dynamic investigation of frame structure with bolted joints *MATEC Web of Conferences* 90 (01043)

[20] Abdullah N A Z, Sani M S M, Hussain N A, Rahman M M and Zaman I 2017 Dynamics properties of a Go-kart chassis structure and its prediction improvement using model updating approach *International Journal of Automotive and Mechanical Engineering (IJAME)* 14 3887-97

[21] Husain N A, Snaylam A, Khodaparast H H, James S, Dearden G and Ouyang H 2009 FE model updating for damage detection–application to a welded structure *Key Engineering Materials* 413 393-400.

[22] Venkat P S N, Chethan M R, Krishna S G G and Chennakeshava R 2017 Dynamic Vibrational Analysis on Areca Sheath fibre reinforced bio composites by Fast Fourier Analysis *Int. J. Mech. Eng* 5(9) 51 – 56.

[23] Mohd Zin M S, Rani M N A, Yunus M A, Wan Iskandar Mirza W I I, Mat Isa A A and Mohamed Z 2017 Modal and FRF Based Updating Methods for the Investigation of the Dynamic Behaviour of a Plate *Journal of Mechanical Engineering* 4 175–89

[24] Ewins D J 2000 Modal testing: theory, practice and application. Research studies press Baldock Vol.2.

[25] Fouzi M S M, Jelani K M, Nazri N A and Sani M S M 2018 Finite Element Modelling and Updating of Welded Thin-Walled Beam *International Journal of Automotive and Mechanical Engineering (IJAME)* 15(4) 5874-89

[26] Mottershead J E, Link M and Friswell M I 2011 The sensitivity method in finite element model updating: A tutorial *Mech. Syst. Signal Process* 25(7) 2275–2296.

[27] Sani M S M, Abdullah N A Z, Zahari S N, Siregar J P and Rahman M M 2016 Finite element model updating of natural fibre reinforced composite structure in structural dynamics *MATEC Web of Conferences* 83 (03007)

[28] Pulungan M A, Sutikno and Sani M S M 2019 Analysis of Bulletproof Vest Made from Fiber Carbon Composite and Hollow Glass Microsphere (HGM) in Absorbing Energy due to Projectile Impact *IOP Conf. Ser. Mater. Sci. Eng.* 206 012001

IOP Conf. Series: Journal of Physics: Conf. Series **1262** (2019) 012014 doi:10.1088/1742-6596/1262/1/012014

Experimental investigation of vibrations and noise characterization for spark ignition engine

Erdiwansyah[1,2], M.S.M. Sani[1,*], R. Mamat[1], J.M. Zikri[1], N.F.D Razak[1] and Munawir[2]
[1]Faculty of Mechanical and Manufacturing Engineering, Universiti Malaysia Pahang, 26600 Malaysia
[2]Fakultas Teknik, Universitas Serambi Mekkah, Aceh, Indonesia

[*]Corresponding author: mshahrir@ump.edu.my

Abstract. An experimental study was carried out for measuring vibrations, noise and combustion characteristics associated correlations using a four-cylinder spark ignition engine. This study aims to investigate the effects of vibrations and noise on spark-polishing machines that are triggered by pure gasoline. This test runs at 15% engine load and 20% with engine speed 1000-2200 rpm at intervals of 300 rpm. The measurement of engine vibration in pistonic uses PCB Piezotronics ICP® Accelerometer (352C22). Overall tests that have been carried out for 20% loads at 1900 and 2200 maximum vibrations, the velocity is 0.214 and 0.234, compared to 0.617 and 0.562 for 15% loads. From the results of engine vibration analysis, it can be reported that increasing engine speed and the load is given engine vibration can significantly decrease. Generally, noise increases for all samples tested; however, reduced noise is recorded when speed and engine load increase. This observation is also consistent with other observations related to vibration and combustion noise, heat release rates, and ignition delays.

1. Introduction

Various automotive industries are now increasingly competing to tie the quality of production, especially on personal vehicle engines and public transportation. All over the world, from cities to remote villages have used both small vehicles and large-bodied vehicles. In recent years the demand for vehicles has increased. Given the ever-increasing demand for the engine, it provides competition for the automotive industry to provide comfort to passengers, be more efficient, less polluting, cheaper, and friendly to the environment. In this context, a comprehensive investigation is important to do both fuel use and repairs to the combustion engine, so that vibration, noise and engine spray characteristics can be reduced. In this study, the relationship between the engine combustion process and its spray characteristics with vibration and noise was investigated.

Various scientists have tried and researched the properties of combustion, sprays, vibrations and noise sourced from internal combustion engines. They are starting from the use of fuel until the addition of engine material has been carried out. Currently, a lot of research into the effects of vibration and engine noise using fuel is done on diesel engines. Very little in the literature of investigation of vibration and noise is carried out on spark ignition engines. Besides, many vibration and noise investigations are carried out of the effects of using various fuels. Several researchers in recent years have studied investigations into the performance of spark ignition engines using alcoholic fuels such as ethanol. However, the effect of vibration and noise on the combustion engine uses very little fuel. Vibration in the engine is a fluctuation in the mechanical system or equilibrium position in the engine. If engine vibration is not controlled, it can cause safety to the occupants[1-2]. Besides, too much vibration also causes noise in the vehicle. The formation of engine vibration is caused by several structures such as engine speed, load, oil viscosity and the properties of the fuel used.

The process for calculating engine combustion optimisation is used from variations in the value of vibration, load, ignition and ignition time. This process, serves to increase fuel consumption, emissions and noise. Therefore, several attempts have been made to reduce engine vibration characteristics. Investigation of the effects of vibration and noise characteristics triggered by diesel-petrodiesel fuel has been carried out [3]. This test is carried out on a four-cylinder engine that has been modified from natural gas-gasoline fuel. The results of tests carried out can reduce vibration and noise and reduce carbon monoxide and hydrocarbon emissions. Investigation of vibrations and noise triggered by biodiesel Karanja fuel has also been carried out by [4]. The highest engine combustion noise is found in the 20% biodiesel mixture. The highest level of vibration is also found in Karanja-biodiesel at 20%. This is due to shorter combustion duration, higher heat release and increased ignition delays. The vibration and noise characteristic effects triggered by ethanol-gasoline blended fuels have been studied by [5]. The results of this experiment show vibrations and noise increased at 1500 and 2500 rpm. This increase is due to oxygen content and latent heat which is higher than the fuel mixture of ethanol-gasoline. While in studies with different vibration and noise characteristics on diesel engines using a diesel-biodiesel blend has been investigated [6]. This experiment was carried out on four-cylinder and four-stroke compression ignition engines.

Furthermore, the effect of vibration on the engine triggered by natural gas fuel is correlated with combustion pressure [7]. The results of this investigation show that the vibrations produced exceed mechanical vibration. Experiments on Genset machines using diesel-n-butanol mixed fuel to measure vibration and noise have been studied [8-10]. The test results can be reported that when the engine load is low, the vibration and noise are slightly better. However, when the engine load is high the vibration and noise increase. Correlation between vibrations, noise with engine combustion is very strong. Testing with methanol-gasoline fuel which is run on a spark ignition engine to evaluate engine vibration has been discussed [11]. This result shows the highest vibration obtained at the engine speed of 1000 and 1300 rpm for gasoline fuel. While the highest vibration of methanol-gasoline is recorded at engine speeds of 1600 and 1900 rpm. Furthermore, research on noise and vibration has been studied by several researchers [12–21]. However, in general, their analysis applies to diesel engines with various fuel mixtures.

The main aim of this study is to investigate the effects of vibrations and noise on spark-ignition engine that are triggered by pure gasoline. This test runs at 15% engine load and 20% with engine speed 1000-2200 rpm at intervals of 300 rpm. Vibration analysis using PCB Piezotronics ICP® Accelerometer (352C22). This tool is used to measure the level of the frequency spectrum and vibration in the spark ignition engine. Analysis of vibration values and engine noise is taken at each engine load and speed.

2. Experimental apparatus and procedure

Spark ignition engine four-cylinder, four-stroke was used as an experimental tool in this study. The specifications of this machine as shown in Table 1. Test parameters for data collection were obtained from engine loads of 15% and 20% with engine speeds of 1000 to 2200 rpm at intervals of 300 rpm. Before the engine data collection is heated first to normalize the vibration and noise on the engine. Schematic machine diagrams in this study are shown in Figure 1.

Table 1. Specifications for engine

Descriptions	Engine
Engine type	Mitsubishi 4G93 SOHC
Bore stroke	81.D mm x 89.0 mm
Number of cylinder	4 (1st cylinder is instrumented)
Fuel injection type	MCI-Multi (Electronically Controlled Multi-point) fuel injection
Piston displacement	1.834 L
CR	9.5:1
Max power	86 kw@5500 rpm
Max torque	161 nm@4500 rpm

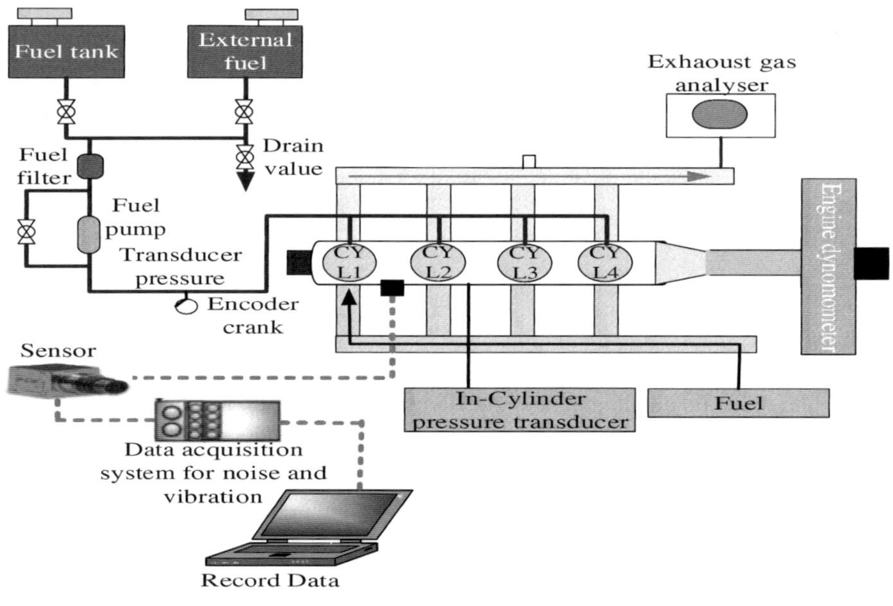

Fig. 1. Schematic diagram for engine setup

Fig. 2. Engine experiment

3. Result and Discussion

As explained in the previous section, the vibration and noise signals measured for engine acceleration are critical parameters for comparing the vibration signals from the spark ignition engine at 15% and 20% loads with different engine speeds. Figure 3 to 7 illustrate the vibration acceleration signal in two cycles for 15% and 20% engine loads at different speeds. Vibration signals generated from the engine have a relationship with the combustion event and valve opening/closing in the engine. From the results of this analysis, it can be noted that engine vibration decreases when the load and engine speed increase is shown in Figure 6 and 7. The maximum value for engine loads is 20% at speeds of 1900 and 2200 rpm at 0.214% and 0.234% lower compared to 15% engine load around 0.0617% and 0.562%. The reason for this is that the engine combustion level is getting higher, so that vibration acceleration is healthier than the lower engine speed. However, the reverse vibration signal increases when the engine speed is lower. Engine load which increases with low engine speed results in increased vibration on the engine as shown in Figure 3-5.

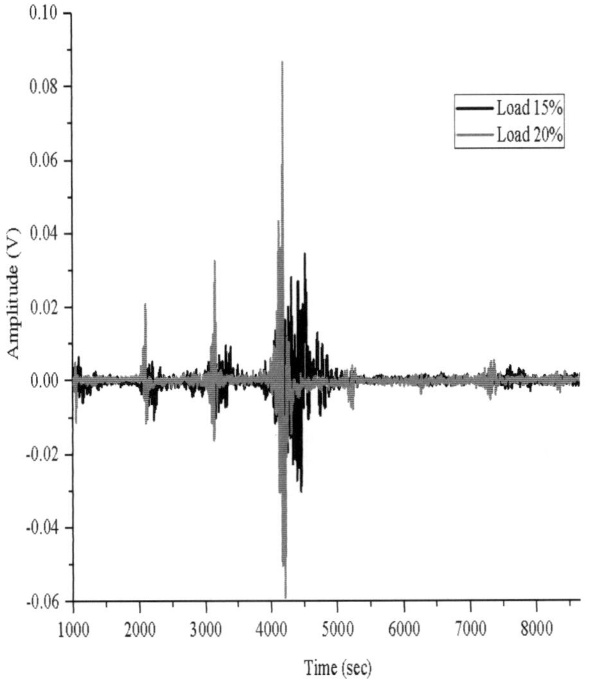

Fig. 3. Time domain signal of vibration acceleration at 1000 rpm

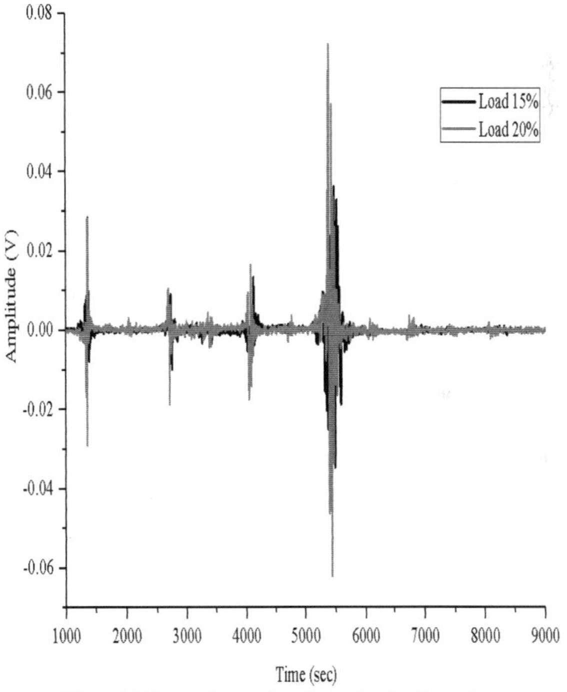

Fig. 4.Time domain signal of vibration acceleration at 1300 rpm

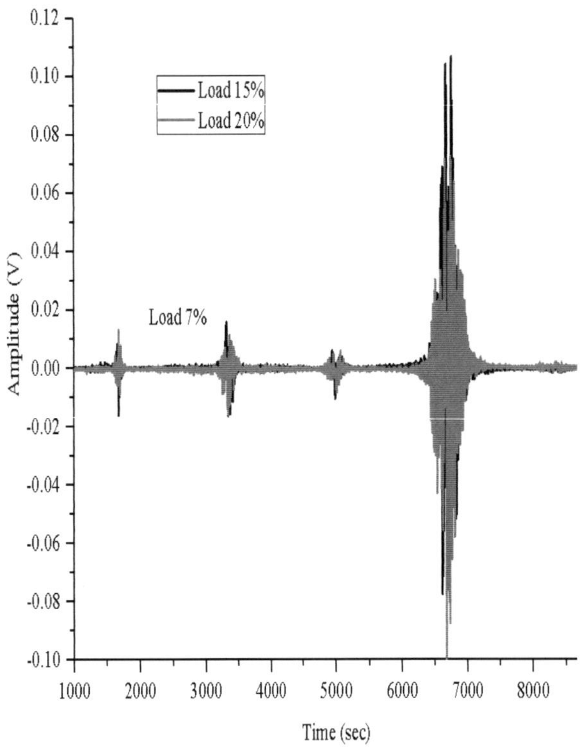

Fig. 5. Time domain signal of vibration acceleration at 1600 rpm

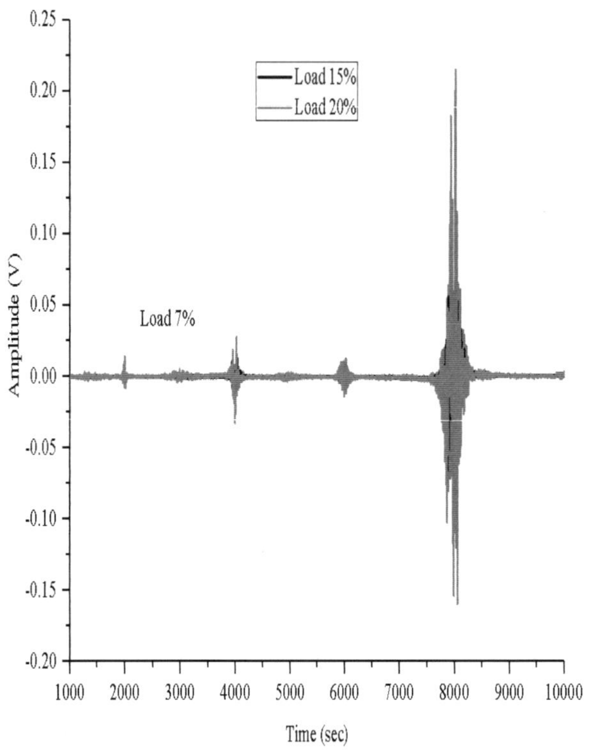

Fig. 6. Time domain signal of vibration acceleration at 1900 rpm

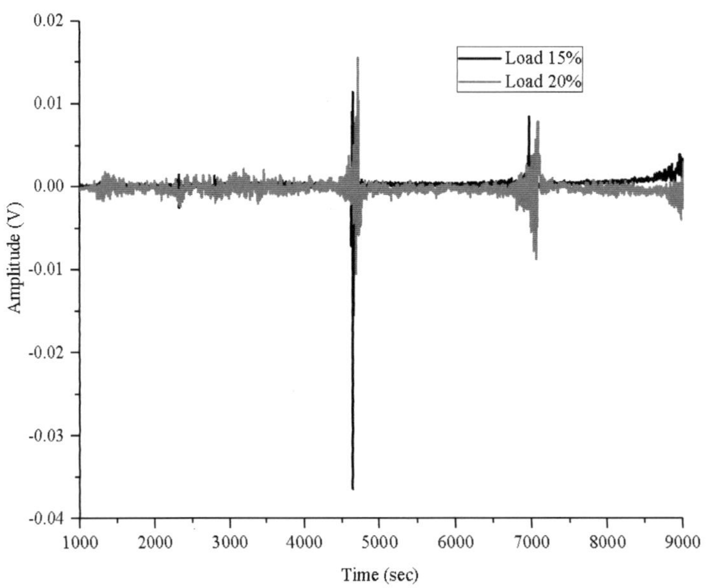

Fig. 7. Time domain signal of vibration acceleration at 2200 rpm

The maximum vibration value for engine speed 1300 and 1600 at a load of 20% is around 0.0812% and 0.174% compared to 0.0392% and 0.157% for a capacity of 15%. So that at the speed of this machine is very vulnerable to the emergence of unwanted engine vibrations. However, when the speed is lower as in Fig. 3, vibration is lower for higher loads than for low engine loads.

IOP Conf. Series: Journal of Physics: Conf. Series **1262** (2019) 012014 doi:10.1088/1742-6596/1262/1/012014

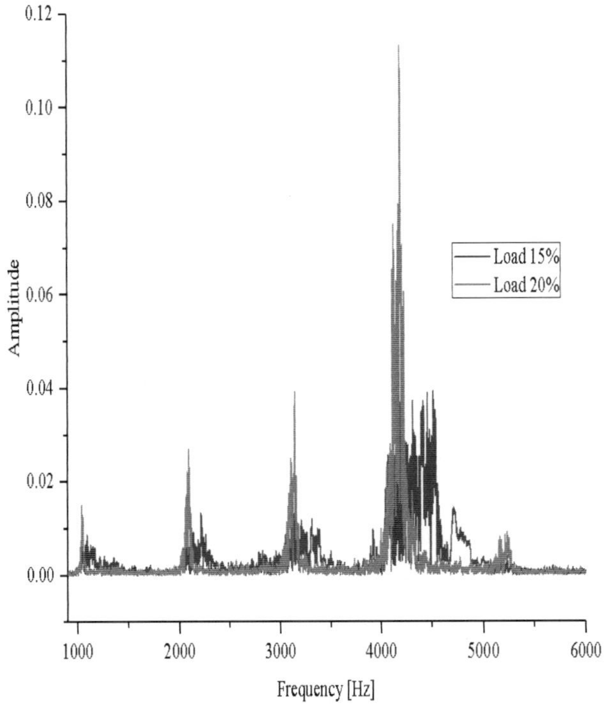

Fig. 8. Frequency spectra of the engine vibrations at 1000 rpm

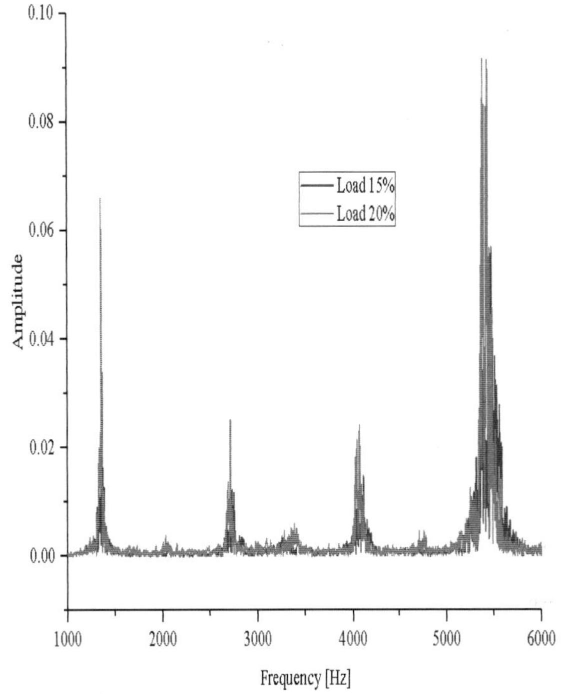

Fig. 9. Frequency spectra of the engine vibrations at 1300 rpm

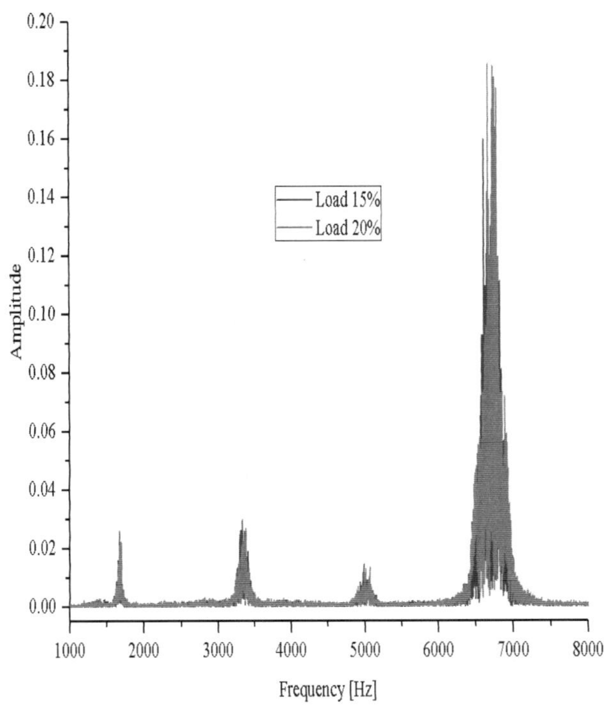

Fig. 10. Frequency spectra of the engine vibrations at 1600 rpm

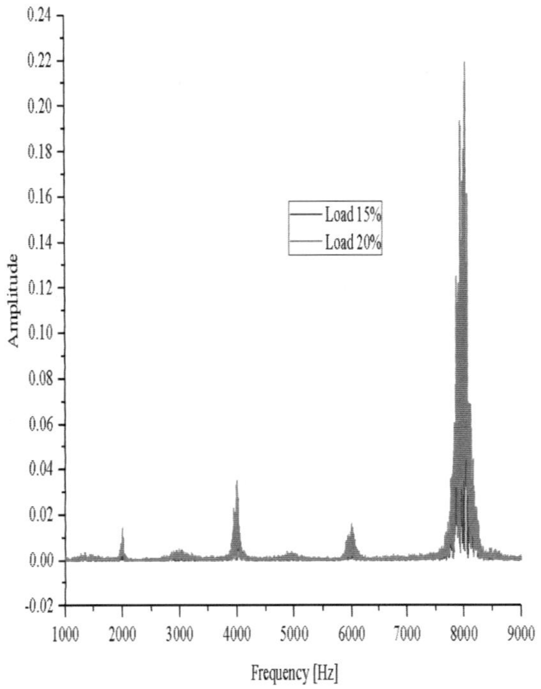

Fig. 11. Frequency spectra of the engine vibrations at 1900 rpm

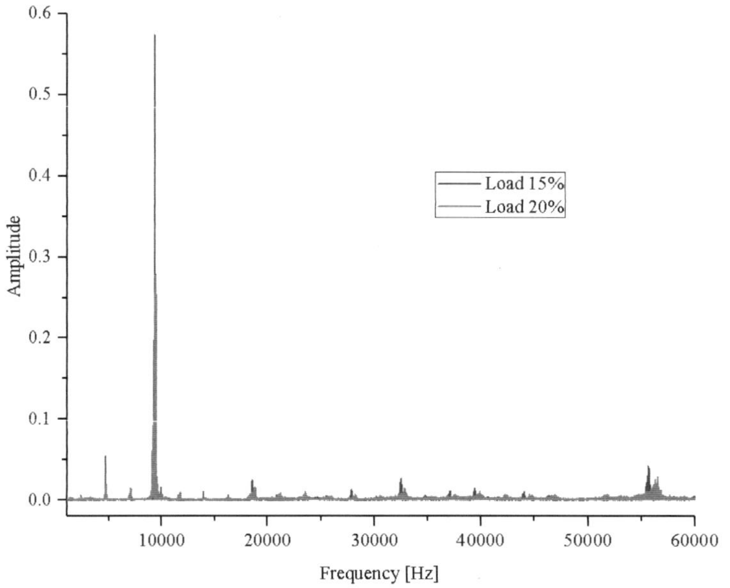

Fig. 12. Frequency spectra of the engine vibrations at 2200 rpm

Fig. 8-12, is the result of the frequency spectra obtained from engine vibrations for engine loads of 15% and 20%. The overall results tested show that the vibration level increases at low and medium engine speeds. However, at high loads and speeds, the vibration generated by the engine decreases as shown in Fig. 12. The characteristics of engine vibration are very significant where at high speeds the maximum frequency value is around 0.3 Hz compared to 0.6 Hz at lower loads. The results of the tests that have been carried out show the highest vibrations of data at engine speeds of 1900 and 1600 rpm with a frequency level of 0.23 Hz and 0.19 Hz respectively.

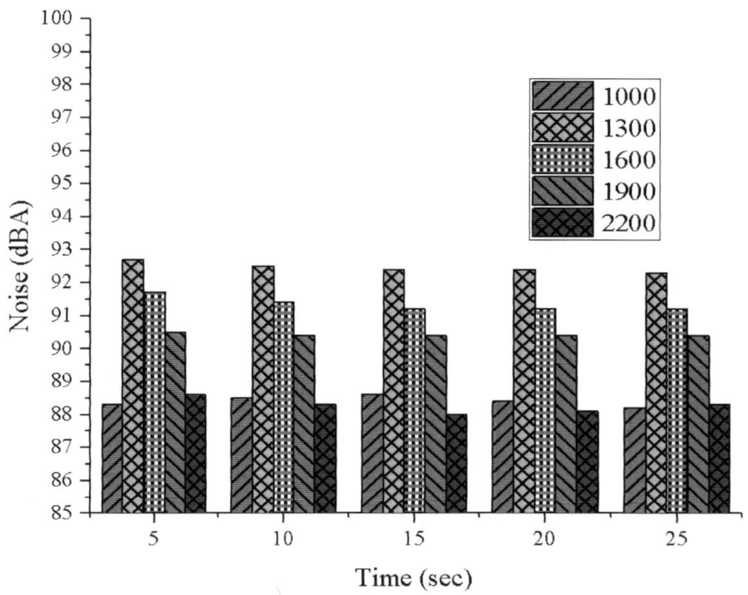

Fig. 13. Average of noise values of different speed at engine load 15%

The engine combustion noise decreases recorded at low speed and high speed shown in Fig. 13. The highest sound is obtained at a speed of 1300 rpm for each second measured. While for 10, 15 and 20 seconds the lowest noise is stared at the speed of 2200 rpm compared to the speed of 1000 rpm. However, at 5 and 25 the lowest noise is obtained at a speed of 1000 rpm. Similar results were found by [5]. Where the results of their tests reported that engine combustion noise for each sample tested experienced an increasing trend.

The increased engine speed and load can reduce engine combustion noise as shown in Fig. 14. The lowest noise is noted when the engine speed reaches 2200 rpm for each second compared to the overall engine speed tested. While the highest noise level is recorded at the engine speed of 1600 rpm. In general, the level of vibration and noise from the results of the analysis carried out on spark ignition engines with a load of 15% and 20% shows an increasing trend. Similar results found by [5]. Where the results of engine operation using ethanol-gasoline fuel increase from the results of the entire sample tested. While the results of experiments conducted by [4], the use of Karanja-Biodiesel 20% can increase engine vibration. Thus, engine vibration increases at low and medium speeds. However, reverse engine vibration decreases when engine speed increases in this study.

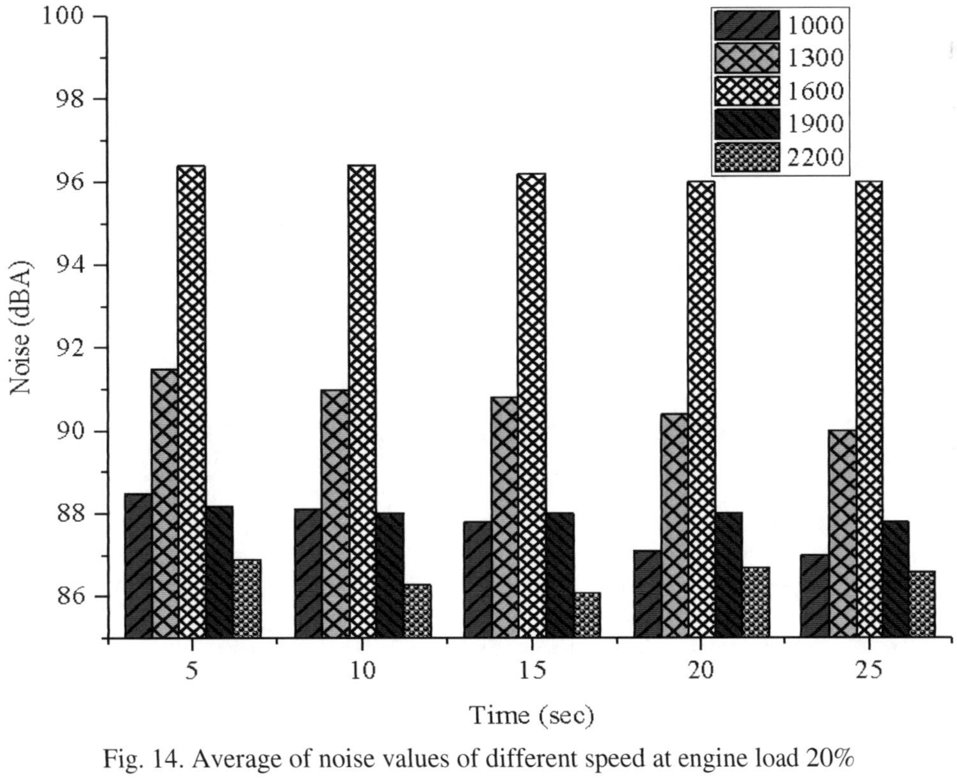

Fig. 14. Average of noise values of different speed at engine load 20%

4. Conclusion

This study focuses on measuring the vibration acceleration rate of combustion spark ignition engines with loads of 15% and 20% and different engine speeds. Based on the results of testing of the two engine loads given, the increased engine speed and load can reduce the vibration level in the engine. Decreasing engine vibration is recorded at speeds of 1900 and 2200 rpm for a load of 20% compared to a load of 15%. Overall tests that have been carried out for 20% loads at 1900 and 2200 maximum vibrations, the velocity is 0.214 and 0.234, compared to 0.617 and 0.562 for 15% loads. From the results of engine vibration analysis, it can be reported that increasing engine speed and the load is given engine vibration can significantly decrease. Meanwhile, engine noise generally decreases when engine speed and load increase. Of the two engine loads tested the lowest sound was recorded at the engine load of 20% for the overall engine speed compared to the engine load of 15%. Moreover, these results indicate a substantial relationship between combustion with vibration and noise originating from the engine.

Acknowledgments
The authors would like to acknowledge for the University Malaysia Pahang (UMP) financial support through short-term research grant scheme (RDU172204 and RDU130131).

References

[1] Hussey M 1983 Fundamentals of mechanical vibrations *Macmillan*
[2] Rao SS and Yap FF 2014 Mechanical Vibrations Ed. 4 *Prentice Hall Upper Saddle River*
[3] Taghizadeh-Alisaraei A, Ghobadian B, Tavakoli-Hashjin T, Mohtasebi SS. 2012 Vibration analysis of a diesel engine using biodiesel and petrodiesel fuel blends *Fuel* 102 414–422.
[4] Patel C, Agarwal A K, Tiwari N, Lee S, Lee C S and Park S 2016 Combustion, noise, vibrations and spray characterization for Karanja biodiesel fuelled engine *Appl Therm Eng* 106 506–517.
[5] Keskin A. 2010 The influence of ethanol-gasoline blends on spark ignition engine vibration characteristics and noise emissions *Energy Sources, Part A Recover Util Environ Eff* 32 1851–60.
[6] Ayd K and Ozcanl M 2016 Vibration, noise and exhaust emissions analyses of an unmodified compression ignition engine fuelled with low sulphur diesel and biodiesel blends with hydrogen addition *International Journal of Hydrogen Energy* 41(26) 11481-11490.
[7] Hunt S 2016 Engine Vibration of a Compressed Natural Gas Engine as an effect of Combustion *Electronic Theses and Dissertations* 5734.
[8] Sani M S M, Rahman M M , Baharom M Z and Zaman I 2015 Sound Intensity Mapping of an Engine Dynamometer *International Journal of Automotive and Mechanical Engineering (IJAME)* **11** 2820-9
[9] Satsangi D P and Tiwari N 2018 Experimental investigation on combustion, noise, vibrations, performance and emissions characteristics of diesel/n-butanol blends driven genset engine *Fuel* 221 44–60.
[10] Razak N F D, Sani M S M, Azmi W H and Zhang B 2017 Noise and vibration analysis for automotive radiator cooling fan *IOP Conf. Ser. Mater. Sci. Eng.* 257 012083
[11] Gravalos I, Moshou D and Gialamas T 2011Vibration Effects on Spark Ignition Engine Fuelled with Methanol and Ethanol Gasoline Blends *Journal of Agricultural Machinery Science* 7(4) 367–372.
[12] Erdiwansyah, Sani M.S.M, Mamat R, Khoerunnisa F, Rajkumar A R and Razak N F D 2018 Vibration Analysis of the Engine Using Biofuel Blends: A Review *MATEC Web Conf 225.*
[13] Heidary B, Hassan-Beygi S R, Ghobadian B, Taghizadeh A 2013 Vibration analysis of a small diesel engine using diesel-biodiesel fuel blends *Agric Eng Int J* 15 117–126.
[14] Yıldırım H, Çınar A, Şayli H and Köylü H 2015 Vibration and noise analysis of an engine fuelled with diesel and biodiesel blends *International Conference on Advances in Mechanical Engineering* Instanbul.
[15] Mirnezami S V, Hassan Beygi S R, Ghobadian B, Abdollahzade R and Ahmadian H 2014 Effect of diesel-biodiesel blends on user vibration transmissibility of riding power tiller through user body *The 21st International Congress on Sound and Vibration* Beijing.
[16] Taghizadeh-alisaraei A, Ghobadian B, Tavakoli-hashjin T 2016 Characterization of the engine's combustion-vibration using diesel and biodiesel fuel blends by time-frequency methods : A case study *Renew Energy* 95 422–432.
[17] Sani M S M, Zaman I and Rahman M M 2015 Analysis of Split Air Conditioner Noise using Sound Intensity Mapping *Journal of Advanced Research in Fluid Mechanics and Thermal Sciences* 9(1) 28-33.

[18] Uludamar E, Tosun E, Tüccar G, Yıldızhan Ş, Çalık A and Yıldırım S, 2017 Evaluation of vibration characteristics of a hydroxyl (HHO) gas generator installed diesel engine fuelled with different diesel–biodiesel blends *Int J Hydrogen Energy* 42 23352–60.

[19] Sani M S M, Rahimi M A F A and Mohamed D 2019 Sound Intensity Mapping of Two Stroke Engine by Using Hemispherical Surface Coordinate Arrangement *IOP Conf. Ser. Mater. Sci. Eng.* 506 012060

[20] Ravi M, Vijaya Kumar K C K and Murugesan A 2015 Certain investigations on the performance of emission, vibration and noise characteristics of C.I engine using biogas and biodiesel as alternate fuel *Int J PharmTech Res* 8 11–9.

[21] Razak N F D, Sani M S M, Azmi W H and Zhang B 2019 The Effect of Change in Flowrate on Power Spectral Density (PSD) of Automotive Radiator System for Flow-Induced Vibration Monitoring *IOP Conf. Ser. Mater. Sci. Eng.* 469 012101.

Experimental Modal Analysis Procedure for a Laser Spot Welded Hat-Plate Structure

M.S Mohd Zin[1], M.N Abdul Rani[1]*, M.A Yunus[1], W.I.I Wan Iskandar Mirza[1] and M.S.M Sani[2]
[1]Structural Dynamics Analysis & Validation (SDAV), Faculty of Mechanical Engineering, Universiti Teknologi MARA (UiTM), 40450 Shah Alam, Selangor, Malaysia.
[2] Advanced Structural Integrity and Vibration Research (ASIVR), Faculty of Mechanical Engineering, Universiti Malaysia Pahang, 26600 Pekan, Pahang, Malaysia

*Corresponding author: mnarani@uitm.edu.my

Abstract. Measuring the dynamic behaviour of a structure made from light thin steel sheets becomes arduous as a result of the double impact, local modes and mass loading issues which lead to inaccurate results. In addition, the attachments of accelerometers, selection of suspension types and methods of excitation adopted to the measurement greatly affect the quality of the experimental results. The aim of this paper is to present detailed procedure for the experimental setup of a welded structure made from thin steel sheets. A hat shaped plate and flat plate connected together by several laser spot welds were used for the demonstration of the experimental modal analysis. The laser spot welded hat-plate structure was tested under free-free boundary conditions using impact testing with roving accelerometers and an LMS Test Lab. This work revealed that the proper procedures used for the experimental setup would help dynamicists enhance and improve the quality of the measured results of frequency response function and the results could be confidently used for the validation purposes and updating analytical models.

1. Introduction

The experimental modal analysis (EMA) has been used for decades by researchers [1-5] to measure the dynamic behaviour of physical test structures. The dynamic behaviour is characterised in terms of natural frequencies and mode shapes of the structures [6]. EMA has helped the researchers to understand experimentally the global behaviour of the structures. Nevertheless, the behaviour of the structures can also be determined using the finite element (FE) method introduced by [7]. The FE method which is one of the most powerful numerical methods requires relatively high-speed computers to solve the engineering problems without using the physical structure [8]. Although solving engineering problems using the FE method, which is highly dependent on valid assumptions about model properties provided by the researchers, is faster than EMA, but the test results obtained from EMA are more reliable than those calculated from the finite element method. However, due to improper techniques of EMA procedure, the measured data is inaccurate and may not represent the real behaviour of the structure [9].

In the automotive industry, the assembled structures are made from thin steel sheets and connected with a thousand numbers of joints [10-11]. Complex structure with a large number of joints can exhibit a local behaviour that is difficult to accurately measure in the experimental modal analysis. Furthermore, EMA has the problems in deciding the measurement points, excitation point and boundary conditions. Hence, the right guideline on the experimental procedure is needed to make sure the measurement data is valid. Accurate EMA data is important to validate the prediction data analysed by finite element software. Numerous research work [12-17] have used EMA data to validate and improve their finite element model close to the measured one. For instance, [18-19] used experimental data to validate the reduce order finite element model of a jointed structure. EMA also used for damage detection on the structure as mentioned by [20-22]. However, no research work presents the guidelines for the experimental procedure for a welded structure made from thin steel sheets.

2. Experimental Set Up of Hat-Plate Structure

In this study, the experimental modal analysis was performed on the hat plate laser spot welded structure with a nominal thickness of 1.5 mm. The schematic diagram of the experimental modal analysis set-up is shown in Figure 1. The structure was discretized into several small squared segments. The purpose of the discretization was to have the appropriate number of the locations of measuring points. The determination of the number of the segments was carried out with the guidance from the results of modal parameters of the structure obtained from the finite element analysis. To simulate free boundary conditions, rubber bands and strings were used.

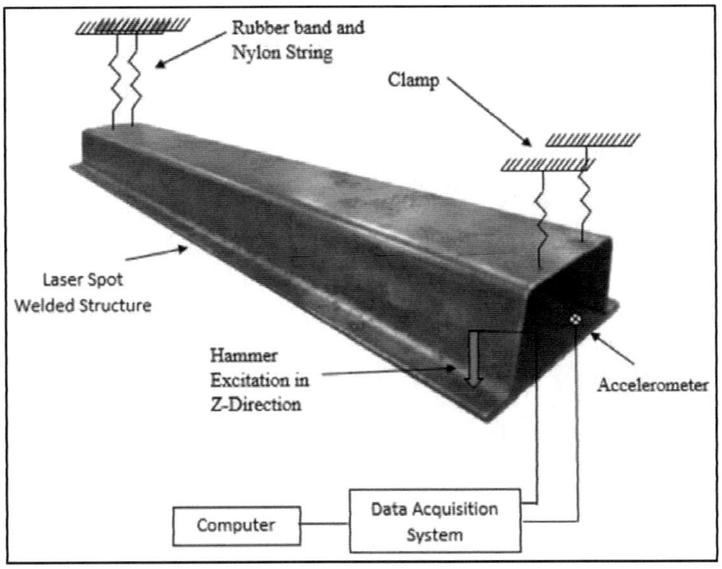

Figure 1: Experimental setup of laser spot welded hat-plate structure

Prior to performing the experimental work, several factors related to the experiments such as the number of accelerometers and measurement points and excitation methods should be considered. In this study, the initial prediction of the dynamic properties of the structure firstly performed to the test physical structure. Furthermore, the calculated natural frequencies and mode shapes of the structure were then used for the selection of the excitation points and the locations of measurement points of the physical test structure. The frequency bandwidth of interest was within 0 to 900Hz.

In the experimental work, an impact hammer and roving accelerometers were used to measure the dynamic behaviour of the physical test structure. A total of three accelerometers was used with one was fixed at the excitation point and the other two were roved to all measurement points.

2.1 Measurement Points

The number and location of points to be used for the measurement of the dynamic behaviour of a structure must be carefully considered and selected. The total number of the measurement points should appropriately be calculated prior to carrying out tests. The smaller the number of measurement points used in the tests, the lower the quality of the mode shapes of the structure measured. Meanwhile, the larger the number of measurement points, the longer the time required for the measurement of the mode shapes. In this regard, the best way is to use the initial FE results for guidance. This is because FE results give an initial overview of the dynamic behaviour of the structure under study.

Furthermore, the distribution and the size of discretization of the measurement points must be equally spread throughout the structure. This will result in a suitable mesh for the mode visualization and also can prevent losing any modes of interest under study. The number of accelerometers used in the experimental measurement must be carefully considered. A large number of accelerometers will increase the amount of mass of the structure and will change the local stiffness of the structure especially for thin steel sheets structure. Therefore, [23] suggested making sure that the total mass of accelerometers must not exceed 10 per cent of the weight of the structure.

In this experiment, one fixed and two roving accelerometers were used in the measurement of the dynamic behaviour of the hat plate spot welded structure. The mass of one accelerometer used in the test was 1.5 grams while the total mass of the assembled structure was 1776 grams. Hence, the total mass of the accelerometers calculated clearly does not give any mass loading issues.

2.2 Selection of suspensions

To accurately generate the experimental frequency response function (FRF) required to fathom out the modal parameters which are the natural frequencies and mode shapes of a structure. The structure must be tested under free-free boundary conditions. In this work, the test on the hat-plate laser spot welded structure was conducted by suspending the structure via four rubber bands and nylon strings as shown in Figure 1. The main reason the test was conducted in that particular set up is to remove the influence of boundary conditions from the test. This is because it is very difficult to incorporate experimental boundary conditions into analytical models. The difficulties have given impetus to researchers to investigate the dynamic behaviour of structures under free-free boundary conditions. The effects of the using of support conditions for free boundary conditions to ensure the accuracy of the measured results were thoroughly discussed [24]–[26].

Free-free boundary conditions do not mean that the structure under tests was completely not being supported by something, but it was sufficient to demonstrate that the structure can move freely within a certain period of time while impact testing was being performed. In addition, the values of rigid body modes measured must be 0 Hz or at least less than 1 Hz. However, the rigid body modes that were much lower than the first elastic mode have an insignificant effect on the accuracy of the measured results. In this regard, the rule of thumb that can be applied and has been proposed by [27], the highest value of rigid body modes must be less than 10 per cent of the first frequency of elastic mode.

In order to simulate a structure in free-free boundary conditions, the structure can be suspended from very soft elastic cords, springs with rubber bands or placed on a very soft cushion. Meanwhile, the selection of suspensions is based on the weight of the structure. In this experiment, four strings and rubber bands were used to simulate a free-free configuration. The selection of the rubber bands was made based on the weight of the laser spot welded hat-plate structure so that the structure can move freely when a force excitation is given. The method of suspending the hat-plate structure is shown in Figure 2.

Figure 2: The structure is suspended using rubber bands and strings

2.3 Methods of excitation

Excitation methods can be categorised by using shakers or impact testing. In the spectral testing method, a shaker is connected to the structure and the shaker provides the necessary excitation force which is based on the specified input voltage. There are three different types of signal inputs produced by shakers such as sinusoidal, random and periodic. There are two main types of shakers that are commonly used by researchers. They are electromagnetic and electro-hydraulic shakers. For the electromagnetic shaker, the force is generated by an alternating current that drives a magnetic coil. The limit of the frequency of the shaker depends on the size of the shaker. On the hydraulic shaker, the force is generated through the use of hydraulics, which provides much higher force levels.

It is worth noting that applying a force excitation through the use of a shaker system is regarded as not particularly suitable for a lightweight structure. This is because the stinger and load cell by which the shaker is attached to the structure will cause mass loading issues. In addition, the way of the shaker being attached to the structure should be properly and carefully handled, especially in the direction of the measurement being carried out. This is to prevent the structure from being damaged. Another disturbing issue, using shaker systems in tests is that the way of the shaker being supported. The main body of the shaker must be isolated from the structure to prevent any reaction forces from being transmitted through the base of the shaker back to the structure. Lastly, the limitation of using shaker systems is the difficulty in adjusting the angle of shaker perpendicular to the face of the structure.

As a result of the limitations of shaker systems, [28] developed a new method of an excitation technique which is called impact testing. Impact testing which offers economical ways in terms of time and cost has become the most popular modal testing method in the structural dynamics community. An impact hammer with a load cell attached to its head used to apply the impulse and to measure the input force. The selection of the size of an impact hammer is dependent on the size of the structure under tests. If the size of the structure is small, then the size of the hammer should be small as well and vice versa.

One of the major setbacks of impact testing is that the difficulty of controlling the input force of the hammer. The force level excitation varies from node to node and therefore, to overcome the varying levels the structure should be excited at least ten times for each of the predefined measured nodes. Then, the average data obtained from the ten excitations is calculated. Another issue in impact testing is to make sure the excitation energy will cover all the measured nodes and frequency of interest during the experiment process. The excitation energy depends on the stiffness of the contacting surface of the structure. If the energy does not cover all the frequency of interest, the tip of the hammer is not suitable for that structure [23].

There are three types of hammer tips which are steel, plastic and rubber that are commonly used in modal testing. The harder the tip, the shorter the frequency content. In this experiment, the hammer tip used was a plastic type, and it can cover the frequency of interest of this study which is from 0 to 900 Hz. The schematic diagram of the modal testing set up using an impact hammer is shown in Figure 3.

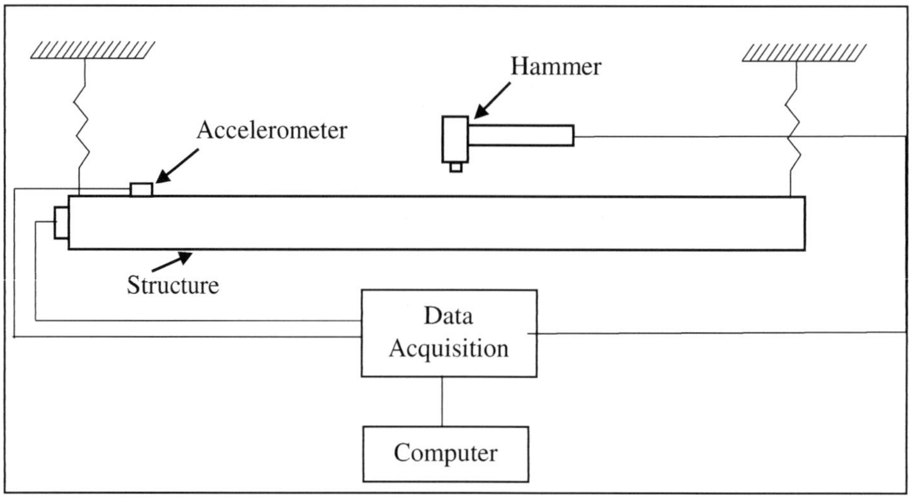

Figure 3: Schematic diagram of hammer excitation test set up

2.4 Attachment of accelerometers

Moreover, the location of the accelerometers attached to the structure is another factor that must be considered. The location cannot be far out from the nodes to minimize the error and missing modes. [29] introduced a new method for accelerometer placement using Generic Algorithm (GA) method. The method can estimate the target mode response with higher accuracy. Besides, the initial prediction from finite element model also can be used to decide the measuring point in the experimental modal analysis. All modes in the frequency range need to investigate the best accelerometers location that covers all mode shapes.

Accelerometers are attached to the structure by using adhesive, stud or magnetic mount at the measured nodes. Petroleum wax is commonly used to mount the accelerometers because it offers quick measurement of vibration. The amount of petroleum wax used to mount the accelerometers also needs to be considered. The excessive amount of wax will affect the accuracy of the measured value and cause mass loading issue. Hence, the usage of wax just enough to attach the accelerometers to the structure.

2.5 Frequency response function (FRF) analysis

Frequency response function (FRF) describes the relationship between excitation input forces and accelerometer output data in the function of frequency [30]–[32]. The dynamic behaviour (frequency, damping and mode shape) of the structure is obtained from a set of FRF measurement. Firstly, the response measured in the experimental modal analysis is in the time domain. Fast Fourier Transform (FFT) is used to transform the measured time domain to the frequency domain. The equation below [33] shows the formula used to calculate the frequencies before plotting in the graph:

$$H(\omega) = \sum_{r=1}^{n} \frac{\emptyset_i \emptyset_j / m}{\sqrt{\left(\omega_\eta^2 - \omega^2\right)^2 + \left(2\xi\omega\omega_\eta\right)^2}} \tag{1}$$

IOP Conf. Series: Journal of Physics: Conf. Series **1262** (2019) 012015 doi:10.1088/1742-6596/1262/1/012015

Where: ω = Frequency, ω_η = Natural frequency, $\varnothing$ = Mode shape, ξ = Damping and m = Mass

3. Results and Discussion

It is important to study the natural frequencies and mode shapes of engineering structures to help engineers to understand and design better structures. The well-accepted method of investigating the dynamic behaviour of the structures is the experimental modal analysis (EMA). However, there are no detailed guidelines, particularly on how to measure the dynamic behaviour of a welded structure made from thin steel sheets via the experimental modal analysis. On that account, the aim of this research was to present detailed guidance on how to accurately measure the dynamic behaviour of the structure using the experimental modal analysis.

Table 1 shows ten natural frequencies and corresponding model shapes of the hat-plate structure. The first mode is 503.832 Hz with the torsional mode shapes, and the last mode is 866.525 Hz. All modes have shown clearly deformation when the excitation frequency equal to the natural frequencies of the structure. The first three modes are shown the global bending and torsional modes while the higher modes tend to the complex in appearance and do not have common names. Luckily, the mode shapes do not have the local issues, and all modes have symmetrical movement.

Table 1: Experimental mode shapes and natural frequencies

Modes (I)	Shape (II)	Frequency Hz (III)
1		503.832
2	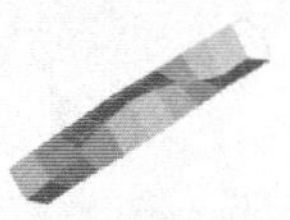	555.534
3		572.479
4		632.315
5		643.033
6		664.890
7		680.817

IOP Conf. Series: Journal of Physics: Conf. Series **1262** (2019) 012015 doi:10.1088/1742-6596/1262/1/012015

8		753.648
9		822.702
10		866.525

Figure 4 shows the frequency response function data of the hat-plate structure. The rigid body mode is lower and in the acceptable range which is 2.5 Hz. According to the rule of thumb, the rigid body modes must be less than 10 per cent of the first elastic mode. Hence, in this case, the rigid body mode was only 0.5 per cent from the first elastic mode. All the peaks in the frequency ranges illustrate that the structure is lightly damped. The sharp peak means that the structure has less than 5 per cent damping and can be negligible [34-35] in solving the equation of motion. The FRF also revealed that the experiment was carried out properly with less amount of noise in the FRF. Hence, this experimental data can be used as reference data in the model updating method.

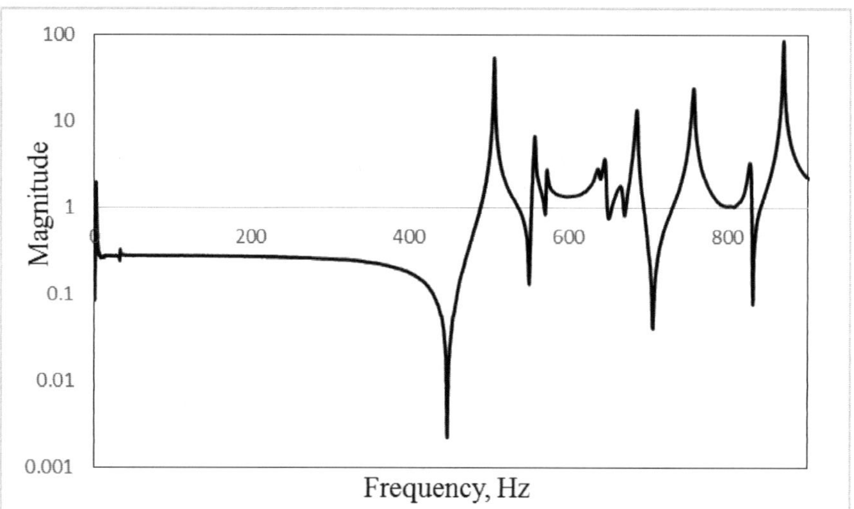

Figure 4: Frequency response function of the hat-plate structure

4. Conclusions

In this paper, the experimental procedure for a laser spot welded hat-plate structure made from thin steel sheets are presented. The five main topics about modal testing of the hat-plate structure which are measurement points, selection of suspensions, methods of excitation, attachments of accelerometers and frequency response function analysis have been thoroughly discussed. The experimental procedure has been successfully used to enhance and improve the quality of the measurement of the frequency response function of the hat-plate structure. The measured results then can be confidently used for validation purposes and also model updating of the finite element model of the hat-plate structure.

IOP Conf. Series: Journal of Physics: Conf. Series **1262** (2019) 012015 doi:10.1088/1742-6596/1262/1/012015

Acknowledgements

The authors wish to acknowledge the Malaysia Ministry of Higher Education (MOHE), Research Management Institute (RMI) and Institute of Graduate Studies of Universiti Teknologi MARA (UiTM) for providing financial support for this study through the fundamental research grant scheme (FRGS) (600-RMI/FRGS 5/3 (0096/2016)). They would also like to extend their sincere gratitude for the support and help given by Mr David Starbuck and Professor Hadariah Bahron and all the SDAV members.

References

[1] M. S. M. Zin, M. N. A. Rani, M. A. Yunus, M. S. M. Sani, W. I. I. Wan Iskandar Mirza, and A. A. Mat Isa, (2008). "Frequency response function (FRF) based updating of a laser spot welded structure", *AIP Conference Proceedings*, pp. 020055.

[2] M. N. Abdul Rani, S. Kasolang, and M. H. Othman, (2016) "Finite Element Modelling and Modal Based Updating for the Dynamic Behaviour of a Laser Spot Welded Structure", *23rd International Congress on Sound & Vibration (ICSV23)*, pp. 1-8.

[3] M. S. Mohd Zin, M. N. Abdul Rani, M. A. Yunus, W. I. I. Wan Iskandar Mirza, A. A. Mat Isa, and Z. Mohamed, (2017). "Modal and FRF Based Updating Methods for the Investigation of the Dynamic Behaviour of a Plate", *Journal of Mechanical Engineering*, Vol. 4(3), pp. 175–189.

[4] A. B. Ahmad Basri, M. N. Abdul Rani, M. A. Yunus, L. Roslan, W. I. I. Wan Iskandar Mirza, and I. Mirza, (2017). "The investigation of dynamic behaviour of a structure using wave-based substructuring method", *MATEC Web of Conferences*, pp. 01011.

[5] W. I. I. Wan Iskandar Mirza, I. Mirza, M. N. Abdul Rani, M. A. Yunus, L. Roslan, and A. B. Ahmad Basri, (2017). "The investigation of the dynamic behaviour of a complex assembled structure using the frequency response function based substructuring method", *MATEC Web of Conferences*, pp. 01012.

[6] P. Avitabile, (2001). "Experimental Modal Analysis", *Sound and Vibration*, pp. 1–11.

[7] M. J. Turner, R. W. Clough, H. C. Martin, and L. J. Topp, (1956). "Stiffness and Deflection Analysis of Complex Structures", *Journal of the Aeronautical Sciences*, Vol. 23(9), pp. 805–823.

[8] S. S. Bhavikatti, (2005). "Finite Element Analysis", *New Age International (P) Limited.*

[9] M. R. Ashory, (1999). "High Quality Modal Testing Methods", *Imperial College of Science, Technology and Medicine London*

[10] A. S. Y. Rashid, R. Ramli, S. M. Haris, and A. Alias, (2014). "Improving the Dynamic Characteristics of Body-in-White Structure Using Structural Optimization", *The Scientific World Journal 2014*, pp. 1–11.

[11] N. A. Z. Abdullah, M. S. M. Sani, N. A. Husain, and I. Zaman, (2017). "Dynamics Properties of a Go-kart Chassis Structure and its Prediction Improvement using Model Updating Approach", *International Journal of Automotive and Mechanical Engineering (IJAME),* Vol. 14(1), pp. 3887–3897.

[12] M. S. M. Fouzi, K. M. Jelani, N. A. Nazri and M. S. M. Sani, (2017). "Finite Element Modelling and Updating of Welded Thin-Walled Beam", *International Journal of Automotive and Mechanical Engineering (IJAME)*, Vol. 15(4), pp. 5874-5889.

[13] R. Omar, M. N. A. Rani, M. A. Yunus, W. I. I. W. I. Mirza, and M. S. M. Zin, (2018). "Efficient finite element modelling for the investigation of the dynamic behaviour of a structure with bolted joints", *AIP Conference Proceedings*, pp. 020082.

[14] A. B. Ahmad Basri, M. N. Abdul Rani, M. H. Othman, W. I. I. Wan Iskandar Mirza, M. A. Yunus, and L. Roslan, (2016). "Experimental Investigation of Dynamic Behaviour of a Structure with Closely Spaced Modes", *Journal of Engineering and Applied Sciences,* Vol. 11(11), pp. 2469–2473.

[15] R. Omar and M. N. Abdul Rani, (2017). "Finite Element Modelling and Updating for Bolted Lap Joints", *Journal of Mechanical Engineering*, Vol. 4(3), pp. 202–222.

[16] W. I. I. W. I. Mirza, M. N. A. Rani, M. A. Ayub, M. A. Yunus, R. Omar, and M. S. Mohd Zin, (2018). "Innovative FRF measurement technique for frequency based substructuring method", *AIP Conference Proceedings*, pp. 020010

[17] M. N. A. Rani, H. Ouyang and M. A Yunus, (2013). "Model Updating for a Thin Steel Sheet Welded Structure", *20th International Congress on Sound and Vibration (ICSV 20)*, pp 1-8.

[18] M. N. Abdul Rani, D. Stancioiu, M. A. M. A. Yunus, H. J. Ouyang, H. Deng, and S. James, (2011). "Model Updating for a Welded Structure Made from Thin Steel Sheets", *Applied Mechanics and Materials*, Vol. 70, pp. 117–122.

[19] W. I. I. W. I. Mirza, M. N. A. Rani, M. H. Othman, S. Kasolang, and M. A. Yunus, (2016). "Reduced order model for model updating of a jointed structure", *Journal of Engineering and Applied Sciences,* Vol. 11(11), pp. 2383-2386.

[20] M. Radzienski, M. Krawczuk, and M. Palacz, (2011). "Improvement of Damage Detection Methods Based on Experimental Modal Parameters", *Mechanical Systems and Signal Processing*, Vol 25, pp. 2169–2190

[21] M. A. Yunus, M. N. Abdul Rani, A. A. Mat Isa, W. M. W. Sulaiman, and R. Hassan, (2015). "Identification of Damage for a Thin Plate Jointed Structure", *Jurnal Teknologi,* Vol 76(8), pp. 33-37.

[22] J. Liu, T. Shi, K. Wang, Z. Tang, and G. Liao, (2012). "Defect detection of flip-chip solder joints using modal analysis", *Microelectronics Reliability,* Vol. 52(12), pp. 3002–3010.

[23] W. Heylen, S. Lammens, and P. Sas, (2007). "Modal Analysis Theory and Testing", *Katholieke Universiteit Leuven, Faculty of Engineering, Department of Mechanical Engineering, Division of Production Engineering, Machine Design and Automation.*

[24] T. G. Carne and D. T. Griffith, (2007). "Support Conditions for Free Boundary-Condition Modal Testing", *IMAC-XXV: A Conference & Exposition on Structural Dynamics*, pp. 1-13.

[25] T. G. Carne and C. R. Dohrmann, (1997). "Support conditions, their effect on measured modal parameters", *Fifth International Congress on Sound and Vibration*, pp. 477–483.

[26] Z. Chen, H. Xu, Z. Zhao, X. Yan, and B. Zhao, (2016). "Investigations on the mechanical behavior of suspend-dome with semirigid joints", *Journal of Constructional Steel Research,* Vol. 122, pp. 14–24.

[27] J. A. Wolf, (1984). "The Influence of Mounting Stiffness On Frequencies Measured In a Vibration Test", *SAE Technical Paper*, pp. 840480.

[28] N. M. M. Maia and J. M. M. Silva, (1997). "Theoretical and Experimental Modal Analysis", *Research Studies Press.*

[29] L. Yao, W. A. Sethares, and D. C. Kammer, (1992). "Sensor Placement for On-orbit Modal Identification of Large Space Structure via a Genetic Algorithm", *American Institute of Aeronautics and Astronautics,* Vol. 14(2), pp. 332–335.

[30] J. T. Wang, C. J. Wang, and J. P. Zhao, (2017). "Frequency response function-based model updating using Kriging model", *Mechanical Systems and Signal Processing*; Vol. 87, pp. 218–228.

[31] M. Arras, (2016). "On the use of Frequency Response Functions in the finite element model updating", *Carleton University.*

[32] F. Shadan, D. J. Inman, and A. Esfandiari, (2016). "Experimental validation of a FRF-based model updating method", *Journal of Vibration and Control,* pp. 1–14.

[33] G. Reynolds, (2000). "The Fundamentals of Modal Testing", *Agilent Technologies.*

[34] V. Arora, (2014). "Structural damping identification method using normal FRFs", *International Journal of Solids and Structures*, Vol. 51(1), pp. 133–143

[35] H. Mevada and D. Patel, (2016). "Experimental Determination of Structural Damping of Different Materials", *Procedia Engineering*, pp. 110–115.

Updating of the Dynamic Behaviour of an Assembled Structure with Bolted Joints in Light of EMA Results

R Omar[1,2], M N Abdul Rani[1*], M A Yunus[1], A A M Isa[3] and Z A Rahman[3]
[1] Structural Dynamics Analysis and Validation, Faculty of Mechanical Engineering, Universiti Teknologi MARA (UiTM), 40000 Shah Alam, Selangor, Malaysia
[2] Machinery Technology Centre, SIRIM Industrial Research, 44200 Hulu Selangor, Selangor, Malaysia
[3] Centre of Excellence of Structural Dynamics & Control, Faculty of Mechanical Engineering, Universiti Teknologi MARA (UiTM), 40000 Shah Alam, Selangor, Malaysia

*Corresponding author: mnarani@uitm.edu.my

Abstract. Bolted joints, which contribute significantly to minimise the magnitudes of resonant peaks of assembled structures and give easy-to-install and uninstall, have been widely used in various types of mechanical structural assembly and. However, modelling of an individual bolted joint within a large assembled structure has always been problematic, especially when it comes to predicting dynamic behaviour. The aim of this paper is to put forward an idea of efficiently and economically modelling of bolted joints of an assembled structure focusing on the bolting and assembly interfaces. The finite element (FE) models of individual components (Plate A and Plate B), bolting and assembly interfaces were developed using the NX11 Simcenter 3D software. Normal modes analysis was carried out for the initial FE models of the individual components and the assembled structure. Experimental modal analysis (EMA) was performed to determine the modal parameters of the individual components and assembled the structure. It was found that a low correlation between the initial FE models and EMA has been recorded. The initial FE models were then tuned in the light of EMA results in order to improve their accuracy and reliability. The proposed updating procedure has been successfully used in improving the low correlation which is largely due to the inaccurate modelling of the bolted joints.

1. Introduction

One of the essential features of bolted joints is easily assembling and dissembling. For this reason, bolted joints have been widely used in assembled structures. In addition, bolted joints play a significant role in the structural, dynamical response. Although bolted joints are important for energy dissipation of structures and alterations in structural damping [1] but efficient and economical representation of an individual bolted joint within a large assembled structure has always been troublesome and difficult, particularly for the prediction of dynamic behaviour.

The dynamic behaviour of an assembled structure with bolted joints can be predicted efficiently with the appropriate FE modelling. There are several factors which can contribute to the development of an efficient FE modelling of the bolted joints such as bolt modelling and joint interfaces [2-5]. However, very little attention has been paid to the development of an efficient FE model of an assembled structure with bolted joints and integrate the proposed method with the FE model updating procedure.

This work is concerned with the idea of bringing forward an efficient and economical modelling method for the bolted joints of an assembled structure focusing especially on the bolting and the assembly interfaces. FE model updating is used to update the initial FE results in the light of experimental data and to improve the accuracy and reliability of the FE model of the assembled structure with bolted joints which could be used for subsequent analyses.

2. FE Modelling and Analysis

The FE modelling and analysis was carried out by using the NX11 Simcenter 3D software. The identical FE models of the steel plates, named as Plate A and Plate B, were developed. The meshes of each plate were created by using 1288 CQUAD8 elements with the element size of 5 mm. The element size was used after several convergent tests were performed. Normal modes analyses for the plates A and B were conducted by using the solution type of SOL 103 Real Eigenvalues in order to characterise the first ten natural frequencies and mode shapes of the plates. The predicted results were used as the initial FE results in this study. The initial FE results were then compared with the measured results of the plates obtained from EMA. The FE model updating procedure was employed to reduce the errors introduced in the initial FE models as a result of invalid assumptions about model properties.

The updated FE models of Plate A and Plate B were assembled together to form a simple lap joint which is called an assembled structure with bolted joints in this work (figure 1). In modelling of the physical stainless-steel bolts and nuts, they were represented by the CBEAM element for the bolts' shank and RBEs for bolts' heads and nuts. The interfaces of the assembled structure were modelled using CELAS elements. The same procedures for FE analysis and model updating were exercised for the FE model of the assembled structure with bolted joints.

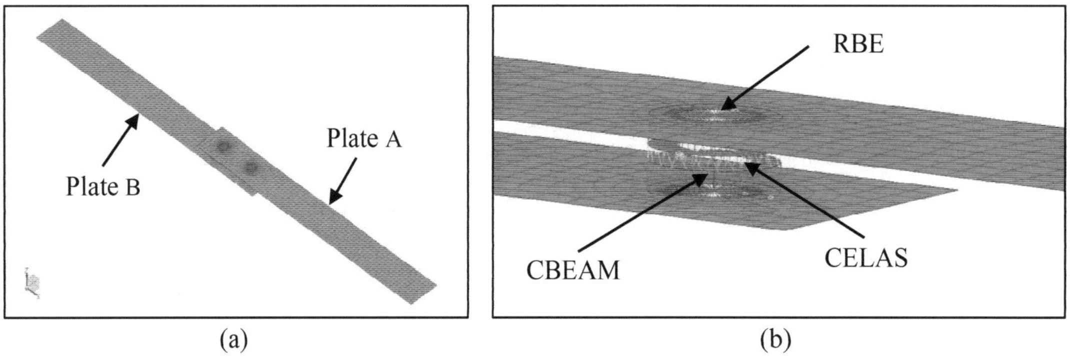

Figure 1. (a) Finite element model of the assembled Plate A and Plate B; (b) Representation of bolt, nut and assembly interface.

3. Experimental Modal Analysis (EMA)

The modal parameters, which are the natural frequencies and mode shapes, of Plate A, Plate B and the assembled structure with bolted joints were characterised by performing the EMA. The experimental setup for the assembled structure is shown in figure 2(a). Plate A and Plate B were joined by using the stainless steel M10 bolts and nuts. The components and assembled structural component under test were suspended to the test rig by using rubber bands to simulate free-free boundary conditions. The equipment used to characterise the modal parameters were the LMS system, LMS Test.Lab software, impact hammer & accelerometers as shown in figure 2(b).

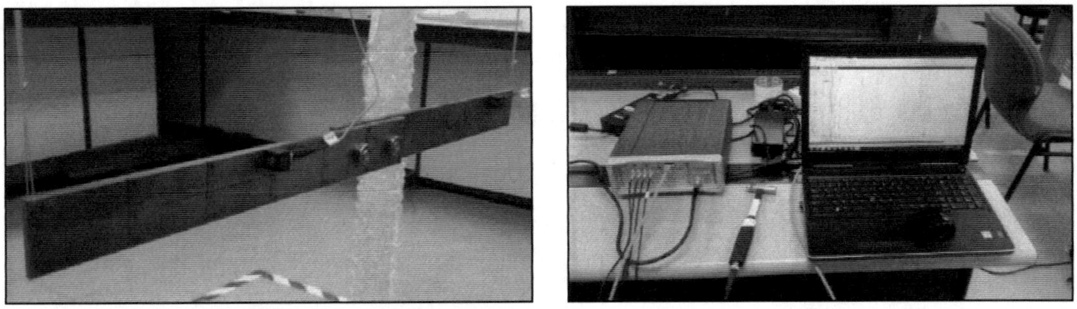

(a) (b)

Figure 2. (a) Experimental set up for the assembled structure with bolted joints; (b) Testing equipment for EMA.

4. FE Model Updating

FE model updating is used to improve the accuracy and reliability of the initial FE model, so that the predicted dynamic behaviour obtained from the updated FE model matches as close as possible to the measured dynamic properties [6-7]. Material and physical properties, such as Young's modulus, Poisson's ratio, shear modulus and mass density, are used as the updating parameters. Those parameters, which are sensitive to the finite element model's stiffness and mass properties [8], shall be systematically adjusted within the identified range to match the experimental results as closely as possible.

In this study, the SOL 200 Model Update of NX 11 Simcenter 3D was used in the attempt to update the FE models of Plate A, Plate B and an assembled structure with bolted joints. The first ten natural frequencies were included in the objective function as in equation (1) [9-10] to minimise the errors between the predicted and measured natural frequencies:

$$min \sum_{i=1}^{m} W_i \left(\frac{\omega_i^n}{\omega_i^e} - 1 \right)^2 \tag{1}$$

where ω_i^n is the i-th predicted frequency, ω_i^e is the i-th measured frequency, the weightage, W_i is set to unity and $i = 1,\ldots,10$.

5. Results and Discussion

In this work, the FE method and EMA were used to obtain the natural frequencies and the mode shapes of Plate A, Plate B and the assembly of both Plate A and Plate B. The FE model updating procedure was carried out in the effort to update the results of the initial FE models of Plate A, Plate B and the assembled structure in the light of the measured counterparts. The comparison of results between the measured, initial FE and updated FE, the updated parameters of Plate A, Plate B and the assembled structure are tabulated in table 1 to table 6.

5.1. FE Model Updating of Plate A and Plate B

The results of the natural frequencies and mode shapes obtained from the EMA and initial FE models for the first ten modes of Plate A and Plate B were compared. The total error for Plate A is 39.417% meanwhile for Plate B is 43.113%. The steepest descent algorithm was used as the optimizer to update the parameters of the FE models of both plates in order to reduce the total errors. The most sensitive parameters identified for the updating are the Young's modulus, Poisson's ratio and mass density. The comparison of results including the updated FE is shown in table 1 for Plate A, and in table 3 for Plate B. For Plate A, it was found that the error in the updated FE model has been successfully reduced from 39.417 to 4.271%. Meanwhile, the error of 43.113% in Plate B has been significantly decreased to 5.23%. The updated values of the parameters are shown in table 2 and table 4 for Plate A and Plate B respectively. It was revealed that there is a slight reduction in mass density and Poisson's ratio while an

IOP Conf. Series: Journal of Physics: Conf. Series **1262** (2019) 012016 doi:10.1088/1742-6596/1262/1/012016

increment was noticed in the Young's modulus for both plates A and B. This process gives an accurate modelling of the Plate A and Plate B before progress to the updating process for the assembled structure.

Table 1. Comparison of experimental, initial FE and updated FE for Plate A.

Mode	I. Experiment (Hz)	II. Initial FE (Hz)	III. Error between I & II (%)	IV. Initial FE MAC	V. Updated FE (Hz)	VI. Error between I & V (%)
1	225.992	215.992	4.425	0.969	225.098	0.396
2	623.044	596.321	4.289	0.915	621.306	0.279
3	1096.832	1045.870	4.646	0.913	1096.350	0.044
4	1221.802	1171.210	4.141	0.772	1219.890	0.156
5	1581.957	1542.180	2.514	0.671	1607.710	1.628
6	2015.090	1935.790	3.935	0.852	2015.480	0.019
7	2216.766	2115.930	4.549	0.938	2217.310	0.025
8	3002.219	2886.310	3.861	0.881	3003.870	0.055
9	3375.650	3225.620	4.444	0.981	3378.800	0.093
10	4041.950	3936.380	2.612	0.840	4105.650	1.576
	Total Error		39.417			4.271

Table 2. Updated parameters for Plate A.

Parameter	Initial Value	Updated Value	Unit
Young's modulus	200000	206400	MPa
Poisson's ratio	0.32	0.302	Unitless
Mass density	7.850E-06	7.458E-06	kg/mm^3

Table 3. Comparison of experimental, initial FE and updated FE for Plate B.

Mode	I. Experiment (Hz)	II. Initial FE (Hz)	III. Error between I & II (%)	IV. Initial FE MAC	V. Updated FE (Hz)	VI. Error between I & V (%)
1	227.041	215.998	4.864	0.958	226.864	0.078
2	626.170	596.143	4.795	0.929	626.046	0.020
3	1099.010	1046.010	4.823	0.899	1099.210	0.018
4	1229.100	1170.520	4.766	0.828	1229.010	0.007
5	1582.200	1542.290	2.522	0.604	1620.060	2.393
6	2027.970	1934.670	4.601	0.877	2030.960	0.147
7	2223.820	2116.590	4.822	0.908	2223.890	0.003
8	3021.030	2885.450	4.488	0.890	3028.390	0.244
9	3390.110	3226.740	4.819	0.913	3389.750	0.011
10	4042.010	3936.400	2.613	0.674	4135.360	2.309
	Total Error		43.113			5.230

Table 4. Updated parameters for Plate B.

Parameter	Initial Value	Updated Value	Unit
Young's modulus	200000	209400	MPa
Poisson's ratio	0.32	0.316	Unitless
Mass density	7.85E-06	7.45E-06	kg/mm^3

5.2. FE Model Updating of Assembled Structure

In this work, the updated FE models of Plate A and Plate B were joined to form a simple lap joint as an assembled structure. In the FE model updating of the assembled structure, the material properties of the bolts and the translational stiffness property of the CELAS elements for the interfaces were found to be the most sensitive parameters and were used as the updating parameters. A genetic algorithm was used as the optimizer for the updating of the assembled structure. The comparison of results including the updated FE is tabulated in table 5. It was found that the total error recorded in the updated FE model of the assembled structure has been reduced from 21.093% to 18.874%. The updated values of the parameters used in the updating procedure are shown in table 6 from which it is revealed that there is a significant increment in the Young's modulus and shear modulus while a reduction in the mass density of the bolts is noted. In addition, there is also a slight increment in the value of the translational stiffness of the interfaces. This achievement discovered that the bolts and nuts used to assemble the plates contributed significantly to the dynamic behaviour of the assembled structure.

Table 5. Comparison of experimental, initial FE and updated FE for assembled plates.

Mode	I. Experiment (Hz)	II. Initial FE (Hz)	III. Error between I & II (%)	IV. Initial FE MAC	V. Updated FE (Hz)	VI. Error between I & V (%)
1	75.075	74.308	1.022	0.982	74.491	0.778
2	200.766	203.724	1.473	0.939	203.785	1.504
3	404.924	396.609	2.053	0.991	397.560	1.819
4	469.304	447.814	4.579	0.950	451.992	3.689
5	630.771	628.602	0.344	0.971	629.301	0.233
6	674.388	649.156	3.741	0.961	651.654	3.371
7	1034.640	1007.230	2.649	0.941	1009.680	2.412
8	1053.230	1060.560	0.696	0.911	1061.550	0.790
9	1276.950	1252.120	1.944	0.948	1255.760	1.659
10	1435.090	1472.260	2.590	0.936	1472.680	2.619
	Total Error		21.093			18.874

Table 6. Updated parameters for assembled plates.

Property	Parameter	Initial Value	Updated Value	Unit
Material	Young's modulus	193000	235619	N/mm^2
(Bolts)	Shear modulus	75000	92150	N/mm^2
	Mass density	7.86E-06	7.49E-06	kg/mm^3
PELAS (Interfaces)	Translational stiffness	300	301.2	N/mm

6. Conclusions

The attempts to model bolted joints using an efficient and economical procedure via the FE model updating procedure in the light of the results of EMA are presented and have been successfully performed. The parameters of the FE models of Plate A, Plate B, bolting elements and interface elements have been successfully used in the updating procedure to determine an efficient FE model of the assembled structure. This study revealed that the bolts' material properties and stiffness values at the joints interfaces have played a crucial role in ensuring the accuracy of the prediction of the dynamic behaviour of the assembled structure with bolted joints. Furthermore, the accuracy of prediction of the dynamic behaviour probably could be improved further if the boundary conditions and damping effects of the interfaces between the assembled plates were included in the FE model of the assembled structure with bolted joints.

Acknowledgements

The authors gratefully acknowledge the Malaysian Ministry of Higher Education (MOHE) and Research Management Centre (RMC) of Universiti Teknologi MARA (UiTM) for providing financial support for this study through Fundamental Research Grant Scheme (FRGS) 600-RMI/FRGS 5/3 (96/2016). They would like to express their appreciation of helpful comments, suggestions and technical support given by Structural Dynamics Analysis & Validation (SDAV) and Machinery Technology Centre (MTC) members.

References

[1] Ibrahim R A and Pettit C L 2005 *J. Sound Vib.* **279** 857
[2] Kim J, Yoon J-C and Kang B-S 2007 *Appl. Math. Model.* **31** 895
[3] Shokrollahi S and Adel F 2016 *J. Theor. Appl. Vib. Acoust.* **2** 65
[4] Ahmadian H, Mottershead J E and Friswell M I 2001 *J. Sound Vib.* **247** 755
[5] Omar R, Abdul Rani M N, Yunus M A, Wan Iskandar Mirza W I I and Mohd Zin M S 2018 *AIP Conf. Proc.* **1952** 020082
[6] Mottershead J E and Friswell M I 1993 *J. Sound Vib.* **167** 347
[7] Sehgal S and Kumar H 2016 *Arch. Comput. Methods Eng.* **23** 515
[8] Mottershead J E, Link M and Friswell M I 2011 *Mech. Syst. Signal Process.* **25** 2275
[9] Abdul Rani M N, Kasolang S, Othman M H, Yunus M A, Wan Iskandar Mirza W I I and Ouyang H 2016 *23rd International Congress on Sound and Vibration* pp 315
[10] Ahmadian H, Mottershead J E, James S, Friswell M I and Reece C A 2006 *Mech. Syst. Signal Process.* **20** 868

Potential of Using Natural Fiber for Building Acoustic Absorber: A Review

Esan Martins Taiwo[1, 2], Khairulzan Yahya[1*] and Zaiton Haron[1]

[1]Department of Structure and Materials, School of Civil Engineering, Faculty of Engineering, Universiti Teknologi Malaysia, 81310 Johor Bahru, Johor, Malaysia
[2]Federal Polytechnic, Ede, Nigeria.

[*]Corresponding author: khairulzan@utm.my

Abstract. Sustainability is one of the significant goals in the evaluation and promotion of building materials in the current society. This paper presents an evaluation of the potential of natural fiber for building acoustics absorber. This was done by comparing all aspects of sustainability including economy, social and environmental needs between the natural and synthetic fibers. Some features of natural fibers such as light weight, bio-degradability, low-cost, carbon-dioxide neutral, low consumption of energy and health aspects has been the major factors that contributed to its use in many construction applications. In addition to that, previous researches showed that sound absorbing materials with natural fibers have good acoustic properties in high frequency range similar with those synthetic fiber. However, high water absorption, prone to termite attack, less fire resistance and low strength are remaining the issues militating against the natural fiber for building acoustic absorber. Despite that, natural fibers have already been proven alternatives to synthetic fiber in building acoustic absorber thereby alleviating some sustainability issues associated with the use of synthetics materials in building acoustics.

1. Introduction

Synthetic fibers such as rock wool and glass wools are practically used as sound absorber in building industries. It is used for many applications in building and construction industry such as partition boards, floors, wall and roof tiles [1]. However, it was found that their negative effects to the environment were high and thus natural and sustainable acoustic materials are now of interest [2]. For these reason, natural fibers including derived from agro-waste, plants and industrial waste were selected. Such natural fibers like coir, pineapple, Sugarcane bagasse, jute, ramie, kapok, rattan etc. were currently investigated. Table 1 shows a few reviews studies on the area of natural fiber as a substitute to synthetic fiber in a building acoustics.

Table 1. Related studies on natural fibers for acoustic absorption

Type of fibers	Acoustic characteristic	Source of reference
Hemp	Higher concentrations and alkalization at higher temperature gives good results in sound absorption coefficients properties.	Yilmaz et al. [3]
Coir	Good acoustic properties at medium and high frequency ranges.	Fouladi et al. [4]
Coconut	The increased in thickness of coconut increases the sound absorption coefficient of coconut fiber	Huang et al. [5]
Date palm	Sound absorption properties of date palm fiber increase with the increases of flow resistivity.	Khidir et al. [6]
Kapok	Hollow structure of kapok fiber help in good acoustic behavior.	Xiang et al. [7]
Bamboo	Good absorption properties mostly at the higher frequency range.	Thilagvathi et al. [8]
Tea-leaf	Better sound absorption properties than polyester.	Ersoy and Kucuk [9]
Wool	Low flow resistivity material resulted good sound absorption properties	Ballagh [10]
Sugarcane bagasse	Resin proportion and the type influenced the sound absorption properties.	Doost-hoseini et al. [11]
Ramie	Superior sound absorption properties over glass fiber reinforced sandwich structures.	Zheng et al. [12]
Jute	Result from modelling showed good agreement with experimental data.	Fatima et al. [13]
Natural fiber mixed nonwoven composite	Excellent absorption properties in the mild to high frequency ranges.	Kucuk et al. [14]
Windmill palm	Good for acoustic absorption materials	Chen et al. [15]
Kapok and milkweed fiber	Good sound absorption and insulation properties.	Liu et al. [16]
Hemp shives	Good absorption properties at lower frequency range.	Balciunas et al. [17]
Cotton and cellulose fiber thermoplastic composites.	Sound absorption of natural fiber in thermoplastic form increases when the density is decrease.	Krucinska et al. [18]
Sisal fiber poly lactic acid bio-composite	Produced highest sound absorption property compared with other composites.	Jayamani et al. [19]
Luffa	Acoustic properties of luffa fiber are higher even for a small thickness when the matrix is not added.	Koruk and Genc [20]
Luffa/epoxy composites	Sound absorption properties of the composites were reduced by chemical treatment.	Jayamani et al. [19]
Bamboo fiber board	Better sound absorption properties more than plywood with same density.	Thilagavathi et al. [8]
Jute composite materials	It shows better acoustical properties than glass fiber	Fatima et al. [22]
Kenaf and hemp	Better acoustical properties than glass fiber.	Berardi and Iannace [23]
Kapok nonwoven fabric	Model of absorption coefficient, thickness and flow resistance.	Liu et al. [7]
Flax and polypropylene fiber	The effects of porosity, air gap and pore diameter on sound absorption at lower frequency.	Merotte et al. [24]
Bark cloth	Large porosity volume fraction is benefit to acoustic absorption property.	Rwawiire eta l [18]
Polyurethane foams loaded with tea-leaf fiber and luffa cylindrical	Good absorption properties at higher frequency range.	Ekict et al. [25]
Pineapple leaves	Acoustic absorption property improves significantly when the tea-leaf was added with the polyurethane.	Putral et al. [26]
Oil palm	Achieve good sound absorption properties at above 1KHz frequency range	Yahya and Sheng [27]

This paper presents an evaluation on the potential of natural fiber for building acoustics absorber. This was done by comparing all aspects of sustainability including economy, social and environmental needs between the natural and synthetic fibers. Some features of natural fibers aspects that have been the major factors contributed to its use in many construction applications are reviewed. In addition to that, previous researches that showed the sound absorbing properties of sound absorber made with natural and issues related with high water absorption, prone to termite attack, less fire resistance are discussed.

2. Comparison of natural fibers with synthetic fibers in building acoustic absorber

Joshi et al. [28] proved that natural fiber composite has much lower environmental impact compared to the synthetic fibers such as glass wool. Thus, natural fibers materials have been recommending as an alternative sound absorber material since its cheaper, lighter and environmentally friendly compare to synthetic materials which can be harmful to human health.

Natural fibers offer a high potential for building material especially acoustic absorber, since they are low-density material, cheap and it's renewable [28]. In addition to that natural fiber are able to recyclability and degradable, and has low consumption of energy, no health risk, and eco-friendly. Due to these aspects, Peter and Narayan [29] reported that natural fibers are very good materials in terms of sustainability. A comparison of various aspects between natural fibers and synthetic fibers is described in Table 2.

Table 2. Comparison between synthetic fibers and natural fibers [34].

Aspects	Natural fiber	Synthetic fiber
Density	Low	High
Cost	Low	Double
Renewability	Yes	No
Recyclability	Yes	No
Consumption of Energy	Low	High
Biodegradability	Yes	No
Carbon dioxide	No	Yes
Health risk	No	Yes
Eco-friendly	Yes	No

Natural fiber is bio-degradable, derived from a renewable resources and require a small energy to process [30]. According to Bismarck et al. [31] on their investigation of carbon dioxide contribution of natural fibers in the environment during the growing and processing, it is concluded that natural fiber production were eco-friendly both at the growing and processing stages. Moreover, Mutnuri et al. [32] found through life cycle energy assessments that natural fiber consume less energy synthetic fiber.

Joshi et al. [28] suggested that there are some major areas that natural fibers can offered in helping and improving of the environment. Such as carbon lower emissions and improve fuel efficiency and recovered energy and carbon credits from end of life incineration. Whereas the synthetic fibers depend on energy for their production. In fact, synthetic fiber causes pollution to the environment especially disposal issues as a result of very large amount of synthetic fiber usage. For example, in US alone disposed over 25 million tons of plastics into the municipal solid waste stream in the year 2001, which accounted for over 11% of the total national waste [33] . Many states in the US have reached their landfill capacity in a shorter period of time and this makes landfill has major issue.

With all the advantages of all aspects that natural fiber can offer to be as a construction material, this has caused the demand for natural fiber increased from 10% to 22% per year (Satyanarayana et al. [35]). The natural fiber market is also predicted to improve in sales from 16% to 30% per year. Thus, Bismarck et al. [31] opined that there is greater opportunity for other regions of the world to join the market in the natural fiber production since there are more market currently for natural fiber production in the United States and Europe which will soon be unable to meet market demands of this product.

3. Ability of Natural Fibers in Absorbing Sound.

Many researches have been carried out on the possibilities of using natural fiber instead of synthetic fiber for sound acoustic absorber due high capability of absorbing sound. The use of natural fiber for sound absorbing materials is very significant in the production of building absorber in the world today. In terms of sound absorption most of the natural materials tends to behave as porous absorbers materials. It is apparent that porosity is one of the factors that determine the absorption characteristics of any fibrous materials. Other factors such are fiber density and thickness. Porous sound absorbing materials have good acoustic properties in high frequency range but with a very weak absorption coefficient at low frequency range. One of the problem with using the natural fibers for building acoustic absorbers is the application of unsustainable material has a binder to bind the fibers together. For a natural fiber to meet the technical need of building acoustic absorber, the fibers has to be characterized and modelled. Natural fiber has to be characterized in order for the improvement in it is properties. This includes characteristics such as fire retardant, anti-fungus, flammability and stiffness.

According to Lee and Joo [36] in their studies observed that the effect of fiber content of non-woven fabrics on the sound absorption properties and found out that the non-woven that are produced with recycled fibers of different diameters having the same length performed better in sound absorption properties, mostly when there is increased in fiber content. The sound absorption coefficient of non-woven that has un-oriented web in the middle layer is very high at higher frequency range than the one with totally oriented web structure. Moreover Ersoy and Kucuk [9] investigated the sound absorption coefficient properties of tea leaves fiber. A layer of woven textile material and three layers of tea leaf fiber waste materials was tested, with and without backing in different configurations were tested. The experimental data indicated that the sound absorption properties of one cm thick tea leaf fiber waste material with backing is almost the same with nonwoven textile material. And also twenty millimeter thick layer of rigidly backed tea leaf fibers and nonwoven textile exhibit almost the same sound absorption coefficient in low frequency range.

Zulkifli et al. [37] observed the sound absorption coefficients of two natural fibers; oil palm fibers and coir fiber in their study. At the fabrication stage, coir fiber sheet was treated with latex and the oil palm fiber sheet was also treated with binder. Both the fibers were compressed under pressure using high precision hydraulic machine for 30 minutes to form the fiber sheets. The coir fiber shows a good sound absorption coefficient for higher frequencies but not too good in the lower frequencies region. The oil palm fiber shows a good absorption coefficient for higher frequency region than lower frequency region when observing the sound absorption coefficient of this material. The structural diversity of synthetic fibers can improve sound absorption coefficient when comparing with natural fiber. The different cross-sections of synthetic fiber such as triangle, hollow and circle are beneficial in improvement of acoustic absorption properties of the synthetic fibers. The sound absorption properties of polyester fiber was studied by Huang et al. [38]. It has been found that the polyurethane composite board can reach a value of 0.69 at lower frequency range. It shows improved sound absorption property in the lower frequency range.

4. Physical and Mechanical Strength of Natural Fibers

Natural fibers are tough, elastic and demonstrate good mechanical strength. The composite from natural fibers is introduced for commercial purpose and becomes a good alternative of synthetic fibers in many applications. The tensile strengths similarly the young's modulus of natural fibers like coir, ramie, flax, hemp and sisal are less than that of synthetic fiber unremarkably utilized in building absorbers. However, the precise strength and modulus of a number of these natural fibers are quite comparable to glass fibers. For example, the superiority of pineapple mechanical properties compared to other natural fibers is due to its high content of alpha-cellulose and low microfibrillar angle. Coir, ramie, hemp, sisal, jute, flax and cotton are kind of natural fiber that possesses high strength and modulus.

The major advantages of natural fiber when comparing it with synthetic is that of good specific mechanical properties due to low energy consumption and density, on the other hand the disadvantages of natural fibers includes high moisture absorption, more inferior fire resistance and lower strength [39]. In general ramie fiber have the best property for structural applications. Sisal, jute and flax offers the best potentials of light weight, low cost and high strength. Overall, natural fibers have low densities and cheaper than synthetic fibers although the strength of synthetic is higher than that of natural fibers. Since natural fibers has specific modulus values, it can be preferable to synthetic fibers in applications where stiffness and weight are primary concern such as in the case of building absorber[39]. A comparison of various mechanical properties between natural fibers and synthetic fibers is shown in Table 3.

Table 3. Comparative mechanical properties of some natural fibers with synthetic fibers [34][40]

Fiber	Young's modulus (Gpa)	Elongation (%)	Tensile strength (MPa)	Density (g/cm3)
		Natural fibers		
Coir	4.0-6.0	15.0-30.0	10	1.2
Ramie	44-128	2.0-3.8	12-17	1.5
Hemp	70	1.6	8	1.48
Sisal	9.0-38.0	2.0-14	11	1.33-1.5
Jute	10-30	1.5-1.8	12	1.3-1.46
Flax	27.6-80	1.2-3.2	7	1.4-1.5
Cotton	5.5-12.6	3.0-10.0	8-25	1.5-1.6
Pineapple	34.5-82.5	2.0	413-1627	1.4
		Synthetic fibers		
Carbon	230.0-40.0	1.4-1.8	4000	1.4
Glass	73	3	2400	2.55
E-glass	70.0	2.5-3.0	2000-3500	2.5

5. Issues on Natural fiber

Although natural fibers have advantage over the synthetic fibers such as low cost and low density, there are many other challenges militating against the advancement of natural fibers into the field of building acoustic absorber. This is due to many reasons such as poor water absorbing capacity, lower mechanical properties, poor fire resistance and low resistant to termite effects. Except of fire resistance, all these issues can be improved with the use of innovative surface treatments, additives such as glue and coatings.

5.1 Higher moisture Absorption

Natural fibers are deliquescent in nature and has poor water absorbing properties, it's mostly absorbed water from air, and this is due to their high cellulose content. The mechanical properties of these fibers can be improved effectively by improving the poor dimensional and environmental stability. Another ways of improving the fiber absorption characteristics is to develop a natural binders to hold the fiber to maintain good shape. Synthetic fibers on the other hand are good moisture resistance.

5.2 Fire Resistance

In most cases, natural fiber shows poor resistance to fire compare to synthetic fiber that could be a major setback of natural fiber for industrial applications such as in building acoustic absorber wherever inflammability and safety are thought-about to be the vital factors. This setback poses challenges for natural fibers to compete with synthetic fibers. Natural fibers have a lower decomposition temperature since they are non-thermoplastic compared to synthetic fiber. Natural fibers primarily composed of hemicellulose, cellulose, lignin and inorganic nonflammable substances. The flammability of natural fiber is often increase by high content of cellulose. The cellulose decomposes at a very high temperature, whereas hemicellulose decomposes at a lower temperature and forms additional fire-retardant gases.

6. Conclusions

The study aims to look at the potential of using natural fiber as an alternative to synthetic fibers in the production of building acoustic absorber. Many researchers have looked on how to replace the conventional synthetic fibers with natural fibers in the production of building absorber because as light weight, bio-degradability, lower cost, carbon-dioxide neutral, low energy consumption and health aspects. The natural fibers such as pineapple, hemp, sisal and jute have impressive mechanical properties. Notwithstanding, with all these benefits of natural fibers, some problem is still impending its application in building absorber such as poor fire resistance properties and higher moisture absorption. This research will contribute to a greater interest in the use of natural fiber for cost-effective building absorber in the construction industry.

Acknowledgements

The study was funded using GUP Q.J130000.2522.19H76, and FRGS R.J130000.7822.4F887 which are supported by the Research Management Centre, Universiti Teknologi Malaysia. Their support is highly appreciated.

References

[1] F. Asdrubali, F. D'Alessandro, and S. Schiavoni, "A review of unconventional sustainable building insulation materials," *Sustain. Mater. Technol.*, vol. 4, pp. 1–17, 2015.

[2] F. Asdrubali, S. Schiavoni, and K. V Horoshenkov, "A review of sustainable materials for acoustic applications," *Build. Acoust.*, vol. 19, no. 4, pp. 283–311, 2012.

[3] N. D. Yilmaz, N. B. Powell, P. Banks-lee, and S. Michielsen, "Hemp-fiber Based Nonwoven Composites : Effects of Alkalization on Sound Absorption Performance," vol. 13, no. 7, pp. 915–922, 2012.

[4] M. Hosseini Fouladi, M. Ayub, and M. Jailani Mohd Nor, "Analysis of coir fiber acoustical characteristics," *Appl. Acoust.*, vol. 72, no. 1, pp. 35–42, 2011.

[5] C. Huang, J. Lin, C. Lou, and Y. Tsai, "The Efficacy of Coconut Fibers on the Sound-Absorbing and Thermal-Insulating Nonwoven Composite Board," vol. 14, no. 8, pp. 1378–1385, 2013.

[6] E. A. Khidir, M. O. M. Ali, M. M. Ali, M. F. B. M. Tahir, and R. Zulkifli, "Analysis of sound absorption of date palm fibers based on flow resistivity," in *Applied Mechanics and Materials*, 2014, vol. 471, pp. 285–290.

[7] X. Liu, X. Yan, and H. Zhang, "Effects of pore structure on sound absorption of kapok-based fiber nonwoven fabrics at low frequency," 2016.

[8] G. Thilagavathi, E. Pradeep, T. Kannaian, and L. Sasikala, "Development of natural fiber nonwovens for application as car interiors for noise control," *J. Ind. Text.*, vol. 39, no. 3, pp. 267–278, 2010.

[9] S. Ersoy and H. Küçük, "Investigation of industrial tea-leaf-fibre waste material for its sound absorption properties," *Appl. Acoust.*, vol. 70, no. 1, pp. 215–220, 2009.

[10] K. O. Ballagh, "Acoustical properties of wool," *Appl. Acoust.*, vol. 48, no. 2, pp. 101–120, 1996.

[11] K. Doost-hoseini, H. R. Taghiyari, and A. Elyasi, "Correlation between sound absorption coefficients with physical and mechanical properties of insulation boards made from sugar cane bagasse," *Compos. Part B Eng.*, vol. 58, pp. 10–15, 2014.

[12] Z. Y. Zheng, Y. Li, and W. D. Yang, "Absorption properties of natural fiber-reinforced sandwich structures based on the fabric structures," *J. Reinf. Plast. Compos.*, vol. 32, no. 20, pp. 1561–1568, 2013.

[13] S. Huda and Y. Yang, "Feather fiber reinforced light-weight composites with good acoustic properties," *J. Polym. Environ.*, vol. 17, no. 2, pp. 131–142, 2009.

[14] Y. Korkmaz and M. Ku, "The effect of physical parameters on sound absorption properties of natural

fiber mixed nonwoven composites," 2012.

[15] C. Chen, Y. Zhang, G. Sun, J. Wang, and G. Wang, "Windmill Palm Fiber / Polyvinyl Alcohol Coated Nonwoven Mats with Sound Absorption Characteristics," vol. 11, no. 2, pp. 4212–4225, 2016.

[16] P. Ganesan and T. Karthik, "Development of acoustic nonwoven materials from kapok and milkweed fibres," *J. Text. Inst.*, vol. 107, no. 4, pp. 477–482, 2016.

[17] G. Balčiūnas, J. Žvironaitė, S. Vėjelis, A. Jagniatinskis, and S. Gaidučis, "Ecological, thermal and acoustical insulating composite from hemp shives and sapropel binder," *Ind. Crops Prod.*, vol. 91, pp. 286–294, 2016.

[18] S. Rwawiire *et al..*, "Investigation of sound absorption properties of bark cloth nonwoven fabric and composites," *Autex Res. J.*, vol. 15, no. 3, pp. 173–180, 2015.

[19] E. Jayamani, S. Hamdan, S. Kok Heng, M. R. Rahman, and M. K. Bin Bakri, "Acoustical, thermal, and morphological properties of zein reinforced oil palm empty fruit bunch fiber bio-composites," *J. Appl. Polym. Sci.*, vol. 133, no. 43, 2016.

[20] H. Koruk and G. Genc, "Investigation of the acoustic properties of bio luffa fiber and composite materials," *Mater. Lett.*, vol. 157, pp. 166–168, 2015.

[21] F. Fibers, "com Characterization of a Material Based on Short Natural Fique Fibers," vol. 9, no. Pérez 1974, pp. 3480–3496, 2014.

[22] S. Fatima and A. R. Mohanty, "Acoustical and fire-retardant properties of jute composite materials," *Appl. Acoust.*, vol. 72, no. 2–3, pp. 108–114, 2011.

[23] U. Berardi and G. Iannace, "Acoustic characterization of natural fibers for sound absorption applications," *Build. Environ.*, vol. 94, pp. 840–852, 2015.

[24] J. Merotte, A. Le Duigou, A. Bourmaud, K. Behlouli, and C. Baley, "Mechanical and acoustic behaviour of porosity controlled randomly dispersed flax/PP biocomposite," *Polym. Test.*, vol. 51, pp. 174–180, 2016.

[25] Y. T. Yang *et al..*, "Monocrystalline silicon carbide nanoelectromechanical systems," *Appl. Phys. Lett.*, vol. 78, no. 2, pp. 162–164, 2001.

[26] A. Putra, K. H. Or, M. Zulke, M. Jailani, and M. Nor, "Sound absorption of extracted pineapple-leaf fi bres," vol. 136, no. February, pp. 9–15, 2018.

[27] M. N. Yahya and D. D. V. Sheng Chin, "A Review on the Potential of Natural Fibre for Sound Absorption Application," *IOP Conf. Ser. Mater. Sci. Eng.*, vol. 226, no. 1, 2017.

[28] S. V. Joshi, L. T. Drzal, A. K. Mohanty, and S. Arora, "Are natural fiber composites environmentally superior to glass fiber reinforced composites?," *Compos. Part A Appl. Sci. Manuf.*, vol. 35, no. 3, pp. 371–376, 2004.

[29] M. Patel and R. Narayan, "How sustainable are biopolymers and biobased products? The hope, the doubts, and the reality," *Nat. Fibers, Biopolym. Biocomposites, AK Mohanty, M. Misra LT Drazl, Ed. CRC Press (Taylor Fr. Group), New York, USA*, pp. 833–853, 2005.

[30] B. C. Suddel and W. J. Evans, "Seventh International Conference on (Wood Fiber)-Plastic Composites," 2003.

[31] A. Bismarck, S. Mishra, and T. Lampke, "Plant fibers as reinforcement for green composites," *Nat. fibers, Biopolym. biocomposites*, vol. 2, pp. 37–108, 2005.

[32] B. Mutnuri, C. J. Aktas, J. Marriott, M. Bilec, and H. GangaRao, "Natural fiber reinforced pultruded composites," *Proc. Compos.*, 2010.

[33] A. Mohanty, M. Misra, L. Drzal, S. Selke, B. Harte, and G. Hinrichsen, *Natural Fibers, Biopolymers, and Biocomposites*. 2005.

[34] P. Wambua, J. Ivens, and I. Verpoest, "Natural fibres: can they replace glass in fibre reinforced plastics?," *Compos. Sci. Technol.*, vol. 63, no. 9, pp. 1259–1264, 2003.

[35] K. G. Satyanarayana, G. G. C. Arizaga, and F. Wypych, "Biodegradable composites based on lignocellulosic fibers-An overview," *Prog. Polym. Sci.*, vol. 34, no. 9, pp. 982–1021, 2009.

[36] Y. Lee and C. Joo, "Sound absorption properties of recycled polyester fibrous assembly absorbers," *Autex Res. J.*, vol. 3, no. 2, pp. 78–84, 2003.

IOP Conf. Series: Journal of Physics: Conf. Series **1262** (2019) 012017 doi:10.1088/1742-6596/1262/1/012017

[37] R. Zulkifli, M. J. M. Nor, A. R. Ismail, M. Z. Nuawi, and M. F. M. Tahir, "Effect of perforated size and air gap thickness on acoustic properties of coir fibre sound absorption panels," *Eur. J. Sci. Res.*, vol. 28, no. 2, pp. 242–252, 2009.

[38] C. Wang and L. Huang, "On the acoustic properties of parallel arrangement of multiple micro-perforated panel absorbers with different cavity depths," *J. Acoust. Soc. Am.*, vol. 130, no. 1, pp. 208–218, 2011.

[39] D. B. Dittenber and H. V. S. Gangarao, "Composites : Part A Critical review of recent publications on use of natural composites in infrastructure," *Compos. Part A*, vol. 43, no. 8, pp. 1419–1429, 2012.

[40] X. Li, L. G. Tabil, and S. Panigrahi, "Chemical treatments of natural fiber for use in natural fiber-reinforced composites: a review," *J. Polym. Environ.*, vol. 15, no. 1, pp. 25–33, 2007.

IOP Conf. Series: Journal of Physics: Conf. Series **1262** (2019) 012018 doi:10.1088/1742-6596/1262/1/012018

A Framework for Green Concrete Noise Barrier Materials using Palm Oil Clinker

Noor Suhaida Galip[1], Zaiton Haron[1,*], Roslli Noor Mohamad[1], Khairulzan Yahya[1], Zanariah Jahya[1], Muhammad Azril Hezmi[1]

[1]School of Civil Engineering, Faculty of Engineering, Universiti Teknologi Malaysia (UTM), 81310 Johor Baharu, Johor, Malaysia

[*]Corresponding author: zaitonharon@utm.my

Abstract. Noise pollution from traffic is an increasing environmental problem worldwide. The most popular mitigation measure is the control of path by means of construction of noise barriers. Current concrete barrier incorporated with perforated surface infilled with non-mineral fibre sound absorbing material on the highway side are the commonly found. The disadvantage of this type is it's easily damaged and the non-mineral fibre is so dreadful to human health. On the other hand, palm oil clinker (POC) is a by-product waste material produced in palm oil mills, are currently dump in open land or landfill sites, which leads to environmental problems. POC is internally porous and if it replaces the aggregate in porous concrete it may absorb sound better than standard porous concrete which is the most important characteristic for noise barrier development. This paper discusses how the palm oil clinker substitution will make concrete noise barrier material to be labelled as green and sustainable. A framework for developing green noise barrier made of palm oil clinker concrete will also be presented and discussed.

1. Introduction

Cities in developing countries faces severe noise pollution problem from traffic. Data collected, by the World Health Organization (WHO) researchers, alongside densely travelled roads were found to have equivalent sound pressure levels for 24 hours of 75–80 dBA [1]. Due to increase in number of vehicles and rapid urbanization of cities, noise level exceeds the permissible value. For many years noise barriers have been erected along major arterials to screen residential areas from high levels of traffic noise. Noise barriers for reduction of noise levels could be warranted for certain portion of road network. Noise barriers can have a substantial effect on the visual environment of a highway and its surrounding environment [2]. They are long continuous structures, and can significantly change the view of the road by blocking view of the roadside and creating a monolithic uniformity of walls instead changing urban scenery. Nonetheless, noise barrier may be highly sought after by neighbourhoods as a means to reduce roads noise level [3].

In Malaysia, the applications of noise barrier become significant after the establishment of its first guideline of Guideline on Noise in 2004 [4]. The most popular mitigation measure in Malaysia is the control of path by means of construction of noise barriers. The requirements for the noise barrier's design and method for installation on highways are documented in Lembaga Lebuhraya Malaysia (LLM) guideline [5]. An obvious benefit from this LLM guideline is the provision of the comfort for residents lining the highway if the traffic noise exceeds the permissible limit given by the DOE 2004 guidelines. Concrete masonry barrier is the most sought after with requirement for sound absorption as much as 60% is suggested to have good acoustic performance. The ability of such absorption can be obtained by providing the porous surfaces or perforated surfaces and in filled consisting of fiberglass or rockwool in hollow concrete masonry. For example, the installation of a 4 metre hollow block concrete masonry noise barrier at Bukit Setiawangsa is to protect the residents from the intense traffic noise coming from the Duta Ulu Kelang Expressway (DUKE) highway [5]. The construction of the DUKE highway has raise the noise level to about 15 or 19 dBA during day and night, and 9 dBA and 7 dBA during day and night. The construction of barriers of sufficient height and length, although relatively expensive (RM 800 to 1000 per square meter) however it could give comfort to residents.

On the other hand, porous concrete made of palm oil clinker has been extensively research for building construction [6-7]. It is well known that POC reduces the compressive strength of concrete in comparison with normal porous concrete due to existing of internal pores in POC aggregate that lead to high porosity. Some researchers coat the POC aggregate [8] to maintain the strength comparable with natural aggregate and some research suggested the high strength cement binder for structural usage. The bad pores in POC can be a good sound absorption enhancer in porous concrete. To the authors' knowledge, none of research so far has focused on POC concrete acoustic performance and its possible application for noise barrier development. Hence, this paper discusses how the palm oil clinker substitution will make concrete noise barrier material to be labelled as green and sustainable. A framework for propose green noise barrier made of palm oil clinker concrete are presented and discussed.

2. Guideline and previous research on the development on noise barrier

Concrete noise barrier can be constructed from concrete panel or masonry block. According to the Malaysian guideline of highway (LLM), basically properties for masonry concrete noise barrier shall be with compressive strength of minimum 7 N/mm^2, transmission loss of 20 to 25 dB and the absorptive surface with sound absorption coefficient of 0.6 [5]. According to FHWA, concrete noise barrier can be precast concrete in the form of concrete panel, can in place barrier and masonry. Structurally, since mostly it will be built lining the road or free standing noise wall, noise barrier are designed to carry self-weight and wind load.

According to [9], one of the most common materials used for concrete noise barrier applications is a combination of two-layer material components consist of sound proof hard backing and porous medium consist of porous concrete for absorbing sound from the source (Figure 1). [9] reported that the standard concrete for hard backing and porous concrete for noise barrier employed in commercial product in Europe is shown in Table 1. It can be seen that porous concrete is developed by eliminating the sand volume in the matrix and aggregate size greater than 3 mm and less than 9 mm with target porosity of 26%.

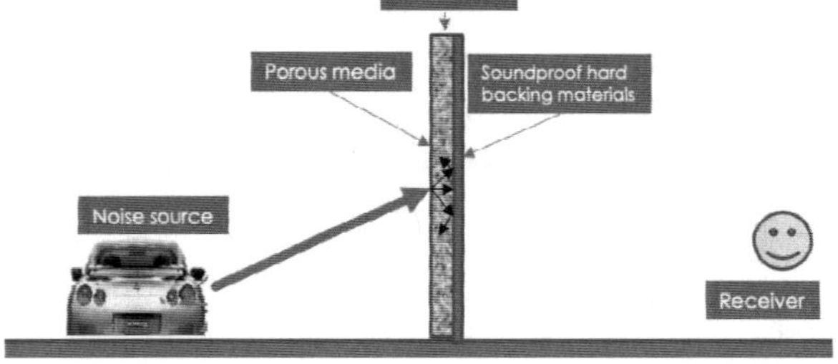

Figure 1. Hard backing and porous concrete materials for noise barrier [11]

Table 1. Mixture (wt %) proportion of employed in commercial product noise barrier [10].

	Porous concrete (PC)	**Hard backing mixture Standard concrete (SC)**
PC II (%)	20	20
Coarse aggregate	80	30
Fine aggregate	–	50
Particle size (mm)	$3 < d_p < 9$	$d_p < 1.25$
Water/solid ratio (wt%)	9.0	12.5
Targeted Open void ratio (%)	26.0	7.3

Few researches have been conducted to substitute the natural aggregate with the waste material especially for noise barrier. [12] studied aggregate replacement with bottom ash from plant into noise barrier with three layer consist of hard backing and two layer porous media called SOL.BA1 and SOL.BA1. Both consist of hard backing (size of aggregate bottom ash<1.25mm) and two layer of medium porous (size particle between 1.25mm and 5 mm and followed by coarser porous medium (size particle of greater than 5mm). The mix proportion is shown in Table 2. Both have the same thickness but SOL.BA1 has greater thickness of coarser porous medium. In comparison with standard porous concrete, it can be seen that three-layer sound barrier SOL.BA1 can absorb higher sound energy given by noise spectrum below 1500Hz (Figure 2).

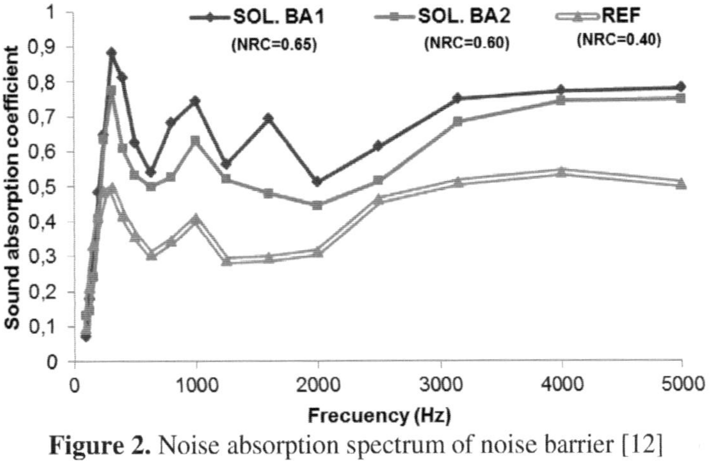

Figure 2. Noise absorption spectrum of noise barrier [12]

Table 2. Mix proportion (% of weight) [12]

Type of material	%BA	Cement	Size aggregate (Dp) mm
Bottom ash-coarse (BA-C)	80	20	Dp>5
Bottom ash-medium (BA-M)	80	20	5>Dp>1.25
Bottom ash-fine (BA-FC)	80	20	Dp <1.25

Further, investigation of the ceramic waste for replacement of both hard backing and porous medium [13]. Again, the recycled porous concrete present better acoustic and mechanical properties than the porous concrete traditionally used to manufacture highway noise barriers. According to [14], the porous medium's compressive strength is acceptable if the minimum compressive strength is 2-3 N/mm^2 with sound absorption coefficient is 0.6. Good sound absorption is described by sound absorption coefficient and also noise rating criteria (NRC). NRC is determined by calculating the arithmetic mean of the absorption coefficients at 250, 500, 1000 and 2000 Hz, in order to have single numeric data of the sound absorption coefficient to clearly identify the effect of the aggregate size of the waste material and the thickness of the specimens.

One of the key important factors for absorbing sound is the porosity of concrete. According to [15] and [16], the size and percentage of pores are important parameter for sound absorption. Interconnected pores are the characterized of good sound absorbing which are typically indicated by void content/porosity typically in the range of 15 to 35% by volume. Investigated the effect of concrete specimen with single size aggregate of various porosity but with the same pore size of 3mm diameter and 2.7mm length [16] (Figure 3). It can be seen that porosity of 15% specimen reached peak absorption at lower frequency compared with those 35% porosity.

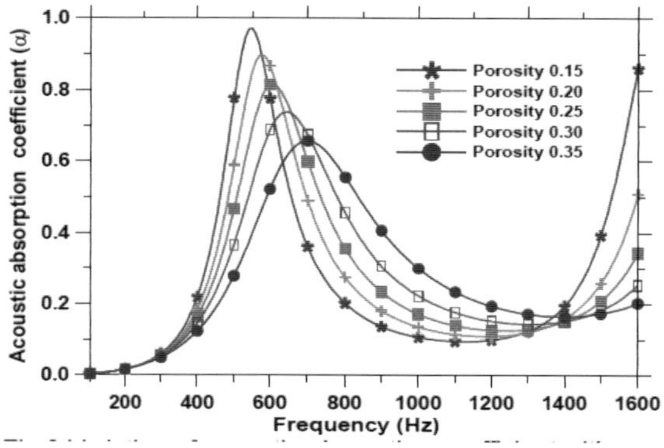

Figure 3. Relationship between Sound absorption and porosity [16]

Thickness also influences the sound absorbing coefficient. The porous concrete with single aggregate size of with similar porosity of 24% it was found that the specimens with different heights reached peak absorption coefficient at a wider frequency range of 300–1,250 [17]. There was no clear relation of the peak absorption coefficient with specimen height, but the 80-mm high specimens had a maximum absorption coefficient of approximately 0.75 (Figure 4).

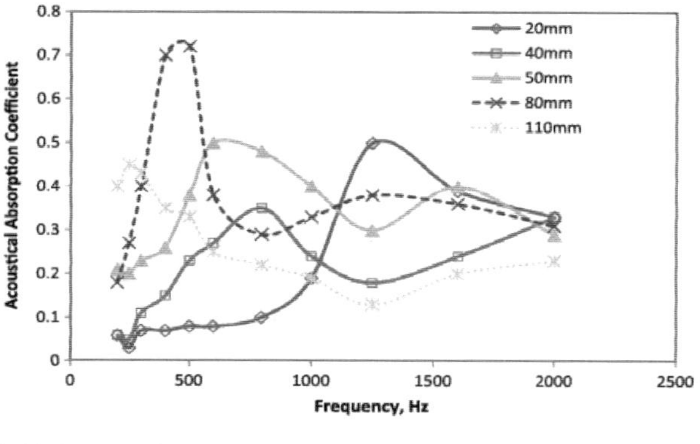

Figure 4. Effect of thickness of sample porous concrete with maximum size of aggregate 9.5mm [17]

3. Potential POC substitution in noise barrier porous medium

3.1 Aggregate replacement

POC is waste derived by combustion of the combustion of palm oil fibre and palm oil shell [18]. Typically, POC is obtained in a large chunk between 20 and 150 mm and it is porous in nature with sharp and rough broken edges, and often flaky with an irregular shape as shown in Figure 5 [19-20]. POC can be grind to required size to serves as possible aggregates in the production of concrete especially for the noise barrier production.

Further, noise barrier material consisting of the entire range of POC particle sizes that offer sound absorptive surface naturally without providing perforated surface infilled with mineral fibres that finally can be utilised in the field of noise barriers. This will overcome the production of standard noise barrier by using ordinary Portland cement, mineral wool and natural aggregate that would emit high CO2 equivalent production. As matter of fact, high amount of natural aggregates, accounting for 70% - 80% of total concrete volume are consumed. Therefore, POC can be used as substitutes to produce an environmentally friendly concrete, thus makes concrete industry to be sustainable, and construction material can be labelled as "green".

Figure 5. Internal pore in POC aggregate [20]

Noise barrier by using POC will also utilized as an alternative material to the current concrete barrier incorporated with perforated surface infilled with absorbing material such as fiberglass/ rockwool on the highway side are the commonly found in Johor Baharu area. The disadvantages of this type are easily damaged and provide unpleasant view (Figure 6). The usage of the non-mineral fiber as infill for sound absorption enhancer also is so dreadful to human health [21]. Many would not know that the fiberglass can cause lung infection, irritation to skin and eyes and the worst case would be cancer [22].

Figure 6. Concrete currently incorporated with absorptive face with perforated surface and absorbing material are easily damaged

3.2 Mechanical Strength

According to [23], 100% replacement of natural aggregate in porous concrete has the compressive strength of 3.43 MPa and density 1238 kg/m3 with porosity of 35.18% which is within an acceptable range stated in ACI 522 (15–35%). In comparison, normal porous concrete has compressive strength of 9.2 MPa and porosity 23% which POC reduced by compressive strength by 62 %. This reduction was due to increased porosity thus caused gradual loss in compressive strength of the concrete [23]. Fortunately, for the noise barrier, the minimum requirement for the porous medium are 2-3 N/mm2, thus 100% replacement would be possible to be used as porous medium in noise barrier. In addition to that, the porosity was also determined under the range for good sound absorbing material. Vermiculite can also be utilised to increase mechanical properties of porous concrete as previously used by [24] for their work on increasing mechanical strength of porous concrete with bottom ash [10].

4. Framework of propose development of noise barrier material by using POC

Based on the previous works of [9], [12], [10], [6], [7] , [23], it is hypothesised that POC aggregate can be utilised as natural aggregate replacement for both the hard backing layer as well as porous layer of free-standing noise barrier. Its internal pores are the main criteria for enhancing sound absorption especially in porous medium of noise barrier. 1st stage of the framework is the characterization of material properties where Palm Oil Clinker (POC) divided into three different particles sizes which are particle sizes dp>5mm, 1.25<dp<5mm and dp<1.25mm for coarse, medium and fine. Environmentally friendly Blended cement (CPII) according to [25] will be used. In order to develop the hard backing, the proportion of mix used by [12] will be utilized (Table 1). This mix design will be compared with standard hard backing material concrete with proportion of cement: Fine aggregate: coarse aggregate of 20:50:30. Second stage of the framework is determination properties of POC concrete as porous sound absorber then determined with medium aggregate ($1.25<d_p<5$mm) and coarse aggregate ($d_p>5$mm). Again, the mixed used by the [12] will be used with the target open void ratio of 26%. The suggested mixtures are shown in Table 3. Each mix proportion is designated by a specific code of "POCC, and POCM" representing palm oil clinker coarse aggregate and, palm oil clinker medium aggregate. The second labels, "20", "40", "50", "70", and "80", denote the amount of POC aggregate. Each specimen type will be compared with other concrete products usually employed in acoustic barriers, a standard porous concrete (PC) as shown in Table 3 [10].

Table 3. Suggested mixture (wt%) proportion of two types of porous concrete

Palm Oil Clinker -Coarse Aggregate Concrete (POCC) Size aggregate > 5 mm With water/solid=9.5%			Palm Oil Clinker -Medium Aggregate Concrete (POCM) 1.25 mm> Size aggregate < 5mm With water/solid=12%		
Mixture	Cement	POC	Mixture	BC	POC
POCC-20	80	20	POCM-20	80	20
POCC-40	60	40	POCM-40	60	40
POCC-60	40	60	POCM-60	40	60
POCC-80	20	80	POCM-80	20	80

The compressive strength of hard backing and porous concrete will be evaluated according to the standard [26]. The sound absorption coefficient and transmission loss will be tested by using impedance tube with 98 mm diameter will be employed for low (50–2000 Hz) and 28 mm diameter for the high acoustic frequency range between 800 and 5000 Hz. The test will be carried out by using impedance tube according to [27]. The physical properties of mixture will also be measured including density, air flow resistivity and porosity.

The best mixture for POCC and POCM will be selected based on specimen that demonstrate the high sound absorption and good compressive strength (>2.0N/mm^2) as suggested by [28]. The best mixture can be potentially used as single layer porous absorber in noise barrier porous medium. High acoustic absorption can be noted by comparing variation of the acoustic absorption coefficient in octave band spectrum for different mix proportions for each type of material together with the reference product i.e normal porous concrete. The relationship with open void ratios, thickness with sound absorption coefficients in the whole range of frequencies will also be studied. Thus, the best mixture in each type of new material will be proceeded to develop multilayer porous concrete by combining hard backing, POCM and POCC. The h_{max} that is thickness of samples that given the highest sound absorption will also be obtained.

Third stage of framework is the multilayer layer material characterisation. At this stage a multilayer material using best material representing each type will be developed. The thickness of material that give highest sound absorption for type I (POCC) , h_{max} will be used as base line for the thickness of combining layers of best type of each type . The arrangement of the layer will depend on the results. Material with the highest sound absorbing coefficient will be put at the first layer of the product as it will be facing the noise source from road. It is hypothesized that POCC called as BT1 will have highest sound absorption followed by POCM called as BT2. Previous research by [12] found that concrete with coarse bottom ash demonstrated highest sound absorption. Thus, several combinations of BT1 and BT2 will be analysed in the top layers (incident noise face). Figure 7 shows the four suggested composite layer of BT1, BT2 and hard backing material (BT3) namely MLC1 to MLC4. Porous and standard hard backing concrete as a conventional product applied in acoustic barriers [28], MLC5 will be casted for comparison. The optimal composite solution would be the product which presented the best acoustic absorption behaviour.

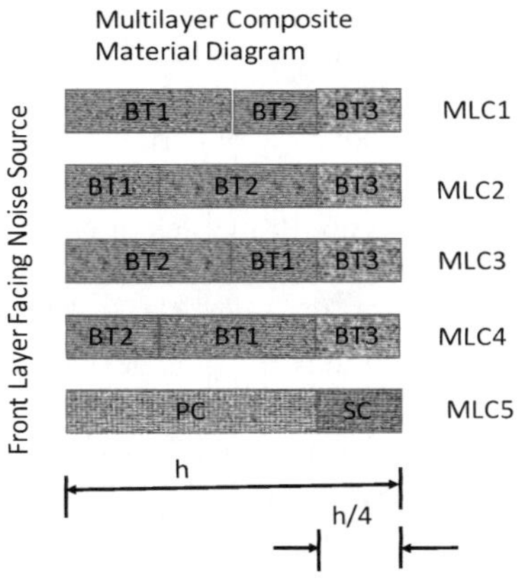

Figure 7. Suggested multilayer material of noise barrier

5. Discussion

A framework for green concrete noise barrier materials by using the entire range of POC particle sizes is proposed. It was based on the previous works by researchers in order to find the ways for reusing and recycling the wastes. Also, it is to help in reducing the environmental noise problem mostly from traffic noise, including motor vehicle, plane and rail noise. However, the most important criteria is that POC itself is internally porous [8], [20] and if it is use in porous concrete it may absorb sound better than standard porous concrete. This will produce new POC noise barrier that would serve as an alternative material to the current concrete barrier incorporated with perforated surface infilled with absorbing material such as fibreglass/ rockwool. It is an environmentally friendly, sustainable, and can be labelled as "green".

6. Conclusions

The replacement of palm oil clinker (POC) in different range of particle sizes for natural river sand and aggregate in the development of noise barrier is believed to improve the acoustic performance based on the properties of the POC. The porous concrete layer/s consists of medium or coarse POC aggregate will enhanced the sound absorption of the noise barrier while the replacement of fine POC in hard backing layer will gives better soundproof.

Acknowledgments

The authors wish to acknowledge FGV Palm Industries Sdn. Bhd. (FGVPISB) for supplying research material palm oil clinker. The authors also would like to extend gratefully acknowledgement to Materials and Structures Laboratory, School of Civil Engineering, University Technology Malaysia for providing equipment's and facilities also to all technicians for their help.

References

[1] Birgitta Berglund, Thomas Lindvall, Dietrich H Schwe, Guidelines For Community Noise, World Health Organization, Geneva, 1999

[2] Arvind Kumar Shukla, An Approach For Design Of Noise Barriers On Flyovers In Urban Areas In India, International Journal For Traffic And Transport Engineering, 2011, 1(3): 158 – 167

[3] Julie Farnham And Edward Beimborn, Techniques For Aesthetic Design Of Freeway Noise Barriers , Transportation Research Record 1312, 2017

[4] Department of Environment (DOE) 2004 Ministry of Natural Resources and Environment, Malaysia

[5] LLM (2012) Lembaga Lebuhraya Malaysia

[6] Jegathish Kanadasan and Hashim Abdul Razak, Utilization of Palm Oil Clinker as Cement Replacement Material, Materials 2015, 8, 8817–8838; doi:10.3390/ma8125494

[7] Jegathish Kanadasan, Auni Filzah Ahmad Fauzi, Hashim Abdul Razak, Paramananthan Selliah, Vijaya Subramaniam and Sumiani Yusoff , Feasibility Studies of Palm Oil Mill Waste Aggregates for the Construction Industry, Materials 2015, 8, 6508-6530

[8] Fuad Abutaha, Hashim Abdul Razak and Hussein Adebayo Ibrahim, Effect of Coating Palm Oil Clinker Aggregate on the Engineering Properties of Normal Grade Concrete, Coatings 2017, 7, 175; doi:10.3390/coatings7100175

[9] Arenas, C., Vilches, L.F., Leiva, C., Alonso-Fariñas, B., Rodríguez-Galán, M., 2016. Recycling ceramic industry wastes in sound absorbing materials, Materiales de Construcción 66(324), http://dx.doi.org/10.3989/mc.2016.10615

[10] Carlos Leiva, Luis F. Vilches, Celia Arenas, Silvia Delgado, and Constantino Fernández-Pereira, Potential Recycling Of Bottom and Fly Ashes in Acoustic Mortars and Concretes, Aci Materials Journal/September-October 2012

[11] Zaiton Haron, Zanariah Jahya, Khairulzan Yahya (2018). Development of Green Soundproof Hard Backing Material, PTSR100621, Patent application, UTM.J.14.01/27.13/1JLD138 (87)

[12] Celia Arenas A, Carlos Leiva A, Luis F. Vilches A, Héctor Cifuentes, Use Of Co-Combustion Bottom Ash To Design An Acoustic Absorbing Material For Highway Noise Barriers, Waste Management, Volume 33, Issue 11, November 2013, Pages 2316-2321

[13] Celia Arenas A, Carlos Leiva A, Luis F. Vilches A, Héctor Cifuentes B, Mónica Rodríguez-Galán, Technical Specifications For Highway Noise Barriers Made Of Coal Bottom Ash-Based Sound Absorbing Concrete, Construction And Building Materials 95 (2015) 585–591

[14] N. Neithalath, J. Weiss, J. Olek, Acoustically efficient concretes through engineered pore structure, Autoclaved aerated concrete: properties and structural design, vol. SP-226, American Concrete Institute, SP (2005), pp. 135-147

[15] Tian, B., Ying, L., Kaimin, N., Sili, L., Jinde, X., Xinjun, Li., 2014. Reduction of tire-pavement noise by porous concrete pavement, J. Mater. Civ. Eng. 26(2), 233-239.

[16] Wai Loan Liew, Mohd. Azraai Kassim, Khalida Muda and Soh Kheang Loh, Feasibility Study On Palm Oil Processing Wastes Towards Achieving Zero Discharge, ARPN Journal of Engineering And Applied Sciences Vol. 11, No. 4, February 2016

[17] Herry Hartono1 , Lee Jin Chai1 and B. C. T. Lee (2016), Lightweight Concrete Using Oil Palm Boiler Clinker (OPBC) – A Review , MATEC conference, 2016

[18] Fuad Abutaha, Hashim Abdul Razak, Hussein Adebayo Ibrahim, Haider Hamad Ghayeb, Adopting Particle-Packing Method To Develop High Strength Palm Oil Clinker Concrete, Resources, Conservation And Recycling Volume 131, April 2018, Pages 247-258

[19] Navy Environmental Health Center (NEHC), "Man-Made Vitreous Fibers," vol. 12, no. October, p.110, 1997.

[20] Hussein Adebayo Ibrahim and Hashim Abdul Razak, Effect of palm oil clinker incorporation on properties of pervious concrete, Construction and Building Materials 115 (2016) 70–77

[21] Iarc, "Iarc Monographs on the Evaluation of Carcinogenic Risks to Humans," *Iarc Work. Gr.Eval. Carcinog. Risks To Humans*, vol. 96, p. i-ix+1-390, 2002

IOP Conf. Series: Journal of Physics: Conf. Series **1262** (2019) 012019 doi:10.1088/1742-6596/1262/1/012019

Correlation of numerical and experimental analysis for dynamic behaviour of a 3 blade propeller structure

P Brapakaran[1], N A Z Abdullah[1], M S M Fouzi[1] and M S M Sani[1,2*]
[1]Advanced Structural Integrity & Vibration Research (ASIVR), Faculty of Mechanical and Manufacturing Engineering, Universiti Malaysia Pahang, 26600 Pekan, Pahang, Malaysia
[2]Automotive Engineering Centre, Universiti Malaysia Pahang, 26600 Pekan, Pahang, Malaysia

*Corresponding author: mshahrir@ump.edu.my

Abstract. In pursuance of deciding the dependability of data gathered by testing a finite element modal in the software version, experimental data is frequently used for validation. On account of finite element analysis, it can sometimes be considered as inaccurate particularly when applied to the complex structure, for example, a propeller blade. This is because of challenges that may happen in the modelling of joints, boundary conditions, and damping of the structure. In this research, a procedure of correlation and validation of the model-based test plan with modal testing results was conducted. Modal properties (normal frequencies, mode shapes, and damping ratio) of a propeller blade structure were resolved by using both test experimental modal analysis (EMA) and finite element analysis (FEA). Correlation of both sets of data was performed for validation. It created the impression that there was a noticeable estimation of error between those two sets of data. Small discrepancies of percentage error of obtained natural frequency for FEA and EMA makes both of the methods can be applied to determine the dynamic characteristic of the propeller structure.

1. Introduction

The dependability of basic mechanical numerical investigation, for example, numerical analysis (FEA) can generally be approved by using experimental analysis data. The difference of qualities accumulated through experiments demonstrates the level of precision for the existing finite element model used in numerical prediction analysis [1-2]. It is critical to guarantee that the existing finite element model that was created during the design process is a dependable and ready to give an exact prediction of structural behaviour and performance before the real structure experiences large scale manufacturing in the assembling zone. As expressed in numerous past experiments, finite element modelling is the most basic instruments used to deride up the complex system or structure [3-7].

One of the most generally used propulsion tool, the propellers plays an essential role on submerged vehicles. For the propellers on the marine vehicles, the working conditions are extremely complicated and numerous off-plan working conditions will be endured like crash back or rapid reversing. Along these lines, the increased efficiency and the thrust of propellers will be useful for their working conditions[8].Other than that, the propeller which have a curvy profile on the structure, can be classified as one complex structure as well. Modelling the exact propeller structure in finite element can be tricky and very hard to achieve. Thus, the goal of modelling in this case to come up with a representative model that is simple enough for mathematical manipulation yet capable for describing, inducing, and reasoning complicated phenomena. By using principle from finite element method, a complicated structure can be assembled using these element matrices by considering connectivity and all the boundary condition [9].

Therefore, in order to validate the existed finite element model whether it is sufficient enough to imitate the response that will be created by real structure, many scientists had proposed the use of modal data obtained both through FEA and experimental modal analysis (EMA) [10-15]. The modal properties of the structure in the subject which have the natural frequencies, damping proportions, and mode shapes will be gathered through both FEA and EMA [16-19]. These two sets of data will be correlated so as to compute the level of discrepancies between them [20-23].

Although discrepancies between prediction analysis and experimental on the real structure are inaccessible because of the presence of many local impacts that are not considered by FE modelling while predicting frequencies and modes, model updating is a technique generally recommended and mostly used when it comes to improving the relationship between the FE model and the experimental data [24]. Therefore, the present research is conducted with the interest of performing dynamic correlation of experimental modal data with the data collected through numerical analysis of a three blade propeller structure which can be considered as a complex structure.

2. Finite element modelling and analysis

Propeller structure, which is the important part of marine vehicle's transportation system, can be considered as a complex structure. Because of this reality, modelling the structure accordingly to the actual structure might be difficult. Along these lines, major simplification in modelling the structure was done while designing the finite element component model of the propeller. The majority of the spaces and curvatures on the surfaces of the propeller structure, just as the accessible joints on the structure, were altogether neglected during modelling process.

Computer-aided design (CAD) model was designed first before being imported into a computer-aided engineering (CAE) software, MSC Nastran/Patran, in order to change the model into a finite element model, the propeller is modelled according to and based on the actual propeller structure. The whole propeller model was modelled as a solid structure where the thickness for the whole structure created will be assigned in properties manager in MSC Nastran/Patran software.

The finite element model of the propeller was created by using 10540 elements of tetrahedral-10 shaped elements while applying Tet Mesh mesher. The propeller model was assigned with the same material properties of aluminium; which details are shown in Table 1. All those properties were assigned according to condition of each surface on actual structure. Since the propeller is a complex structure, few properties are neglected for the free-free boundary condition testing.

Table 1. Nominal value of material properties (aluminium) assigned to propeller model.

Properties	Nominal Value
Young's modulus	69 GPa
Poisson ratio	0.3
Density	2800 kg/m^3

Neither boundary condition nor external forces were applied to the model as the model was let to be in free-free boundary condition for calculation of modal properties using SOL 103 in MSC Nastran/Patran, which is the solution for the normal modes analysis. The arrangement of the condition of movement for natural frequencies and typical modes requires a special decreased type of the equation of motion. The reduced equation of motion in matrix form is as shown in Equation (1):

$$[M]\{u\}+[K]\{u\}=0 \tag{1}$$

where [M] is the mass matrix, [K] is the stiffness matrix and u is the vector of the variable describing the motion. Equation 1 was solved by assuming a harmonic solution of the form as Eq. (2)

$$\{u\}=\{\Phi\}\sin\omega t \tag{2}$$

where $\{\Phi\}$ is the eigenvector or mode shape and ω is the circular natural frequency. This harmonic solution is the key to the numerical solution of the problems and indicates that all the degrees of freedom of the vibrating structure move in asynchronous manner. The structural configuration does not change its basic shape during motion and the only thing that changes is its amplitude. If the differentiation of the assumed harmonic solution is performed and substituted into the equation of motion, the simplification of the equation will be represented as Eq. (3).

$$([K]-\omega^2 [M])\ \{\Phi\}=0 \tag{3}$$

As Eq. (3) was reduces to an eigenvalue problem, it will be represented as Eq. (4)

$$[K]-\omega_i^2\ M\{\Phi_i\}=0 \tag{4}$$

where Φ is the eigenvector (mode shape) corresponding to its eigenvalue and the eigenvalue Z_i is related to the natural frequency. The constructed finite element model as viewed in graphic interface of MSC.Nastran/Patran software is as shown in Figure 1. Meanwhile, the computed first 5 mode shapes obtained in FEA is shown in Figure 2.

Figure 1. Finite element model of propeller structure

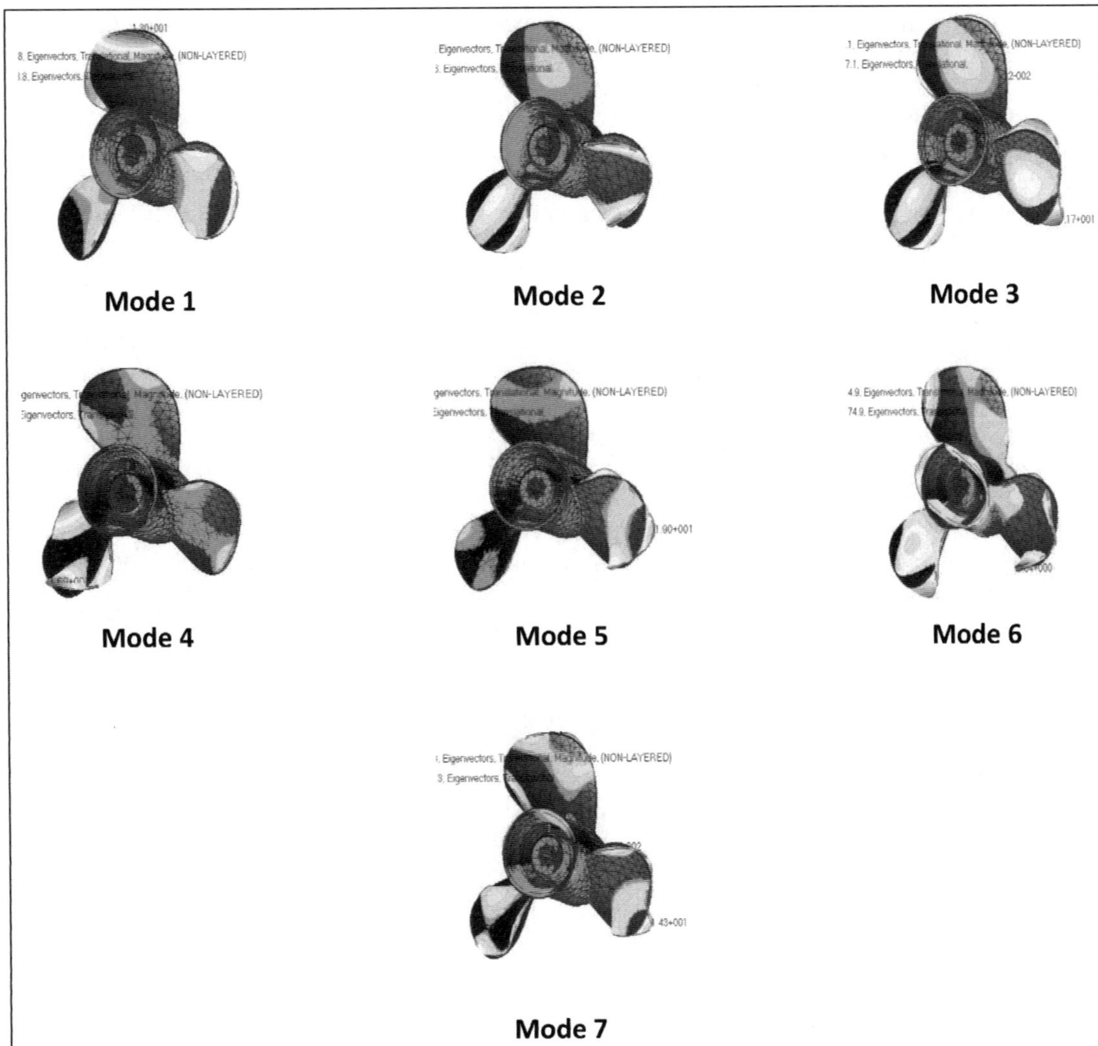

Figure 2. FEA mode shapes of propeller structure.

3. Experimental modal analysis

Experimental modal analysis (EMA) or once in a while called as modal testing, is the way toward separating dynamic qualities of a system, hardware or structure experimentally. Completing EMA on a structure or system has the advantage of having modal characteristic defined from actual measurements. A vital property of modes or the measured frequency response functions from modal testing can be used to describe the structure's dynamic or modal properties, which are the natural frequencies, damping ratios, and mode shapes.

There are a few kinds of response domain (regardless of whether frequency or time domain) that can be accumulated by EMA so as to extract those modal properties. On the other hand, the most normal information utilized for parameter extraction, and was utilized in this investigation, are the accumulated frequency response functions (FRFs), which use excitation input and the corresponding output of the test structure. The technique for excitation is the impact hammer test. Impact hammer testing is one of the most frequently used methods in modal testing. Aside from the capacity to figure FRF estimations in an FFR analyser, impact hammer testing is likewise a quick, advantageous, and minimal effort method for finding the methods of machines and structure.

Before conducting the impact hammer testing, an experimental model of the propeller was created by using post-processing software. The model consists of lines and points which virtually represent the geometric shape of the propeller structure (refer Figure 3). Data gathered through numerical analysis was used as a guide to performing the experiments in a way of indicating the best location for accelerometers on the propeller structure. In order to perform the impact hammer test, the propeller structure was hanged from a test rig by using the elastic cable in order to put the structure under the free-free boundary condition (see Figure 4). The measurements were made using modal analysis software and several other types of equipment such as PCB 086D20 impact hammer with medium soft tip attached, 4-channel NI DAQ device, and a tri-axial PCB accelerometer. Roving accelerometer method was adopted for the testing procedure, where one knocking point and 36 measurement points were assigned on the structure. Roving accelerometer test was done by creating an initial disturbance on the propeller structure at one fixed position while the single tri-axial accelerometer was roved around other measurement points. Nodal points were ensured to be avoided when choosing the impact and measurement point. The vibrational response was measured by using the 4- channel NI DAQ device. Curve fitting procedure available in the modal analysis software was used to extract the modal properties of the propeller structure from the computed FRFs.

The computed FRF based on all measurement points is shown in Figure 5. The estimated modal parameters were extracted by applying a curve fitting method to the FRF graph in order to obtain a set of experimentally derived data. The outcome of curve fitting, which is a set of modal parameters, which consist of natural frequencies, damping, and residues, for each of the identified modes within the range of frequency of interest, is shown in Figure 6. After the curve fitting process was completed, the modal parameters are stored in a Shape Table as residue mode shapes. The mode shapes were then displayed through the created experimental model. The computed modal parameters are shown in Figure 6. According to the computed FRF graph, the natural frequency for the first mode is 1530 Hz.

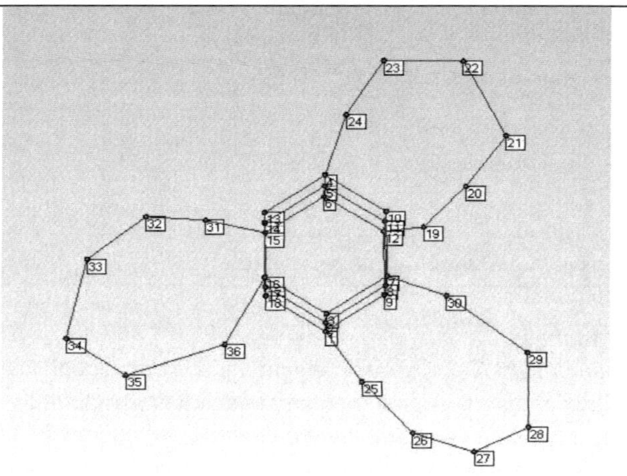

Figure 3. Experimental model of propeller structure.

IOP Conf. Series: Journal of Physics: Conf. Series **1262** (2019) 012019 doi:10.1088/1742-6596/1262/1/012019

Figure 4. Hanging propeller structure under free-free boundary condition.

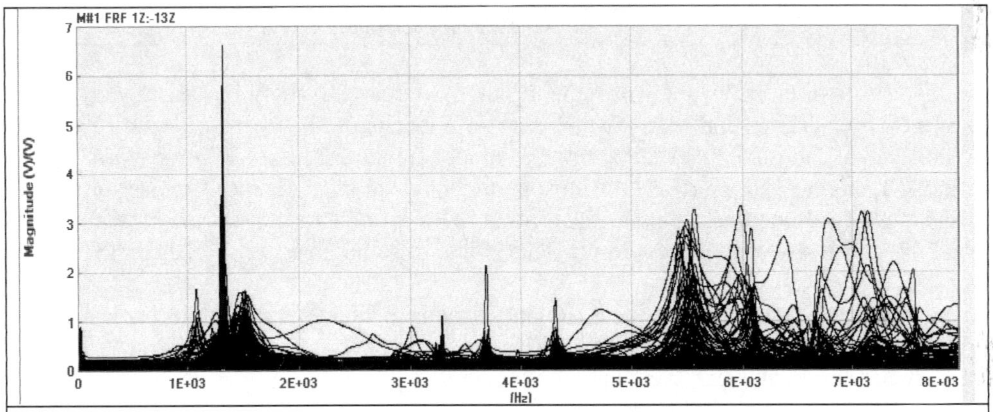

Figure 5. Computed FRF for all measurement points.

	Select Shape	Frequency (or Time)	Damping	Units	Damping (%)	Label	MPC
1	Yes	1.53E+03	38.9	(Hz)	2.54	G-PLY-PLY	0.0317
2	Yes	3.29E+03	4.68	(Hz)	0.142	G-PLY-PLY	0.104
3	Yes	3.36E+03	0.224	(Hz)	0.00666	G-PLY-PLY	0.61
4	Yes	4.34E+03	21.5	(Hz)	0.495	G-PLY-PLY	0.0253
5	Yes	5.54E+03	36.7	(Hz)	0.663	G-PLY-PLY	0.528
6	Yes	6.1E+03	34.2	(Hz)	0.56	G-PLY-PLY	0.302
7	Yes	7.59E+03	25.3	(Hz)	0.333	G-PLY-PLY	0.395

Figure 6. Computed modal data from FRF graph.

IOP Conf. Series: Journal of Physics: Conf. Series **1262** (2019) 012019 doi:10.1088/1742-6596/1262/1/012019

4. Result and Discussion

Correlation of data that was obtained through finite element analysis and modal testing was conducted in order to analyze the discrepancies existed between those two sets of data. In addition, correlation of data is essential in order to have accurateness estimation on the existing propeller model. Table 2 represent the correlation of natural frequencies of propeller structure that was gathered via experimental and numerical (finite element) analysis. Value of discrepancies between those two sets of data was calculated by accepting the value obtained through experiment as the actual value.

Table 2. Correlation of natural frequencies between EMA and FEA.

Mode	Experimental natural frequency (Hz)	Numerical natural frequency (Hz)	Percentage of error (%)
1	1530	1933.8	26.39
2	3290	3284.6	0.16
3	3360	3387.6	0.82
4	4340	4351.8	0.27
5	5540	5440.8	1.79
6	6100	5974.9	2.05
7	7590	7576.3	0.18
		Average Error	4.52

Since there is a discrepancy between the result of numerical analysis and modal testing, the percentage error is calculated and been tabulate at table 2. The highest percentage error for portal frame structure obtained is about 26.39% while the other modes (mode 2,3,4,5,6 & 7) were just below 10%. Mode 2, mode 3, mode 4 and mode 7 is the most perfect mode among all since the error just below 1%. Mode 5 and mode 6 in moderate range as the error is below 10%. Mode 1 has the highest error which more than 10% and it shows the worst result. This is due to some errors encounter during performing the FEA and EMA analysis.

There are few errors resulting from the assumptions made to characterize the mechanical behavior of the physical structure. Such errors typically arise from simplifications of the structure, inaccurate assignment of mass properties, incorrect modeling of boundary conditions, incorrect modeling of joints, incorrect geometrical shape assumptions. Other than that, when the finite element formulation neglects particular properties, Errors in the connectivity of the mesh i.e. some elements are not connected or are connected to the wrong node.

In order to reduce the existing discrepancies, model updating procedure can be applied on the finite element model of the propeller, thus improving the model to have better correlation with the actual structure. Several updating parameters (Young's modulus and all the thicknesses used in finite element analysis) can be considered to be included in model updating procedure.

Besides that, during performing modal analysis using impact hammer test and run analysis in ME'Scope VES, the noise may occur because during the knocking process, it is possible to wait longer until the structure perfectly stops being moving or vibrate before the next knocking process can be done. There may be internal vibrate inside the structure which invisible to be seen. This may contribute to longer time to complete the testing since there are 36 node points on the structure. Moreover, by improving higher number of node points, it can result the more accurate mode shape.

5. Conclusion

This study was undertaken to correlate the experimental modal data to the data gathered from the finite element analysis of the propeller structure. A finite element model of the propeller structure was produced and the percentage of errors between those two sets of data was obtained. Defined that the outcome obtained from the numerical analysis is close to the modal testing and because of some existing error, will affect the imperfection of both results.

Issues of each close mode frequently occur in engineering designing practice because of structure symmetries with small damping and exact determination of the model parameters. The percentage differences between finite element analysis and experimental analysis are within 10% except for mode 1 by cause of the factor been discussed previously. Small discrepancies of percentage error of obtained natural frequency for FEA and EMA makes both of the methods can be applied to determine the dynamic characteristic of the propeller structure. It is recommended that further research be undertaken by performing modal updating in order to diminish the percentage of error. The updating procedure is observed as parameter identification which means to bring the finite element prediction to be as close as possible to the actual test subject.

Acknowledgement

The authors of this paper would like to acknowledge a great support and encouragement by focus group of Advanced Structural Integrity of Vibration Research (ASIVR), Universiti Malaysia Pahang (UMP) for providing all the equipment used for this work Fundamental Research Grant Scheme (FRGS/1/2017/TK03/UMP/02-19) – RDU 170123.

References

[1] Kishore M L P, Behera RK, Pradhan SK, Parida PK 2016 Effect of Material Behavior on Dynamic Characteristics Determination of Marine Propeller Blade Using Finite Element Analysis *Procedia Engineering* 144 767-774.

[2] Li J, Rao Z, Su J, Qu Y and Hua H 2018 A numerical method for predicting the hydroelastic response of marine propellers *Applied Ocean Research* 74 188-204.

[3] Rakshit S. and Khare S R 2019 Symmetric band structure preserving finite element model updating with no spillover *Mechanical Systems and Signal Processing* 116 415-431.

[4] Arjmandi M and Ramezani M 2019 Finite element modelling of sliding wear in three-dimensional textile hydrogel composites *Tribology International* 133 88-100.

[5] Nioi M, Pinna C, Celotto S, Swart E, Faruggia D, Husain Z and Ghadbeigi H 2019 Finite element modelling of surface defect evolution during hot rolling of Silicon steel *Journal of Materials Processing Technology* 268 181-191.

[6] Panwar V, Gupta P, Bagha A K and Chauhan N 2018 A Review on studies of Finite Element Model Updating and Updating of Composite Materials *Materials Today: Proceedings* 5(14, Part 2) 27912-27918.

[7] Ádány S 2019 Modal identification of thin-walled members by using the constrained finite element method *Thin-Walled Structures* 140 31-42.

[8] Chen F, Liu L, Lan X, Li Q, Leng J, Liu Y 2017 The study on the morphing composite propeller for marine vehicle *Part I: Design and numerical analysis. Composite Structures* 168 746-757.

[9] Zienkiewicz O C and Taylor R L 2005 The finite element method for solid and structural mechanics *Butterworth-Heinemann Elsevier*.

[10] Sani M S M and Ouyang H 2016 Identification of Nonlinearities in Joints of a Wing Structure *MATEC Web of Conference* 83 (03006)

[11] Ewins D 2000 Model validation: Correlation for updating *Sadhana* 25(3) 221-234.

[12] Jelani K M, Razip M H M, Nazri N A, Sani M S M and Yasar M 2018 Dynamics investigation on motorcycle chassis based on Finite Element (FE) modelling and updating *IOP Conf. Ser. Mater. Sci. Eng.* 469 012103

[13] Reynders E 2012 System identification methods for (operational) modal analysis: review and comparison *Archives of Computational Methods in Engineering* 19(1) 51-124.

[14] Zahari S N, Sani M S M, Husain N A, Ishak M and Zaman I 2016 Dynamic analysis of friction stir welding joints in dissimilar material plate structure *J. Teknol.* 78

[15] Friswell M I, Mottershead J E and Ahmadian H 2001 Finite–element model updating using experimental test data: parametrization and regularization. *Philosophical Transactions of the Royal Society of London. Series A: Mathematical, Physical and Engineering Sciences* 359(1778) 169-186.

[16] Sani M S M, Abdullah N A Z, Zahari S N, Siregar J P and Rahman M M 2016 Finite element model updating of natural fibre reinforced composite structure in structural dynamics *MATEC Web of Conferences* 83 (03007)

[17] Kim J, Yoon J C and Kang B S 2007 Finite element analysis and modeling of structure with bolted joints *Applied Mathematical Modelling* 31(5) 895-911.

[18] Fouzi M S M, Sani M S M, and Muchlis Y 2019 Finite Element Modelling and updating of welded joint for dynamic study of exhaust structure I*OP Conference Series: Materials Science and Engineering* 469 1-10

[19] Narasaiah P and Basarkod S S 2014 Dynamic Analysis of Composite Propeller of Ship using FEA *International Journal of Scientific Engineering and Technology Research* 3(27) 5467-5472.

[20] Prasad A S, Ramakrishna S and Madhavi M 2018 Experimental Investigations on Static and Dynamic Parameters of Steel and Composite Propeller Shafts with an Integrated Metallic Joints. *Materials Today: Proceedings* 5(13, Part 3) 26925-26933.

[21] Abdullah N A Z, Sani M S M, Hussain N A, Rahman M M and Zaman I 2017 Dynamics properties of a Go-kart chassis structure and its prediction improvement using model updating approach *International Journal of Automotive and Mechanical Engineering (IJAME)* 14 3887-97

[22] Fouzi M S M, Jelani K M, Nazri N A and Sani M S M 2018 Finite Element Modelling and Updating of Welded Thin-Walled Beam *International Journal of Automotive and Mechanical Engineering (IJAME)* 15(4) 5874-89

[23] Mohd Zin M S, Rani M N A, Yunus M A, Wan Iskandar Mirza W I I, Mat Isa A A and Mohamed Z 2017 Modal and FRF Based Updating Methods for the Investigation of the Dynamic Behaviour of a Plate *Journal of Mechanical Engineering* 4 175–89

[24] Abdullah N A Z, Sani M S M, Rahman M M, and Zaman I 2015 A review on model updating in structural dynamics *IOP Conference Series: Materials Science and Engineering* 100 1-6

IOP Conf. Series: Journal of Physics: Conf. Series **1262** (2019) 012020 doi:10.1088/1742-6596/1262/1/012020

Normal mode finite element analysis of aerofoil wing structure with different materials

M N A M Asri[1], N A Z Abdullah[1], M S M Fouzi[1] and M S M Sani[1,2]*
[1]Advanced Structural Integrity & Vibration Research (ASIVR), Faculty of Mechanical and Manufacturing Engineering, Universiti Malaysia Pahang, 26600 Pekan, Pahang, Malaysia
[2]Automotive Engineering Centre, Universiti Malaysia Pahang, 26600 Pekan, Pahang, Malaysia

*Corresponding author: mshahrir@ump.edu.my

Abstract. There are a lot of study regarding on the aircraft components. One of the critical components for an aircraft would be the wings. The wing structure of an aircraft is one of the complex structures of a designed aircraft. This paper is about identifying the modal properties which are the mode shape and the corresponding natural frequencies of the aircraft wing structure. The modal properties of the wing structure would be compared for two different materials applied towards the wing structures which are the aluminium alloy AA-7075-T6 and AA-2024-T3 which currently being widely used by the aircrafts. The study for this aircraft wing structure would be using the approach of finite element analysis (FEA) method. The 3D model is design by using SolidWorks. The modal properties are identified with the help of MSC PATRAN and MSC NASTRAN. The results obtained for both materials for 10 mode shapes are compared and it could be observed that AA-7075-T6 are much lower compared to AA-2024-T3. Hence, from the result, AA-7075-T6 is more suitable to be used for the wing structures.

1. Introduction

There are several types of aircraft. A typical aircraft that are easily to be seen would be on the fixed wing aircraft or known as airplane, aeroplane or planes. The airframe of the aircraft consists of five major components which are fuselage, wings, stabilizers, flight controls surfaces and landing gears [1]. There are many studies that had been done related to the aircraft itself as well as its components especially on the wing structure of the aircraft itself [2].

A major portion of the lift for an aircraft is developed from the wings itself. It could be found that the wing structures itself do carry some of the loads found by the aircraft structure especially when the aircraft are flying [1]. The components that are affected the most due to the loads are at the fuselage as well as the wings where the stresses of bending and compression would occur the most for both components either when it is statics on the ground or when it is flying or landing [1]. In addition to that, for both components' fuselages and wings, it could be distorted in two ways – bending and twisting if it were subjected to flutter [3]. When flutter occurs, resonance is one of the contribution towards the structural failure [4]. The wings could also face vibrational problem due to the rotation of the engine rotor for the aircraft with attached engines below the wings.

As for this paper, the only component that would be analyse would be on the wings structure of the aircraft. And the main concern in analysing this structure is to identify the dynamic properties of the structures which are the natural frequencies and the mode shape. By identifying the dynamic properties of a system, it could help in term of identifying the reasons of the occurred vibration that may cause damage towards the system components as well as it could reduce the noise emitted from the system to the surrounding [5]. Plus, in achieving an effective design and gain control of the vibrations of the structural components, it do depends on those vibrational behaviour [6-10].

A typical aircraft wing structures – in a simplest form does made up from a framework of spars, ribs and covered with metal [5, 11-12]. Figure 1 shows a typical aircraft wing structure. Generally, for a typical aircraft – metallic, non-metallic and composites materials are used throughout all the components of the aircraft. Originated from wood, and nowadays replaced by alloys and the latest technology enable composites to be used, the most important criteria to be considered for an aircraft wing are high strength and low in weight [13-14]. Although composites are much more reliable compared to aluminium alloys, but aluminium alloys are currently used widely in the civil aircraft industries as its low cost, lightweight and easy to be manufactured and treated and this material (aluminium alloy) would be the materials to be analyse for this paper [15]. The objectives for this study are to find the difference in the modal properties of the wing structure with different material used by the wing structure itself.

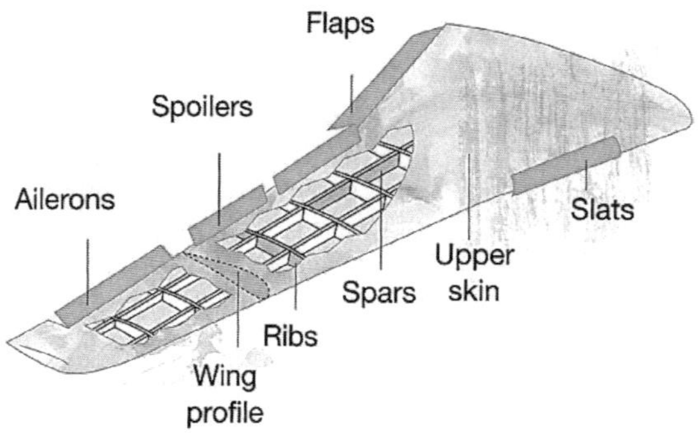

Figure 1. Aircraft wing structure.

2. Methodology

The method that is used in this study is finite element analysis (FEA). The fundamental steps in any finite element analysis (FEA) method would consists following steps as shown in Figure 2:

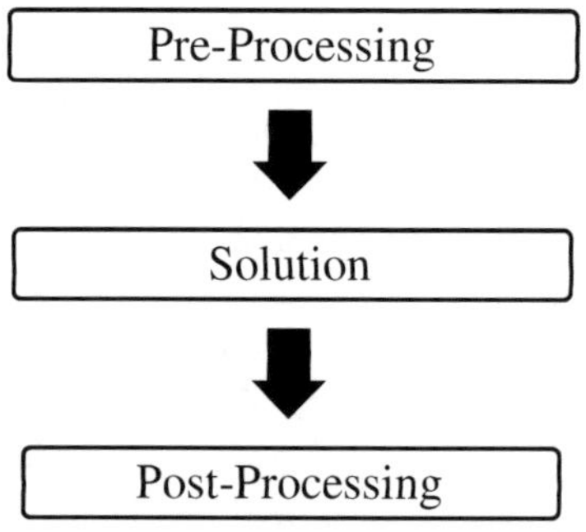

Figure 2. FEA process

2.1 Modal Analysis

Modal analysis determines the modal properties of a structure in form of natural frequencies and mode shapes due to vibrational excitations. For this study, damping ratio is not to be considered as well as any applied loads. For aircraft wing structure, vibration does occur due to the rotating engine mounted under the wing. This analysis is important in order for a structure to be identified for the dynamic loading conditions [11, 16].

2.2 Finite Element Analysis (FEA)

FEA approach was used in term of normal mode analysis in order to obtain both modal properties of the wing structure which are the natural frequencies and the mode shape. The software to be used for FEA would be MSC PATRAN for both pre-process and post-process. The analysis was done by using the solver software which are MSC NASTRAN. The 3D model was designed in SOLIDWORKS 2017 before it was imported into MSC PATRAN. The consistency of the analysis is then validated by refining the mesh size. The mesh size was reduced until convergence was achieved for the magnitude of the natural frequencies during post-process as in Figure 4.

2.3 Simplification

Aircraft wing structures are complex structures. Typical aircraft wing structure would have the framework made up from several structural components which are ribs, spars, stringers, skin, flaps, fuel tanks, aileron and wing tips. And for every type of aircrafts would have different type of wing designs and configurations depending on the uses of the aircraft [3]. Some of the simplification was made for this model based on the following assumption:

- The material used is homogenous and isotropic.
- The material is elastic.
- The complex structure was simplified to only the spars and the rib as the study would cover for only the wing skeleton structure.

2.4 Pre-Processing

The 3D model was imported from SOLIDWORK 2017 in term of parasolid (.x_t) file. As the simplification was applied for this model, the wing skeleton structure was design as a whole instead of assembling it part by part. The design for the wing structure is as in Figure 3. The geometric parameter as in Table 1. There are total 16 ribs in the wing skeleton structure based on literature [8]. The model was then assigned with the material properties as tabulated in Table 2. The boundary condition for the wing structure model was set to free-free in order to obtain the static modal properties. Once the material, boundary condition and constraint were set up, the model was proceeding to mesh. For meshing, a 3D solid meshing type was used with tetrahedral element shape, TetMesh mesher with Tet10 topology. Based on the mesh convergence study graph, the ideal mesh global edge length is 0.03mm as could be seen in Figure 4.

Table 1. Geometric parameter of the wing design

Airfoil	NACA 64215
Root Chord	2400 mm
Tip Chord	700 mm
Span Length	5500 mm
Front Spar	18% to 25% of chord
Rear Spar	62% to 70% of chord

IOP Conf. Series: Journal of Physics: Conf. Series **1262** (2019) 012020 doi:10.1088/1742-6596/1262/1/012020

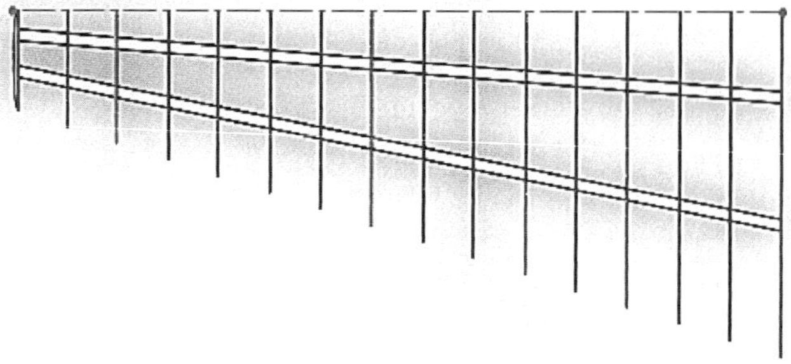

Figure 3. The plan view of the wing skeletal structure

Table 2. Material properties of wing structure

Class of material	Material	Elastic Modulus (GPa)	Density (kg/m^3)	Poisson Ratio
Aluminium (reference)	$7075 - T6$	71.70547	2810	0.33
Aluminium	$2024 - T3$	73.08443	2780	0.33

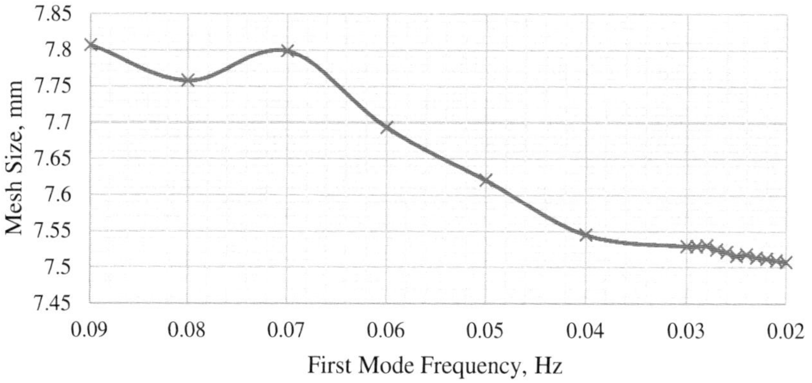

Figure 4. Mesh Convergence Graph

2.5 Post-Processing

The acceptability of the model design of the wing structure are depending on the result of the modal analysis. The generated results of the modal properties were cross checked with other findings from the literature. In the comparison, the most important finding to be checked would be the mode shape of the structure in which it should have the same mode shape for every number of the mode shape. The intended mode shape and natural frequencies would be only 10 mode shapes.

3. Results and Discussion

The normal mode analysis allows computing normal mode of the input structure and visualization of the computed modes by animating the displacement of the structures. It is critical in obtaining the natural frequencies as well as the corresponding mode shape in order to identify the dynamic behaviour of a system. The results obtained for the modal properties that was run by using the MSC PATRAN/NASTRAN are as displayed in Table 3, Table 4, Table 5 and Table 6.

Table 3. Modal properties of AA-7075-T6

Mode	Frequency (Hz)	Maximum Deformation (m)	Description
1	7.5294	0.206	Bending
2	10.564	0.357	Torsion
3	28.629	0.392	Bending
4	39.286	0.429	Torsion
5	71.832	0.531	Torsion
6	91.388	0.853	Bending
7	130.80	0.516	Torsion
8	144.46	1.060	Torsion
9	168.85	0.348	Torsional Bending
10	205.44	0.772	Torsion

Table 4. Modal properties of AA-2024-T3

Mode	Frequency (Hz)	Maximum Deformation (m)	Description
1	7.6424	0.207	Bending
2	10.723	0.359	Torsion
3	29.059	0.394	Bending
4	39.875	0.431	Torsion
5	72.910	0.534	Torsion
6	92.759	0.858	Bending
7	132.76	0.519	Torsion
8	146.63	1.064	Torsion
9	171.38	0.350	Torsional Bending
10	208.52	0.776	Torsion

Table 5. The first 10 mode shapes for AA-7075-T6

Mode	Mode Shape	Mode	Mode Shape
1	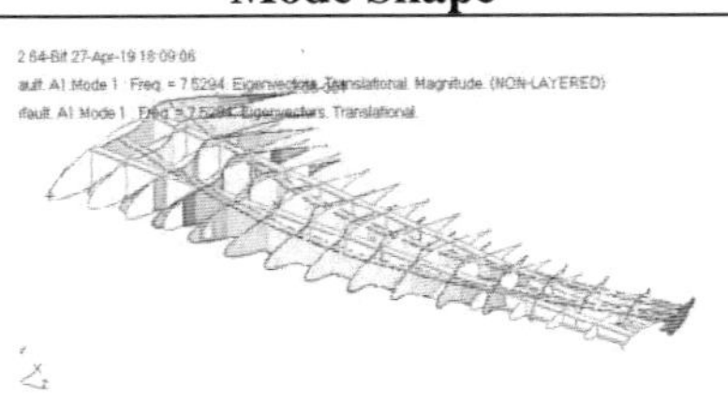	2	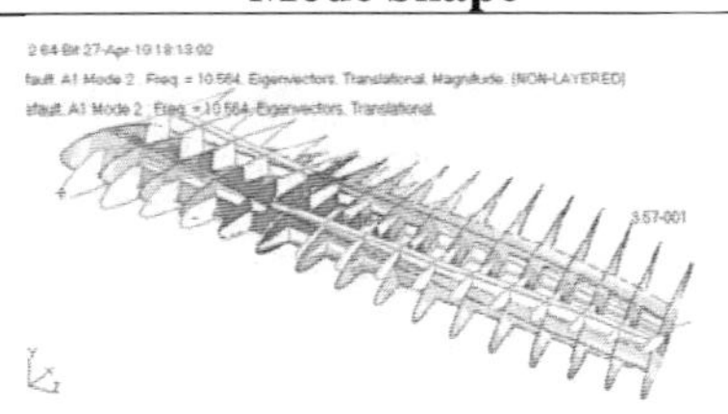
3	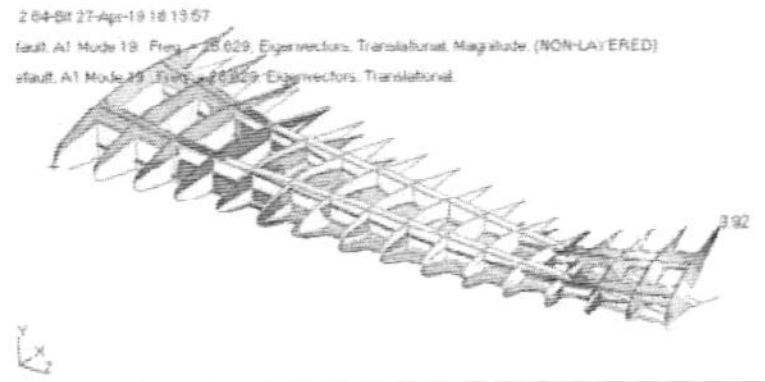	4	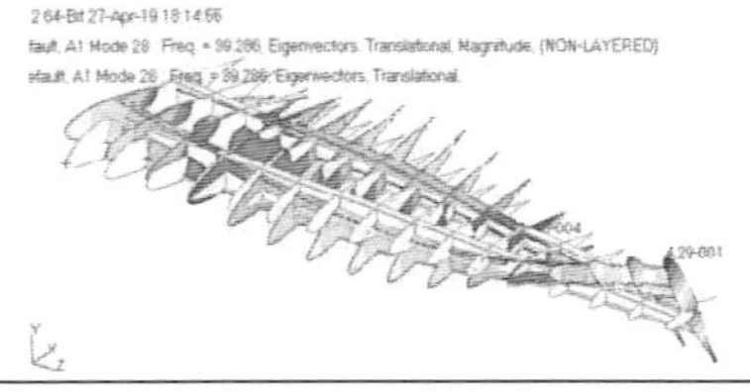

IOP Conf. Series: Journal of Physics: Conf. Series **1262** (2019) 012020 doi:10.1088/1742-6596/1262/1/012020

5	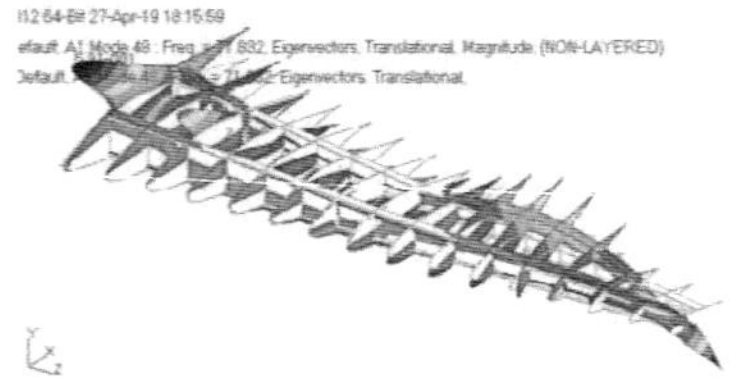	6	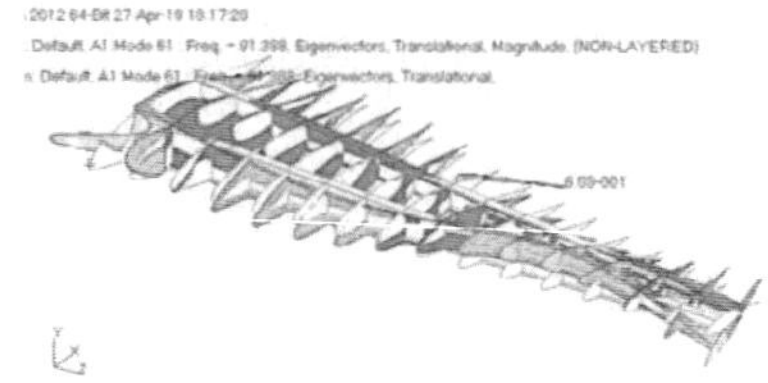
7	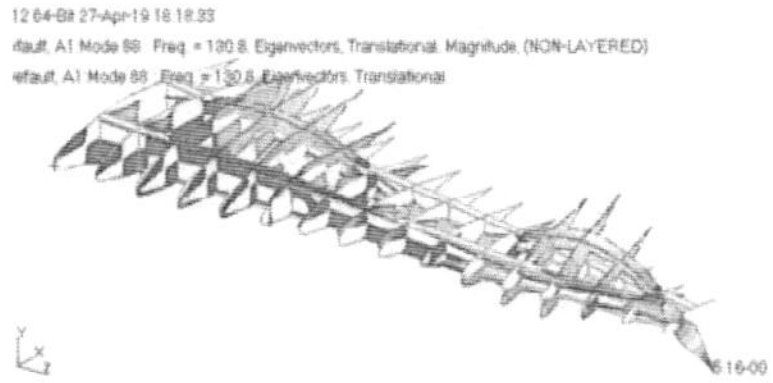	8	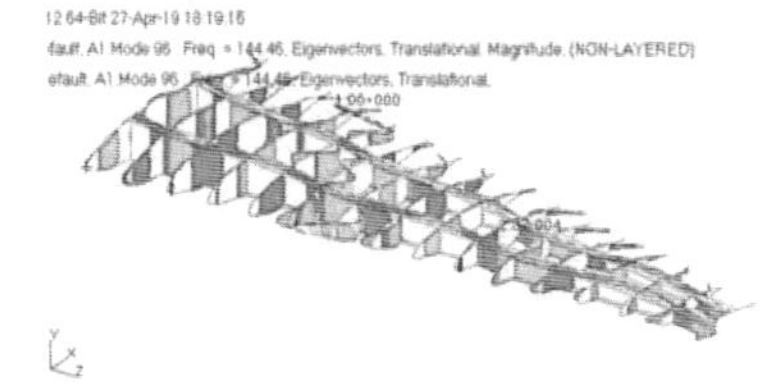
9	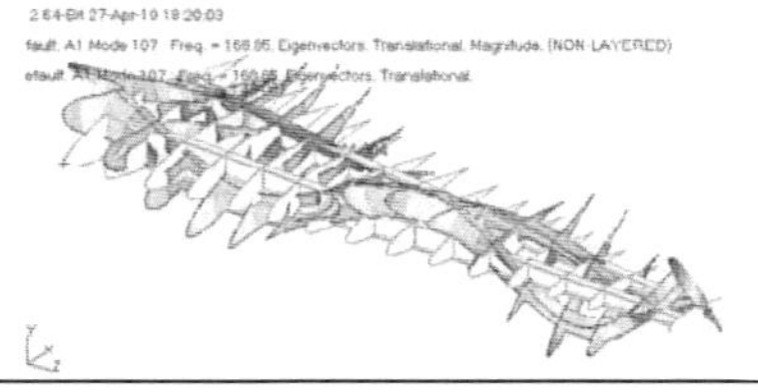	10	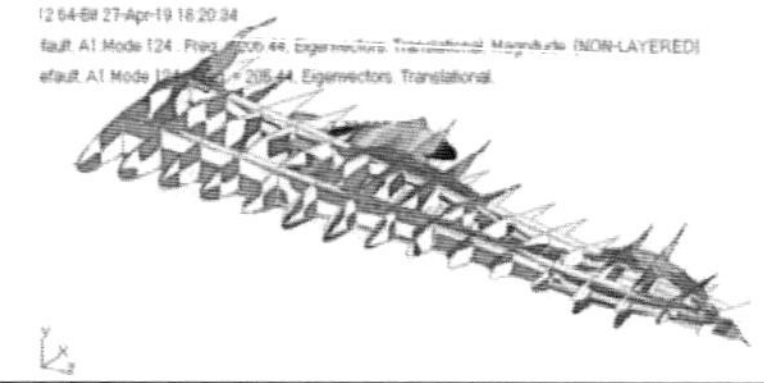

Table 6. The 10 mode shapes for AA-2024-T3

Mode	Mode Shape	Mode	Mode Shape
1	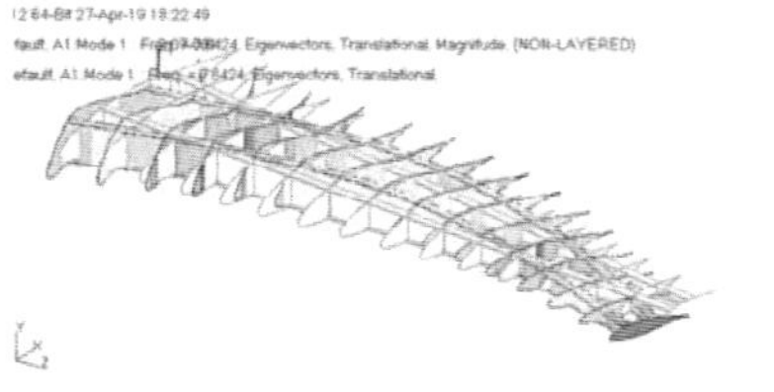	2	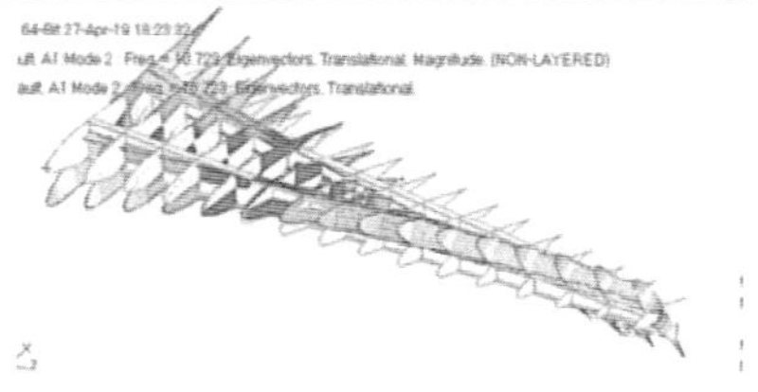
3	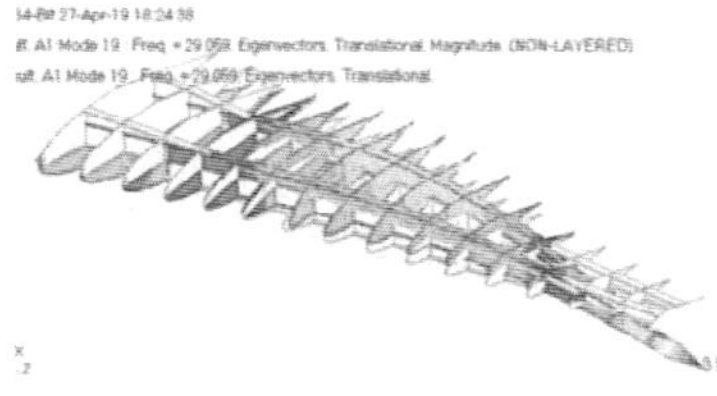	4	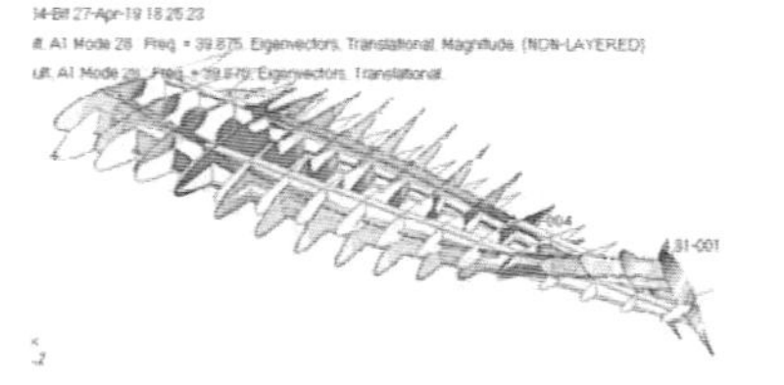
5	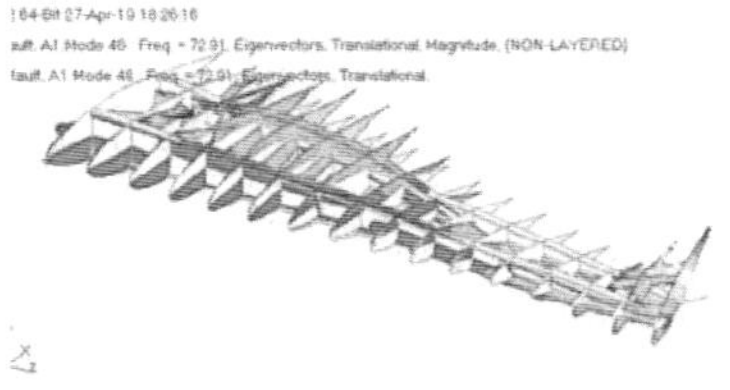	6	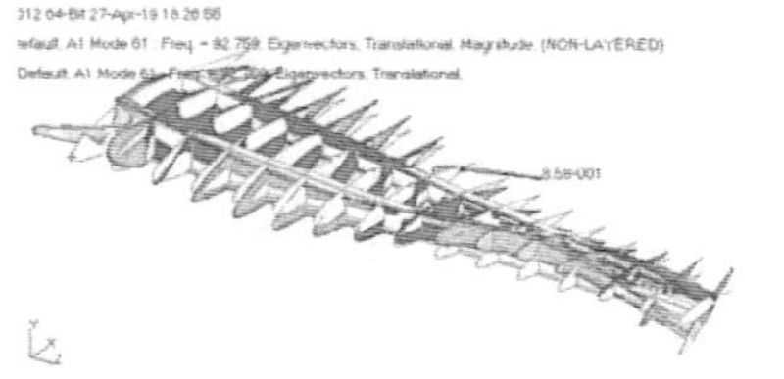
7	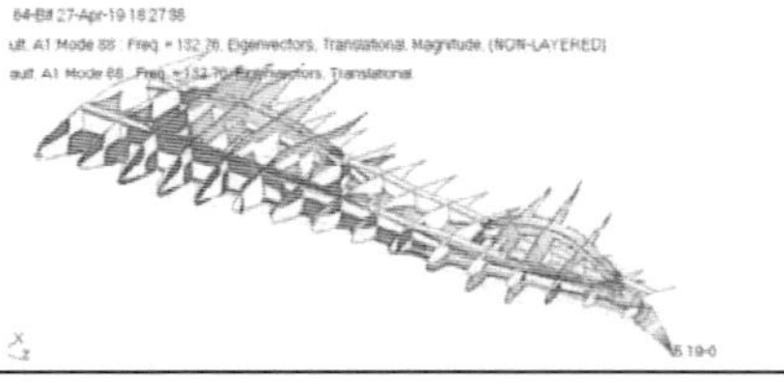	8	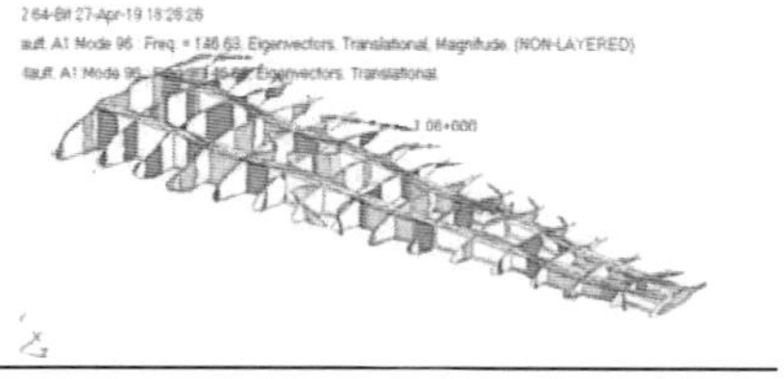

9	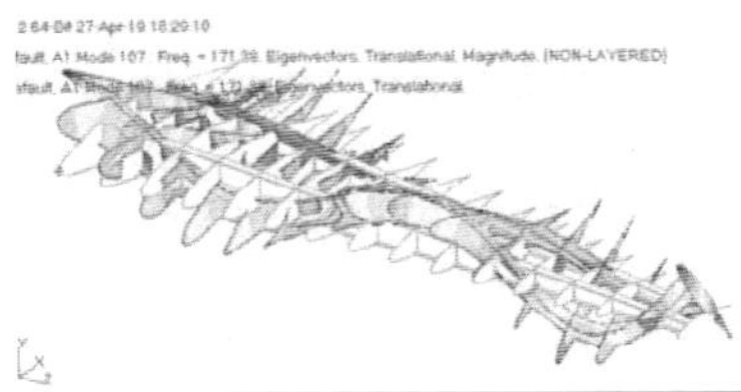	10	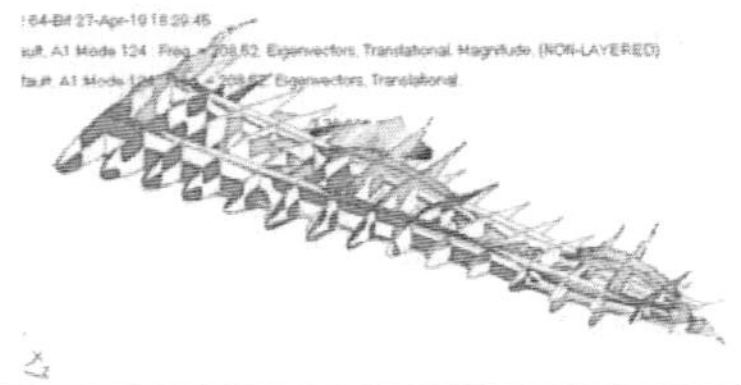

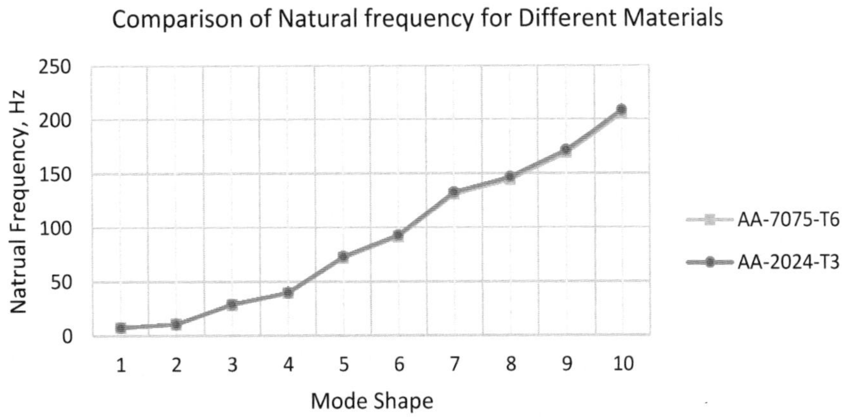

Figure 5. Graph for comparison of the natural frequency for the different materials

Based on the obtained results, it could be observed that the maximum deformation for AA-7075-T6 was 1.060 m which was correspond to the natural frequency of 144.68 Hz while for AA-2024-T3 was 1.064 m correspond to the natural frequency of 146.63 Hz. The fundamental frequency could be considered at 7.5294 Hz and 7.6424 Hz for both AA-7075-T6 and AA-2024-T3 respectively which the deformation value was the lowest which are at mode 1. The fundamental frequency was the lowest frequency obtained from the structures when being excite and the fundamental frequency would be the frequency to be concerned the most [17-19].

4. Conclusion

From the above results, it could be sum up that the difference of magnitude of natural frequencies and the mode shape of AA-7075-T6 and AA-2024-T3 are minimal. The result obtained are validated and verified. As the difference between the two results of the modal properties are minimal, it could be observe that AA-7075-T6 and AA-2024-T3 are both suitable to be used in building the wing structures as both do have alike properties but in term of strength levels, AA-7075-T6 are quite higher compared to AA-2024-T3 so the uses of AA-7075-T6 in aircraft components generally are much more accepted as in the finding [12]. In detail, although the modal properties were look alike, it could be observed that the natural frequencies for AA-7075-T6 was much lower compared to AA-2024-T3 which make AA-7075-T6 much more suitable material for aircraft wing structure. The resonance of the wing structure could be calculated by identifying the natural frequencies of the wing structure itself. Once the resonance is determined, it could prevent the failure in design of the aircraft component generally.

Acknowledgements
The authors of this paper would like to acknowledge a great support and encouragement by focus group of Advanced Structural Integrity of Vibration Research (ASIVR), Universiti Malaysia Pahang (UMP) for providing all the equipment used for this work Fundamental Research Grant Scheme (FRGS/1/2017/TK03/UMP/02-19) – RDU 170123.

References
[1] METU 2014 Aircraft Basic Construction *METU Aerosp Eng* 1–22
[2] Kuntoji N and Kuppast D V V 2017 Study of Aircraft Wing with Emphasis on Vibration Characteristics *Int. J Eng Res Appl* 7(4) 1-8
[3] Kermode A C 2006 *Mechanics Of Flight* 11th Edition Pearson Prentice Hall
[4] Vulovic Z 2015 Modal Parameters of Light Aircraft Wing
[5] Khadse N A and Zaweri S R 2015 Modal Analysis of Aircraft Wing using Ansys Workbench Software Package *IJERT* 4(7) 225–230
[6] Sani M S M, Nazri N A, Zahari S N, Abdullah N A Z and Priyandoko G 2016 Dynamic Study of Bicycle Frame Structure IOP Conf. Ser. Mater. Sci. Eng. 160 012009
[7] Jelani K M, Razip M H M, Nazri N A, Sani M S M and Yasar M 2018 Dynamics investigation on motorcycle chassis based on Finite Element (FE) modelling and updating *IOP Conf. Ser. Mater. Sci. Eng.* 469 012103
[8] Sani M S M and Ouyang H 2016 Identification of Nonlinearities in Joints of a Wing Structure *MATEC Web of Conference* 83 (03006)
[9] Zahari S N, Sani M S M, Husain N A, Ishak M and Zaman I 2016 Dynamic analysis of friction stir welding joints in dissimilar material plate structure *J. Teknol.* 78
[10] Fouzi M S M, Sani M S M, and Muchlis Y 2019 Finite Element Modelling and updating of welded joint for dynamic study of exhaust structure I*OP Conference Series: Materials Science and Engineering* 469 1-10
[11] Kumar T S V, Basha A W, Pavithra M and Srilekha V 2015 Static and Dynamic Analysis of A Typical Aircraft Wing Structure Using MSC Nastran *Int J Res Aeronaut Mech Eng* 3(8)1–12.
[12] Sureka K and Meher R S 2015 Modeling and Structural Analysis on A300 Flight Wing By Using Ansys *IJMERR* 4(2) 123–130
[13] Lopes J C O 2008 Material Selection for Aeronautical Structural Application *Ciência Tecnol dos Mater* vol 2044(3) 78–82
[14] Sruthi K, Kishore T L and Rao M K 2017 Design And Structural Analysis Of An Aircraft Wing By Using Aluminium Silicon Carbide Composite Materials *IJEDR* 5(4) 949–959
[15] P Rambabu N E Prasad and V V Kutumbarao 2017 Aerospace Materials and Material Technologies *Springer*
[16] Naghate S and Patil S 2012 Modal Analysis of Engine Mounting Bracket Using FEA *Int J Eng Res Appl* 2(4) 1973–1979
[18] Fouzi M S M, Jelani K M, Nazri N A and Sani M S M 2018 Finite Element Modelling and Updating of Welded Thin-Walled Beam *International Journal of Automotive and Mechanical Engineering (IJAME)* 15(4) 5874-89
[19] Abdullah N A Z, Sani M S M, Hussain N A, Rahman M M and Zaman I 2017 Dynamics properties of a Go-kart chassis structure and its prediction improvement using model updating approach *International Journal of Automotive and Mechanical Engineering (IJAME)* 14 3887-97

Removal of Dust from the Solar Panel Surface using Mechanical Vibrator

Osam H. Attia [1, 2,*], Nor Mariah Adam [1], Azizan As'arry [1], Khairil Anas Md Rezali[1]
[1]Mechanical and Manufacturing Engineering, Faculty of Engineering, Universiti Putra Malaysia (UPM), Malaysia.
[2] Engineering Affairs Department, University of Baghdad .Iraq

*Corresponding author: osamhsattai@yahoo.com

Abstract. Soiling and its effect on the performance of solar modules are generally of high concern for regions with a high deposition of dust and low frequency and less intensity of rainfall. The procedure of removing dust using traditional methods is capital and labour intensive. Additionally, most of the cleaning methods consumed power from the energy produced by the solar system. Therefore, the main objective of this study is to investigate the effect of vibration magnitude on the dust removal index of solar panel. In this work, wind energy was transformed into mechanical energy i.e. vibration. The mechanical vibrator attached to a panel produced harmonic excitation force to overcome the adhesive force between the dust particles and the surface of the solar panel. The generated vibration force has a linear relationship with the air velocity. This new designed and fabricated system was able to remove 3.5 gram of dust out of 5 grams on the panel with a vibration force of 3.128 N at a tilt angle of 15°. The new system has effectively proven that wind energy if being converted into vibration force can be used for dust removal from the solar panel surface.

1. Introduction

Solar energy as one of the sources of energy is environmentally friendly and does not in any way contribute to the global warming [1-2]. The amount of energy radiation from the sun in one second is more than what the people have used since the beginning of time [3]. Solar energy most often is called "alternative energy" when compared to coal and oil [4]. The solar radiation varies from one country to others [5]. Solar panels (PV) are ideally located in regions with high insolation levels such as desert, remote desert areas, and un-shaded areas [6]. The locations of the installation and environmental parameters (ambient temperature, wind, rainfall, and water scarcity) have a direct bearing on the performance of the solar panels [7- 8]. The performance was usually affected by soiling and high cell temperatures. In nutshell, the performance is highly dependent on installation conditions [9].

Dust deposited on the solar panels resulted in energy losses [10]. Therefore, identifying a specific techniques or methods to remove dust becomes necessary. This will further ameliorate the performances of the solar system [11]. The technique or method to adopt depends on the site of the solar panels, environmental parameters, and mechanical parameters [12]. Several methods were employed to remove dust from PV system via the main concept; i.e. ability to overcome the adhesive force between dust particles and panel surface. The various methods usually fall as either natural, mechanical, or electrical technique [13]. Some of the techniques are combinations of mechanical and electrical like the one saw in the work of Mazumder, Zahn, Sharma, Zhang, Calle, Immer and Mardesich [14], which focused on electrodynamics system to remove the dust. The technique doesn't require water and manual labour and it continuously keeps the solar collectors clean [15]. In some situation with help of wind and presence of cover glass, the PV panels can be free from falling dust particles [16]. Also, Piliougine and his Nanophysics group at the Institute for Nano Energy developed a self-cleaning nano-hydrophobic material which will be used to coats the solar panel in order to maintain peak efficiency over long periods of time [17]. While [18] used piezoceramic actuators to remove the dust from solar panel. All these methods above consuming power and increasing the cost of electricity that produced. Therefore, this study produces a conversion of renewable energy (wind energy) into mechanical vibration for dust removal from PV surface without consuming power from the solar system and thereby improving its performance.

2. Experimental Procedure

2.1 Experimental rig set up
The experimental rig fabricated for converting wind energy into vibrating force, which used to remove dust from the solar panel surface. The experimental rig fabricated from the locally available materials and it has two major parts. The first part consists of a ducting box and mechanical vibrator (MV) with its rotating mechanism, while the second part comprises of solar panel and its frame structure with measurement instruments attached.

2.1.1 Air ducting box
A ducting box fabricated in order to obtain a streamline flow of air for the effective rotation of the impeller. The fabricated ducting box was shown in 'figure 1' and its comprises steel table fabricated using steel right angle (75×75×3mm) and a box fabricated from 12 mm thick plywood with a dimension of 1200×1010×900mm. A leather tarpaulin used to connect the wooden box and the variable speed fan. An opening on the box created for measurement of some environmental parameters, which can be open and close by sliding Perspex cover (850 ×1300 x 4mm).

2.1.2 Mechanical vibrator and its mechanisms
This assembly comprises two parts, namely rotating parts and vibrating parts. The rotating parts in this experimental rig are impeller, shafts and gear assembly for a change of direction of rotation. Four (4) rotating impellers assembly was fixe to a steel frame with a total weight of 1646 g. The rotational movement of the impeller was transmuting with the aid of long shaft 1320 mm that shown in 'figure 2'. An adjustable structure was also providing with four ball bearings to support the shaft. A bevel gear assembly, which was fabricating by using CNC machine, was also use to change the direction of shaft rotation and transmit the rotational movement down to the vibrating parts.

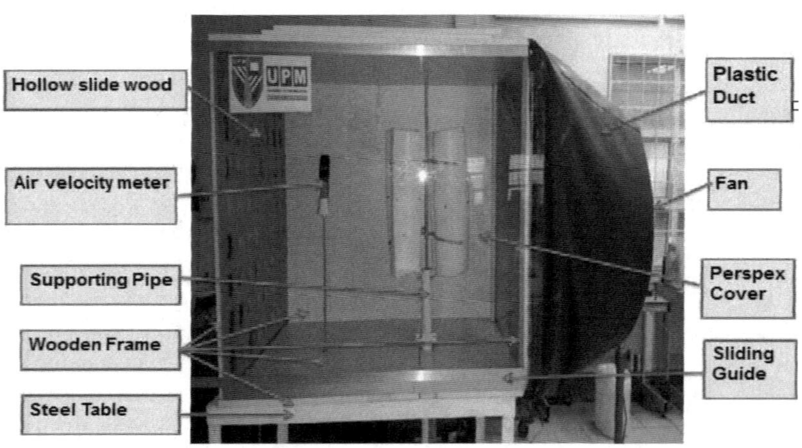

Figure 1. Air ducting box

Vibrating part of the rig contains a rotating shaft (stainless steel pipe SSt304, 12, 0.7 mm for diameter & thickness respectively, and length of 1100 mm) used to transfer the rotational speed from rotating parts to vibrating parts. Rotating cams (initially, a steel type made using laser cutting machine from 2 mm plate with outer diameter of 300 mm, later, it was changed to plastic type which was improvised from plastic pulley, specifically to reduce noise, with 8 cams. It has an external diameter of 240 mm, Adjustment structure: The adjustment structure made from 2 mm thick steel plate. In all, three (3) plates used, one is having square cross-section with dimension of 400 × 400 mm and it is serving as base plate. Another two (2) H cross-section plate with dimension 400 × 100 mm placed on top of the former from both ends. This structure will serve as support for ball bearing to guide the rotating shaft. Lastly, another plate with dimension 400 × 200 mm serving as support for the moving parts assembly was at the top. All the plates arranged and joined together with aid of M10 stud, Moving parts consists of 310 mm long stainless steel pipe with 12 mm and 11.3 mm external and internal diameter respectively. This shaft welded to a 40 mm diameter circular disc with thickness of 2 mm. The disc side is encapsulated into a box and spring loaded to allow free return after it was being strike by the cam. The weight of whole assembly is 150 g and Nylon stinger (silicone type with diameter d = 12 mm) was used to translate the motion from moving parts to solar panel. This material used to minimize damage of the panel due to impact as shown in 'figure 3'.

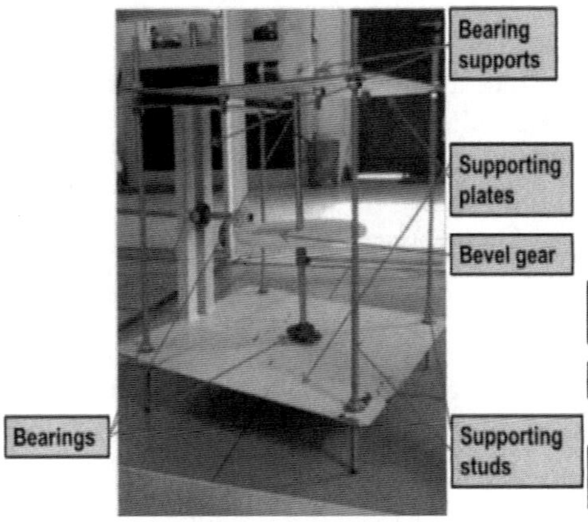

Figure2. Vibrating Parts 2

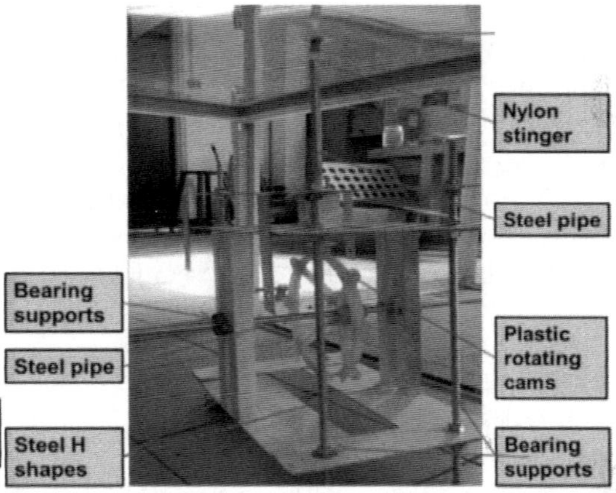

Figure 3. Vibrating parts 3

2.2 Solar Panel

Rectangular steel frame with dimensions of 690×1200 mm was fabricated using 25 mm square steel section to support the solar panel. The solar panel's plain surface was representing by Perspex plate (1050×540×2 mm) which fixed to the aluminum frame (right angle section 25×25×1 mm) using adhesive. This arrangement was fixed to the rectangular frame with 6 supporting pads to prevent damage to the panel. The whole assembly has a total weight of 1500 grams.

3. Dust Removal Index (I)

Dust material, size and deposition density has a strong effect on loss of power output from PV system. Dust not only reduces the incoming radiation on the solar cell but also changes the dependence on the angle of incidence of such radiation. Dust removal index (I) which simply represents the ratio of the difference between the final index of dust after applying the effect of any parameter (I_a) and the index of dust before applying the effect of the same parameter (I_b) to the index of dust before applying the effect of the same parameter. The index of dust removal is just a number that represents the amount of dust removed from panel surface. So, the dust removal index can simply as follow:

IOP Conf. Series: Journal of Physics: Conf. Series **1262** (2019) 012021 doi:10.1088/1742-6596/1262/1/012021

$$I = \frac{I_b - I_a}{I_b} \quad (1)$$

The value of dust removal index is constrained to vary from zero to one. If the amount of dust removed increases, the value of dust removal index approaches to one.

4. Frequency of vibration and acceleration

Wind energy have the ability to rotate the mechanical vibrator, from the output rotating shaft with 8-rotating cams produced a reciprocating motion for the stinger that transfer the energy to panel surface. With the aid of tachometer, rotational velocity of output rotating shaft measured to calculate the frequency according to air speed. . The measurement of the acceleration was done by with the aid of accelerometer (DeltaTron, type 4514-001, which fixed on the upper surface of the panel, with sensitivity 100.5), cDAQ-9171 and personal computer. With the aid of LabVIEW project build and DAQ assistance, acceleration data was acquired for a period of 10 sec. Maximum and minimum input range of 5 and -5 with 5 dB references. Then, the value of the acceleration can be obtained by 10 second continuous running for mechanical vibrator and choose the value of arithmetic mean as final value of acceleration. The surface vibration response is a characteristic, which is unique to each structure and depends on the structural properties. This vibration phenomenon can be modeled as a sinusoidal curve.

5. Testing the Mechanical Vibrator

After fabrication of the experimental rig, the MV tested by measuring the acceleration generated due to the vibration received by the solar panel, by fixing the accelerometer on the solar panel top surface. Data from all the measurements acquired with the project build in LabVIEW software 2013 as shown in 'figure 4'. Various readings observed by varying the fan speed so that the air velocity inside the ducting will change. Indoor experimental conditions were noted which are; air velocity closed to zero, temperature equal to 22 ^{0}C and relative humidity 50 %.

6. Results and Discussion

Wind energy converted to mechanical energy, which intended to use for dust removal from the solar panel. This relationship is proportional to each other as depicted in Table 1. Furthermore, the 'figure 5' shows the linear relation between air velocities with frequency, acceleration. If the system assumed to have acted as a single degree-of-freedom system, the increase in the magnitude of acceleration will increase the vibration force subjected to the panel.

To test the effectiveness of utilizing acceleration on panel surface to remove dust, a 5 g of dust particles with a size of $75 \leq \Phi < 150$ μm dusted on the solar panel surface. The solar panel was tilt to an angle of 15° and the experiment was run for ten seconds with the air velocity vary between 1 to 6 m/s as shown in table 1.The effect of air speed via MV has resulted changes in dust removal index. This relatioship was depicted in 'figure 6', which shows that with the increase in air speed via mechanical vibrator, the dust removal index was increased. Additionally, the produced energy will cause dust particles to removed or change re-suspension position.

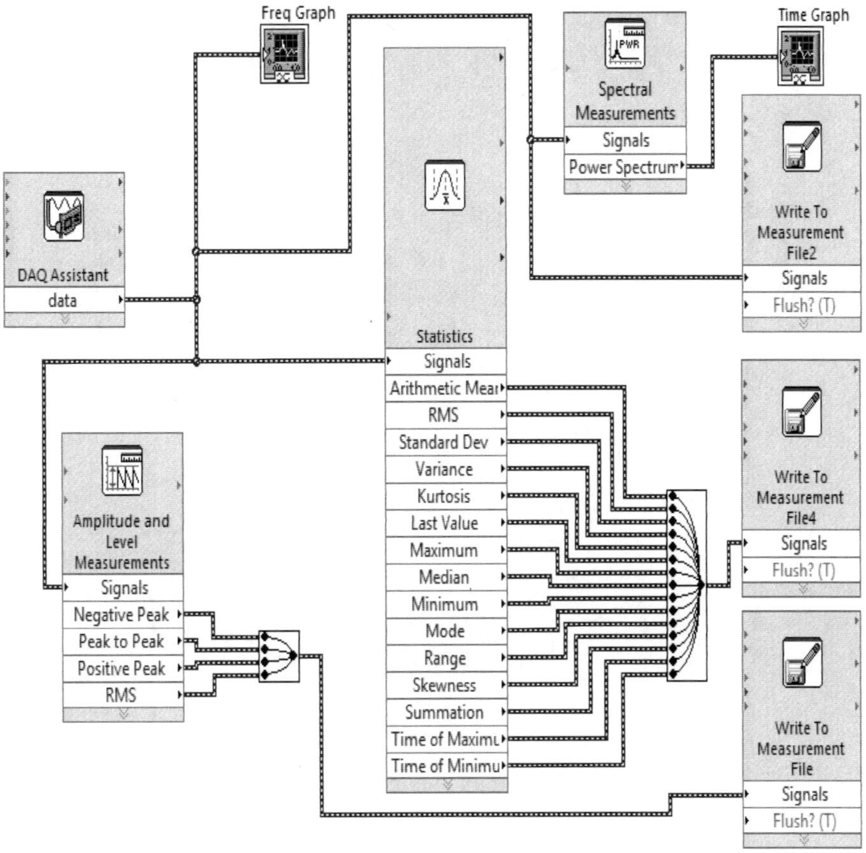

Figure 4. LabVIEW project for data acquisition.

The photographs of the solar panel before and after dust removal are shown in 'figure 7'. From the photographs it means that the dust moves away from the region of high energy and migrates to region with low energy as also found in the work of [18].

Table 1. The performance Mechanical vibrator

Run No.	Air Speed m/sec	Output Shaft RPM	Frequency Hz	Acceleration m/s^2
1	0.98	22	9.219	0.297
2	2.11	51	21.372	0.630
3	2.98	80	33.525	1.155
4	4.01	112	46.935	1.559
5	5.02	147	61.603	2.086
6	5.95	165	69.146	2.264

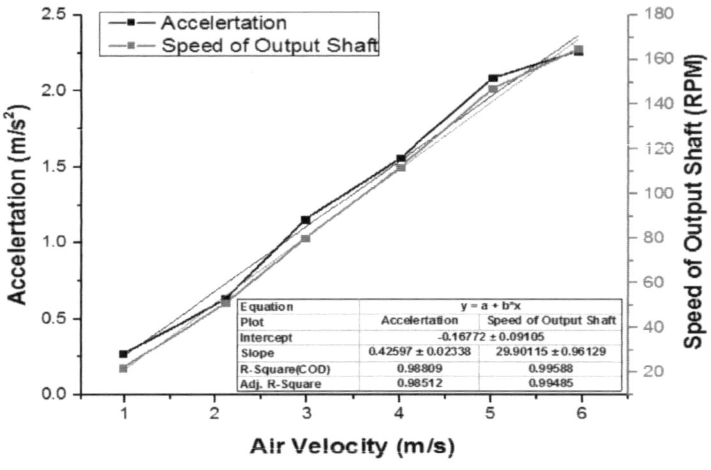

Figure 5. The performance of mechanical vibrator.

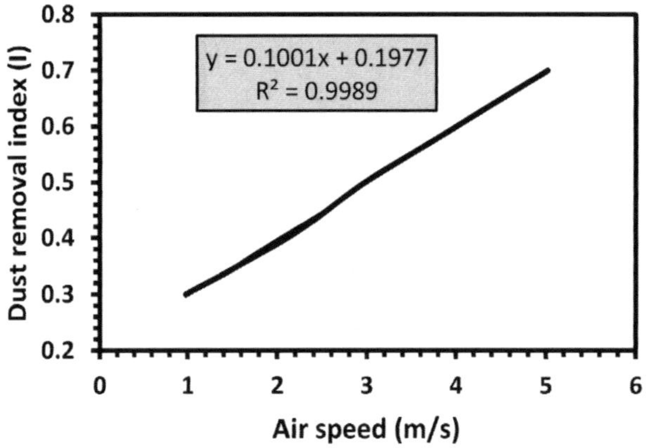

Figure 6. Effect of air speed on Dust Removal Index via mechanical vibrator.

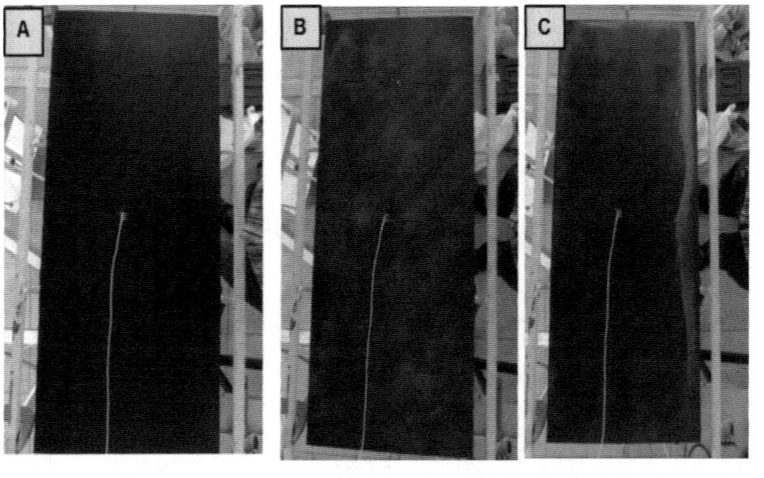

Figure 7. Photography of dust removal from solar panel surface at 3.128 N; (A: cleaned surface, B: surface accumulated 5 grams of dust, C: dust remaining on the surface).

7. Conclusion

Wind energy have a significant effect on dust removal from solar panel surface that achieved by using via a mechanical vibrator without consuming any part of power produced also reducing manpower used for solar panels maintenance. Where wind is available on the solar panel site, this method produce a continuous operation of mechanical vibrator, where no chance for the dust to be accumulated hence reducing the operation cost. The dust removal index increased with increasing the excitation frequency at specific excitation amplitude. This increment is very logical because increasing the excitation force frequency leads to increasing the dynamic response at each point on the panel surface. The increasing of magnitude of vibration leads to the increase in the vibration inertial force, which was transform into kinetic energy for the dust particles and overcome the adhesion force.

Acknowledgements
The authors would like to express their gratitude for the support provided by the Universiti Putra Malaysia. The first author also acknowledges a scholarship program from the Ministry of Higher Education, Iraq.

References

[1] Subbarao S and Dhananjaya K 2017 Role of Renewable Energy and Sustainable Technologies in Building an Eco-friendly and Sustainable Anti-poaching Unit in a Forest *DHARANA-Bhavan's International Journal of Business* **9** 51-5
[2] Chu S, Cui Y and Liu N 2017 The path towards sustainable energy *Nature materials* **16** 16-22
[3] Modi A, Bühler F, Andreasen J G and Haglind F 2017 A review of solar energy based heat and power generation systems *Renewable and Sustainable Energy Reviews* **67** 1047-64
[4] Solangi K, Islam M, Saidur R, Rahim N and Fayaz H 2011 A review on global solar energy policy *Renewable and sustainable energy reviews* **15** 2149-63
[5] Kannan N and Vakeesan D 2016 Solar energy for future world:-A review *Renewable and Sustainable Energy Reviews* **62** 1092-105
[6] Mokheimer E M, Dabwan Y N and Habib M A 2017 Optimal integration of solar energy with fossil fuel gas turbine cogeneration plants using three different CSP technologies in Saudi Arabia *Applied Energy* **185** 1268-80
[7] Adinoyi M J and Said S A 2013 Effect of dust accumulation on the power outputs of solar photovoltaic modules *Renewable Energy* **60** 633-6
[8] Saidan M, Albaali A G, Alasis E and Kaldellis J K 2016 Experimental study on the effect of dust deposition on solar photovoltaic panels in desert environment *Renewable Energy* **92** 499-505
[9] Semaoui S, Arab A H, Boudjelthia E K, Bacha S and Zeraia H 2015 Dust effect on optical transmittance of photovoltaic module glazing in a desert region *Energy Procedia* **74** 1347-57
[10] Piedra P and Moosmüller H 2017 Optical losses of photovoltaic cells due to aerosol deposition: Role of particle refractive index and size *Solar Energy* **155** 637-46
[11] Ahmed O K 2016 Effect of the Dust on the Performance of Solar Water Collectors in Iraq *International Journal of Renewable Energy Development* **5** 65-72
[12] Mazumder M, Horenstein M, Stark J, Hudelson J N, Sayyah A, Heiling C and Yellowhair J 2014 Electrodynamic removal of dust from solar mirrors and its applications in concentrated solar power (CSP) plants. In: *Industry Applications Society Annual Meeting, 2014 IEEE*: IEEE) pp 1-7
[13] Atten P, Pang H L and Reboud J-L 2009 Study of dust removal by standing-wave electric curtain for application to solar cells on Mars *IEEE Transactions on Industry Applications* **45** 75-86
[14] Mazumder M, Zahn M, Sharma R, Zhang J, Calle C, Immer C and Mardesich N 2006 Development of self-cleaning transparent dust shields using low-power electrodynamic fields for solar panels on Mars. In: *Proceeding of the ESA/IEEE-IAS/IEJ/SFE Joint Conference on Electrostatics 2006,* pp 177-204
[15] Mazumder M, Horenstein M, Stark J, Erickson D, Sayyah A, Jung S and Hao F 2013 Development of self-cleaning solar collectors for minimizing energy yield loss caused by dust

deposition. In: *ASME 2013 7th International Conference on Energy Sustainability collocated with the ASME 2013 Heat Transfer Summer Conference and the ASME 2013 11th International Conference on Fuel Cell Science, Engineering and Technology*: American Society of Mechanical Engineers) pp V001T11A13-VT11A13

[16] Gaier J R and Perez-Davis M E 1991 Effect of particle size of Martian dust on the degradation of photovoltaic cell performance

[17] Piliougine M, Cañete C, Moreno R, Carretero J, Hirose J, Ogawa S and Sidrach-de-Cardona M 2013 Comparative analysis of energy produced by photovoltaic modules with anti-soiling coated surface in arid climates *Applied Energy* **112** 626-34

[18] Williams R B, Tanimoto R, Simonyan A and Fuerstenau S 2007 Vibration characterization of self-cleaning solar panels with piezoceramic actuation. In: *48th AIAA/ASME/ASCE/AHS/ASC Structures, Structural Dynamics, and Materials Conference*: S) pp 512-20

IOP Conf. Series: Journal of Physics: Conf. Series **1262** (2019) 012022 doi:10.1088/1742-6596/1262/1/012022

Joining Strategy Analysis of Structural Dynamic in Top Hat Plate of Dissimilar Materials

N.A. Nazri[1] and M.S.M. Sani[1,2*]
[1]Advanced Structural Integrity & Vibration Research (ASIVR), Faculty of Mechanical and Manufacturing Engineering, Universiti Malaysia Pahang, 26600 Pekan, Pahang, Malaysia
[2]Automotive Engineering Centre, Universiti Malaysia Pahang, 26600 Pekan, Pahang, Malaysia

*Corresponding author: mshahrir@ump.edu.my

Abstract. Structural dynamic can be described as a study regarding analysis of model under exertion of dynamic loading. Joining in structure such as riveting, welding and soldering were strongly effects the structural dynamic properties. The aim of this paper is to conduct various joining strategy analysis in finite element software by using top hat plate structure in car as a model and result is compared with experimental data. Normal mode analysis with free-free boundary condition is applied. Resistance spot welding (RSW) technique is selected to weld flange part in top hat plate between dissimilar materials which is mild steel 1010 and stainless steel 304 with specified parameters applied. Modal testing via impact hammer test is carried out at the welded top hat plate for experimental analysis purpose. Correlation in between FEA and EMA results need to be conducted. By selecting identified parameters to execute model updating on the structure using sensitivity analysis method, the discrepancies in natural frequencies were reduced between FEA and EMA. CWELD element was chosen to represent weld model for spot weld joints due to its accurate prediction of frequencies and mode shapes compare to other weld modelling. As a closure, connection in structure parts play significant role in structural dynamics and model updating is one of successful method to lessen the discrepancies between FEA and EMA.

1.0 Introduction

The complex behaviour of joining elements holdings an important role in the whole dynamic characteristics, such as natural frequencies, mode shapes, and non-linear response features to external excitations [1]. It is necessary to evaluate the vibrational behaviour of the structure in order to prevent resonance fracture because of external dynamic loads [2]. Selecting proper joining method is a fundamental strategy in assembling structure with good performance. Various joining methods are existed in industry which has their own pros and cons. In between so abundance of manufacturing technologies, welding has been observed and confirmed as a key enabling technology to innovative and sustainable manufacturing [3] and resistance spot welding is one of the most popular method of welding metal sheets [4]. Pressure, heat and clamping duration are three main observed parameters to conduct resistance spot welding which are vary among different materials due to dissimilarities in mechanical properties. Previous study has reported that the time taken reflects how long current will flow in the joint, that is totally depends on the material type and also the thickness of material that needed to be weld [5].

This paper is attempted to conduct joining strategy of top hat structure using MSC Nastran Patran software and determine which joining is capable in resembling resistance spot welding joining. Modal parameters of the structure which is natural frequency and mode shapes are required to be analyzed from FEA and EMA and comparison will be carried out.

2. Finite Element Analysis

Finite element analysis can be defined as a computer simulation technique for modeling and analyzing the effect of the part or model built. This tool is very crucial to rectify the structure failure before manufacturing and test is carried on [6]. MSC Patran is utilized to conduct simulation for top hat structure. Completed geometry design is declared with respective material types and the properties can be seen in Table 1. Boundary condition is set to be free-free with normal mode analysis using SOL 103.

Table 1. Dimension and material properties of frame structure

Properties		Value
Material properties (SUS304L)	Young's modulus, E (GPa)	190
	Density (kg/m^3)	8000
	Poisson ratio	0.29
Physical properties	Length (mm)	1004
	Width (mm)	510
	Depth (mm)	510

Four different joining strategies were implemented on top hat structure which is rigid body element (RBE), bar element (CBAR), weld element (CWELD) and beam element (CBEAM). Figure 1 shows how the joining methods were depicted. However, it is difficult to establish a sufficiently accurate dynamic FE model using FE modelling techniques and thus, lead to the development of modal updating [7- 8].

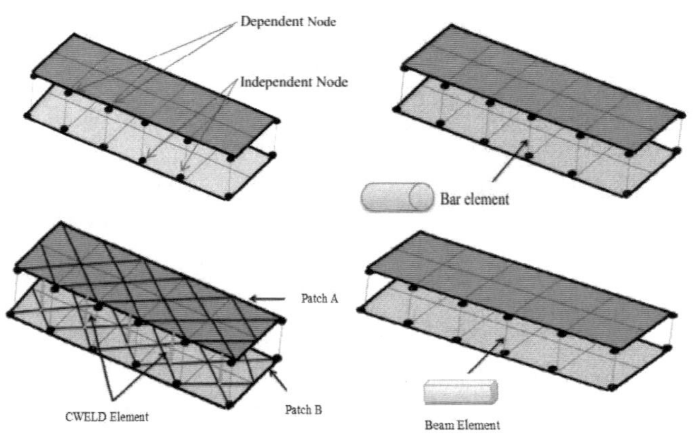

Figure 1. Joining methods in FEA [9].

3. Experimental Modal Analysis

Impact hammer testing is one of method to obtain modal parameter by giving excitation to the structure in the form of forces. Selected part in the structure is excited and the output response will be measured by accelerometers with the Fast Fourier Transform (FFT) analyzer and thus allow modal parameters to be extracted from the frequency response function (FRF) [10-11]. The selection of hammer tips is essential as the input excitation range of frequency is dominantly controlled by the hardness of the tip selected [10]. Springs were used to hang the structure in order to ensure the free-free boundary condition could be applied with 90° angle in between structure and hanger so that the free motion of vibration is not prevented. Figure 2 depicted the structure in hanging condition and the experiment setup. Roving accelerometer method is conducted while hammer is set to be fixed at one point. Frequency response function (FRF) and coherence graphs were observed for each excitation to ensure the data obtained is reliable.

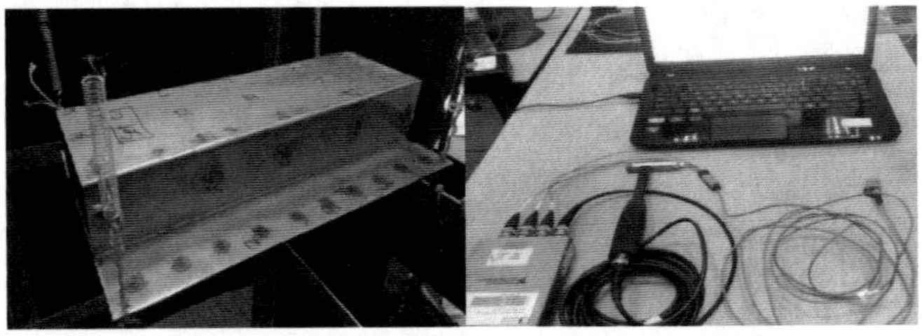

Figure 2. Structure in hanging position and experiment setup

4. Model Updating

Model updating is a process of adjusting certain parameters of the finite element model [12-13]. The purpose of modifying the parameter is to meet a better correlation between finite element model and experimental analysis. There are numerous methods of updating, and for this paper, optimization algorithm (SOL200) in NASTRAN software was utilized to perform model updating. For updating, sensitive parameters need to be identified through sensitivity analysis that was performed earlier before updating process. Model updating is such a procedure that updates the uncertainty parameters in the initial finite element model based on the experimental results so that a more refined model can be extracted [14].

5. Result and Discussion

Among four joining methods, one need to be selected which gives the lowest percentage of error when compared with EMA, as the joining seems to be the most reliable to represent spot weld in real tested structure. Data from Table 2 shows apparently the differences in percentage of error for natural frequencies when compared with EMA while Figure and Table 3 demonstrates the mode shapes for EMA and FEA with joining strategies respectively.

From the result, it can be seen obviously that CWELD contributes the least percentage of error while RBE experienced overestimation of natural frequencies. CBAR and CWELD both were capable of reducing the error. However, for mode shape prediction, CBAR gives the worst illustration of deflection while RBE gives the best demonstration of mode shapes when compared with EMA. Both CWELD and CBEAM experience a little failure in estimating the mode shape in fifth and sixth modes.

Table 2. Percentage of error between EMA and FEA for 4 types of modelling

Modes	Natural Frequency									
	EMA	RBE	Error %	CBAR	Error %	CWELD	Error %	CBEAM	Error %
1	212	224	5.66	210	0.94	219	3.30	201	5.19
2	261	302	15.71	271	3.83	265	1.53	251	3.83
3	334	371	11.08	330	1.20	371	11.08	369	10.48
4	383	385	0.52	353	7.83	378	1.31	382	0.26
5	418	396	5.26	387	7.42	424	1.44	409	2.15
6	443	462	4.29	455	2.71	436	1.58	457	3.16
7	583	543	6.86	544	6.69	549	5.83	543	6.86
8	715	659	7.83	658	7.97	650	9.09	658	7.97
9	791	823	4.05	812	2.65	767	3.03	777	1.77
10	944	986	4.45	980	3.81	984	4.24	983	4.13
Total Average Error			6.57		4.51		4.24		4.58

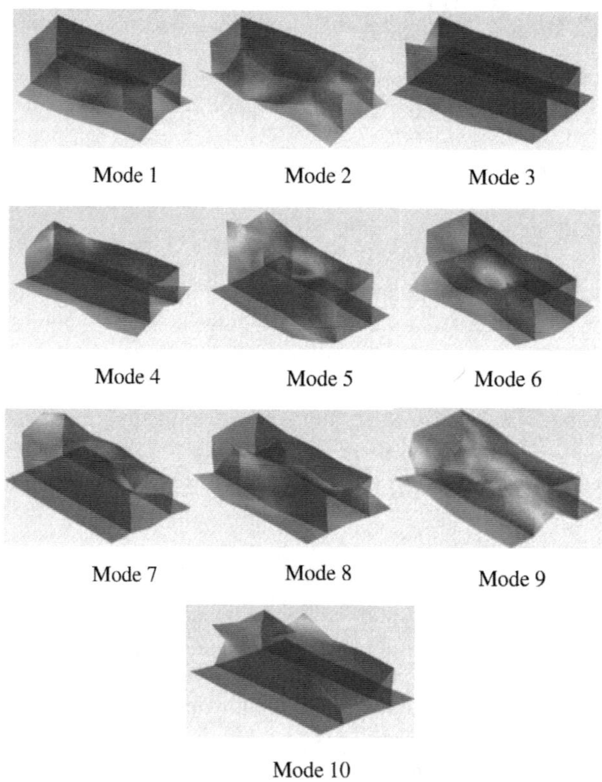

Figure 3. Mode shape of EMA

IOP Conf. Series: Journal of Physics: Conf. Series **1262** (2019) 012022 doi:10.1088/1742-6596/1262/1/012022

Table 3. Mode shapes of FEA

Modes	Type of modelling			
	RBE	CBAR	CWELD	CBEAM
1				
2				
3				
4				
5				
6				
7				
8				
9				
10				

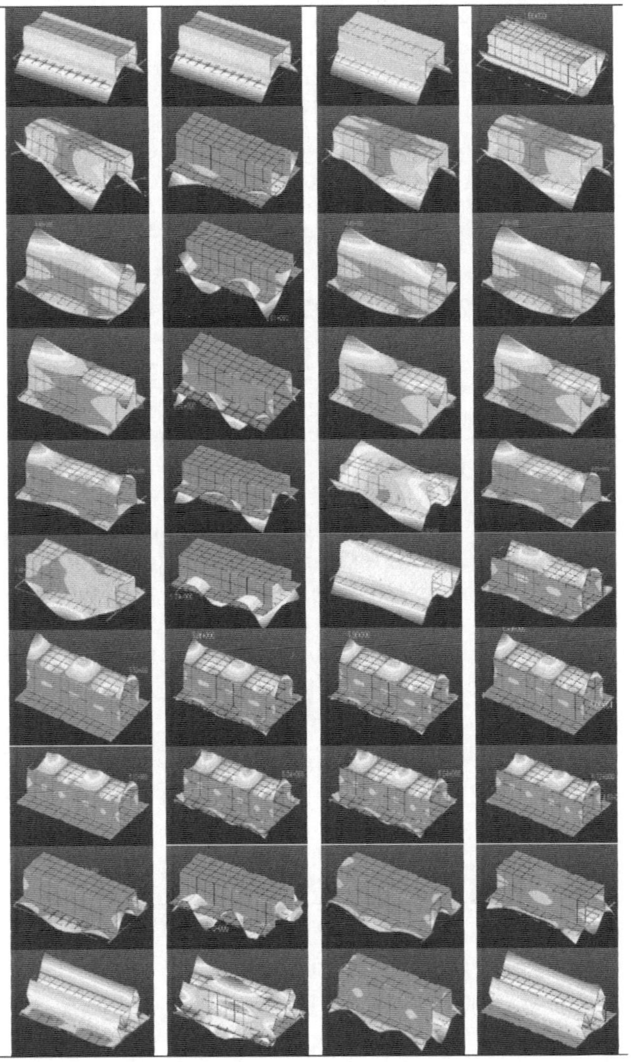

By considering all joints modelling for spot weld, it can be stated that generally all modelling strategy can be modelled when matching mesh was implemented to the structure [15]. For model updating purpose, RBE that represent weld joints do not have both geometrical and material properties which disallowed the updating purpose compare to CBAR, CWELD and CBEAM that contains significant parameters in modelling of weld that can be used to be update. From these comparisons, it is found that the natural frequency of CWELD is more tempting to approach EMA compare to other two models, so the CWELD element is selected to represent spot weld joining and will be proceed to model updating method to improve the correlation between numerical and experimental counterparts.

5.1 Sensitivity Analysis and Model Updating Result

It is essential to choice those updating parameters which will be most effective in producing a genuine improvement in the modelling of the structure [16-18]. Sensitivity analysis is a one step behind model updating which allowing the procedure of selecting sensitive parameters for updating. Sensitivity of parameters can be identified from F06 file in NASTRAN by giving huge values for the most sensitive parameters while the lowest values depicted the least sensitive and not suggested for updating purpose. Initially, six parameters were chosen to undergo sensitivity analysis, and the sensitivity of top five natural frequencies were tabulated in Table 4 below.

Table 4. Sensitivity analysis for six parameters

Output Type	Young Modulus (Mild steel)	Young Modulus (Stainless steel)	Young Modulus (Weld Joint)	Density (Mild steel)	Density (Stainless steel)	Diameter (Weld Joint)
NF1	7.05	108.6	-5.01E-7	-18.7	-102.4	-5.03E-7
NF2	93.8	35.8	3.28E-7	-85.8	-31.3	-5.87E-8
NF3	132.5	56.3	-2.15E-7	-104.5	-73.2	6.54E-7
NF4	8.63	198.2	-5.98E-8	-13.7	-209.7	-3.82E-8
NF5	166.7	45.9	-1.67E-7	-122.5	-76.7	7.23E-7

From the table tabulated, it can be seen clearly that parameters related to weld joints does not sensitive enough to contribute in updating. Based on the table, by neglecting the negative signs, Young modulus and density for both materials were seems to gives effect on updating strongly as they were considered sensitive. To be clear, four parameters were selected for updating which are Young Modulus for mild steel, Young Modulus for stainless steel, density for mild steel and last but not least, density for stainless steel. Table 5 presented the initial and updated result of FEA.

Table 5. Percentage of error between FEA and EMA after updating

Modes	EMA (Hz)	Initial FEA of CWELD		Updated FEA of CWELD	
		FEA (Hz)	Error (%)	FEA (Hz)	Error (%)
1	212	219	3.77	214	0.94
2	261	265	1.53	261	0.00
3	334	371	10.48	344	2.99
4	383	378	0	367	4.18
5	418	424	0.72	416	0.48
Total Average Error			2.75		1.72

Regards to the table displayed, percentage of error was successfully deducted from 2.75% to 1.72% and this result proved the ability of model updating to reduce the discrepancy between FEA and EMA. It is also verified that four selected parameters significantly affect the behaviour of model. The changes in value for updated parameters are shown in Table 6.

Table 6. Updated value of parameters

Parameter	Initial value (i)	Updated value (ii)	Deviation $\lvert(ii-i)/i\rvert$
Young Modulus of mild steel (GPa)	200	224	0.12
Young Modulus of stainless steel (GPa)	190	161.5	0.15
Density of mild steel (kg/m^3)	8000	9200	0.15
Density of stainless steel (kg/m^3)	8030	8110	0.01

6. Conclusion

As a summary, this study has deal with joining strategy of FEA on modelling spot weld connection for forecasting the structural behaviour. Four types of joining has been simulated which is RBE, CBAR, CWELD and CBEAM and each of them shows dissimilarities in the results of natural frequency and mode shapes. CWELD joining has been agreed to be selected as the most reliable model of spot weld as the natural frequencies of CWELD give the lowest error to EMA besides the prediction of mode shapes does not differ in many modes. Model updating was carried out by updating four sensitive parameters which is Young Modulus of mild steel and stainless steel, and density for both materials. After model updating is run, the result shows that the percentage of error was successfully lessen and achieve the goal of updating.

Acknowledgments

The authors of this paper would like to acknowledge a great support and encouragement by focus group of Advanced Structural Integrity of Vibration Research (ASIVR), Universiti Malaysia Pahang (UMP) for providing all the equipment used for this work Fundamental Research Grant Scheme (FRGS/1/2017/TK03/UMP/02-19) – RDU 170123.

References

[1] Ibrahim R A and Pettit C L 2005 Uncertainties and dynamic problems of bolted joints and other fasteners, *Journal of Sound and Vibration* 279 (3) 857-936.

[2] Lee Y S, Yang M S, Kim H S and Kim J H 2002 A study on the free vibration of the joined cylindrical–spherical shell structures *Computers & Structures* 80 (27) 2405-2414.

[3] Martinsen K, Hu S. J and Carlson B E 2015 Joining of dissimilar materials *CIRP Annals - Manufacturing Technology* 64 (2) 679-699.

[4] Ambroziak A and Korzeniowski M 2010 Using Resistance Spot Welding for Joining Aluminium Elements in Automotive Industry *Archives of Civil and Mechanical Engineering* 10 (1) 5-13.

[5] Aslanlar S 2006 The effect of nucleus size on mechanical properties in electrical resistance spot welding of sheets used in automotive industry *Materials & Design* 27 (2) 125-131.

[6] Sani M S M, Nazri N A, Zahari S N, Abdullah N A Z and Priyandoko G 2016 Dynamic Study of Bicycle Frame Structure IOP Conf. Ser. Mater. Sci. Eng. 160 012009

[7] Guo N, Yang Z, Jia Y and Wang L 2016 Model updating using correlation analysis of strain frequency response function *Mechanical Systems and Signal Processing* 70 (Supplement C) 284-299.

[8] Abdullah N A Z, Sani M S M, Hussain N A, Rahman M M and Zaman I 2017 Dynamics properties of a Go-kart chassis structure and its prediction improvement using model updating approach International Journal of Automotive and Mechanical Engineering (IJAME) 14 3887-97

[9] Nazri N A and Sani M S M 2017 Finite element normal mode analysis of resistance welding jointed of dissimilar plate hat structure IOP Conf. Ser. Mater. Sci. Eng 257(1) 012059

[10] Avitabile P 2001 Experimental Modal Analysis J Sound Vib. 35(1) 20–31.

[11] Izham M H M, Abdullah N A Z, Zahari S N and Sani M S M 2017 Structural dynamic investigation of frame structure with bolted joints MATEC Web of Conferences 90 (01043).

[12] Mottershead J E, Link M and Friswell M I 2011 The sensitivity method in finite element model updating: A tutorial Mech. Syst. Signal Process. 25 2275–96

[13] Sani M S M, Abdullah N A Z, Zahari S N, Siregar J P and Rahman M M 2016 Finite element model updating of natural fibre reinforced composite structure in structural dynamics *MATEC Web of Conferences* 83 (03007)

[14] Ren W X and Chen H B 2010 Finite element model updating in structural dynamics by using the response surface method *Engineering Structures* 32 (8) 2455-2465.

[15] Zahari S N, Sani M S M and Ishak M 2017 Finite element modelling and updating of friction stir welding (FSW) joint for vibration analysis MATEC Web of Conferences 90 (01021)

[16] Zahari S N, Zakaria A A R, Sani M S M and Bujang I 2016 A review on Model updating of joint structure for dynamic analysis purpose MATEC Web of Conferences 74 (00023)

[17] Friswell M I and Mottershead J E 1995 Finite element model updating in structural dynamics, *1st ed. (Springer Science & Business Media, The Netherlands).*

[18] Fouzi M S M, Jelani K M, Nazri N A and Sani M S M 2018 Finite Element Modelling and Updating of Welded Thin-Walled Beam International Journal of Automotive and Mechanical Engineering (IJAME) 15(4) 5874-89

Model Updating of Frame Structure with Bolted Joints

M.H.N. Izham[1] and M.S.M. Sani[1,2*]
[1]Advanced Structural Integrity & Vibration Research (ASIVR), Faculty of Mechanical and Manufacturing Engineering, Universiti Malaysia Pahang, 26600 Pekan, Pahang, Malaysia
[2]Automotive Engineering Centre, Universiti Malaysia Pahang, 26600 Pekan, Pahang, Malaysia

[*]Corresponding author: mshahrir@ump.edu.my

Abstract. Experimental analysis is often viewed as an important source of reference compared to finite element due to the accuracy of giving reliable data. Currently, finite element analysis is widely used as an early adoption in development, hence trusting the finite element data is crucial for the user. Finite element analysis often shows discrepancies to the test result. The complexity of the joints (bolted) might cause the discrepancies to occur. This study aims to reduce the discrepancies between the experimental and numerical analysis on a frame structure with bolted joints by model updating. Model updating is a process of making adjustment to certain parameters of finite element model to reduce discrepancy between analytical predictions of finite element analysis (FEA) and experimental results. Modal properties (natural frequencies, mode shapes, and damping ratio) of a frame structure with bolted joints are determined using both experimental modal analysis (EMA) and finite element analysis (FEA). Both data obtained is correlated before optimising the properties with sensitivity analysis. Joint strategy of this paper is focusing on RBE2, CBAR and CELAS element. CELAS was selected to represent the bolted modelling due to its lowest percentage average of 2.03% compares to CBAR 6.55% or RBE2 3.56%. Selected parameters were identified by performing a sensitivity analysis and the discrepancies was reduced by performing model updating procedure.

1.0 Introduction

Structural systems or space frame structure is build up or assemble of many components. Components are commonly being axial element, with circular or rectangular section which all the components either can resist tension or compression. The joint or mechanical joint on structures plays an important role where it connects two or more components. Joint has a big impact to the structure related to strength which consuming 20-30% of the overall weight of the structure, which making the design is critical to the structure [1]. Engineers commonly used bolted joints as the solution to joint two or more components with the idea of low cost, easy to assemble/dissemble and the most importantly is suit to the design of using bolted joints. At the same time, some engineers are trying to reduce welding joints as good welding joints will need a special skill worker to execute a good joint and to maintain it consistency. Figure 1 shows the frame structure with bolted joints.

IOP Conf. Series: Journal of Physics: Conf. Series **1262** (2019) 012023 doi:10.1088/1742-6596/1262/1/012023

Figure 1. Frame Structure with bolted joints

As an early stage of investigation of dynamic characteristics finite element analysis (FEA) is commonly used method in current lifestyle. However, the results of numerical analysis vs experimental analysis have discrepancies [2] is a common topic to be discussed among researchers or engineers. Few ideas of why discrepancies happens on FEA, modelling structures and bolted joints [3] can be a big thing that impacting the results. The frequency of a structure can be significantly affected by the mass that might contribute by the inconsistency of the structure thickness due to manufacturing process. Common mistake by user is using a fixed connection at the bolted joints which causing a modelling error impacting the results [3]. There is other study on few most common errors in numerical studies, such as model structure errors, model parameters errors and model order errors [4].

Modelling a bolted joint is a challenging task where few researchers is focusing on the bolted joints modelling/strategy itself. A common mistake assuming a fixed connection on the bolted joints will lead to modelling error. There are two types of joints strategies which are by 3D modelling or assigning the joints using the FEA element that comes with the FEA software. Few types of modelling techniques are investigated on structure with bolted joints modelling. Consists of solid belt model, couple bolt model, spider bolt model and no bolt model. Experiment was conducted on large marine engine diesel which parts connected with long stay bolts. Among all four modelling techniques, solid belt model is recommended since it gives the best result and looking at computational time which it can save 62% computational time, 21% memory usage compares to solid belt model without scarifying the results badly, coupled bolt model is also recommended [5]. Couples of bolted joint modelling such as solid elements, line element and no bolt are discussed with pro and cons of the joint modelling. Result explains Solid always give a better accuracy, and line elements doesn't produce a bad result either. No bolt or fixed joints has a worse displacement with no properties of bolt been able to define as there is no modelling [6]. Joint element strategy, CFAST element is used for bolted joints, this technique allows the local effect to be modelled in a simple way [7]. A review paper focusing of model updating on joint structure discussed on bolt can be model as spring element (CELAS and CBUSH), connecter element (CBAR and CBEAM) and rigid connection element (RBE2) via NASTRAN software or it can also be modelled as thin layer [8]. Studies also shows, other researchers applied the element as part of their joints strategy with conclusion such as, CBAR and CBEAM are most suitable to use in welded joints compare to CTUBE [9] and spot weld joint modelling is best works with CWELD compare to CBAR and RBE2 [10].

Experimental modal analysis (EMA) has been used since early 1970's [11]. Experimental Modal Analysis (EMA) or also known modal testing or modal analysis is a process to determine the modal properties or vibration characteristic (natural frequency, damping ratio and mode shape) that permits a

IOP Conf. Series: Journal of Physics: Conf. Series **1262** (2019) 012023 doi:10.1088/1742-6596/1262/1/012023

dynamic mathematical model to be formed. Structure is excited by a force using tool called impact hammer and the output response will be measured by accelerometers with the Fast Fourier Transform (FFT) analyzer, modal parameters will be extracted from the frequency response function (FRF) [11–13].

Model updating is a process to reduce the discrepancies between FEA and EMA. When the discrepancy is high, the predictions are often discussed which either the numerical or experimental setup that cause us a high difference in results. There are few model updated technique been proposed as a model updating method such as Bayesian method [2], sensitivity base model updating [14], Stochastic method by [15] and reduced order characteristic polynomial (ROCP) [16]. One of the iterative method is design sensitivity and optimization algorithm called SOL 200 by NASTRAN [17] and a study is found to utilize SOL 200 performing model updating with the CBAR element to represent the FSW joint [18]. SOL 200 is used as well to update the discrepancies on go kart chassis by updating the sensitive parameters which are young modulus and diameter of tube [19].

This paper aims to reduce the discrepancies between FEA and EMA with the most suitable joint element such as RBE2, CBAR and CELAS taking into considerations when simulating the joint on frame structure. Once suitable joint element is selected, model updating will be carried out using the sensitivity analysis.

2. Frame Structure with Bolted Joints

The structure used in this paper is divided into two which are L bars (4 pcs), straight/flat bars (16 pcs) together with 32 pcs of M6 screws and Nuts. M6 screws are tighten with control value of 8-9Nm. Dimension details of frame structure with the material properties is tabulate in table 1. Stainless steel (SUS304L) is used for this study. Frame structure is bolted to the ground making it as a fixed-free boundary condition. Frame structure model in MSC Nastran/Patran is shown in Figure 2.

Figure 2. Frame structure of FEA 3D model

Table 1. Dimension and material properties of frame structure

	Properties	Value
Material properties (SUS304L)	Young's modulus, E (GPa)	190
	Density (kg/m^3)	8000
	Poisson ratio	0.29
Physical properties	Length (mm)	1004
	Width (mm)	510
	Depth (mm)	510

IOP Conf. Series: Journal of Physics: Conf. Series **1262** (2019) 012023 doi:10.1088/1742-6596/1262/1/012023

3. Finite Element Analysis

Boundary condition and external forces were assigned to the model as the model was let to exist in fixed-free boundary condition. Calculation of modal properties in MSC.Nastran/Patran was done by using SOL 103, which is the solution for normal modes analysis. By using the normal modes analysis, the natural frequencies and mode shapes of the structure can be computed. In normal modes analysis, the equation of motion is stated as Eq. (1).

$$[M]\{\ddot{u}\}+[K]\{u\}=0 \tag{1}$$

where K and M are stiffness and mass matrices respectively. These system matrices are computed automatically by MSC Nastran/Patran, based on the geometry and properties of the FEA model. By assuming a harmonic solution, the equation from above can be reduced to an eigenvalue problem which is stated as below

$$[K-\lambda_i M]\{\phi_i\}=0 \tag{2}$$

where $\{\phi\}$is the eigenvector (mode shape) corresponding to the eigenvalue λ (natural frequency). Each eigenvalue is proportional to a natural frequency and is corresponding to eigenvector. The eigenvalues are related to the natural frequencies as

$$f_i=\sqrt{(\lambda_i)}/2\pi \tag{3}$$

4. Experimental Modal Analysis

Experimental modal analysis is also known as modal testing or modal analysis. This is the process where dynamic characteristic will be collected. The excitation method use in this paper is impact hammer with roving tri-accelerometer. Below Figure 3 is showing the equipment used for impact hammer.

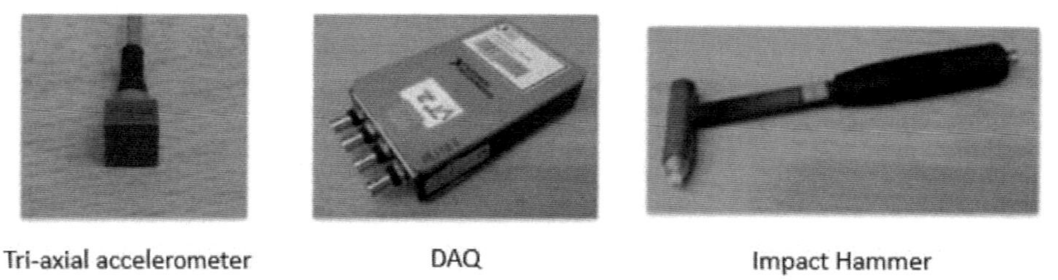

Figure 3. Impact hammer equipment

The portal frame structure is assigned and labelled into 60 grids points of excitation. VES. Figure 4 displays the points assigned on the frame structure with MEscope

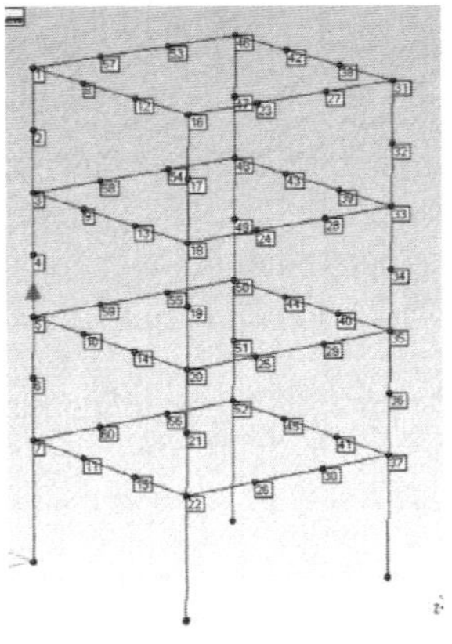

Figure 4. Frame structure with labelled grid points.

5. Data Correlation

Data obtained from the FEA and EMA is correlated to analyze the discrepancies, where discrepancies between both results will be compared and the percentage error will be captured. It has been discussed before, the results of FEA often have discrepancies between the experimental [2]. Three joints elements have been taken into considerations to select the most reliable element to represent bolted joints on frame structure. Table 2 summarized the average percentage of error for CBAR, RBE2 and CELAS.

Table 2. Percentage of error between EMA and FEA for 3 types of joint modelling

Modes	Natural Frequency (Hz)						
	EMA	FEA					
		CBAR	Error (%)	RBE2	Error (%)	CELAS	Error (%)
1	39.6	38.69	2.30	38.99	1.54	39.45	0.38
2	71.7	68.21	4.87	68.25	4.81	70.65	1.47
3	84.4	74.06	12.25	87.49	3.66	80.47	4.66
4	90.4	84.28	6.77	94.24	4.24	91.84	1.59
Average Error (%)			6.55		3.56		2.03

The results reveal that, RBE2 would not be suitable for updating since RBE2 does not present any material properties. Unlike CBAR and CELAS, both elements contain material properties allowing the updating process to be carried on. Average percentage error for CBAR and CELAS are respectively 6.55% and 2.04%. The objective of this study is to reduce the discrepancies between FEA and EMA, looking at the percentage error, by fact CELAS average percentage error is more attractive hence CELAS is chosen to represent the bolted joint with frame structure to perform updating procedure.

6. Finite Element Model Updating

Model updating is a process adjusting selective parameters in finite element to reduce the results discrepancies between FE and EMA [20]. The target of correlation is to minimize the percentage of error in between FE and EMA. This can be completed by performing the model updating. Approximation subject to the simple first-order Taylor series expansion was used in SOL 200 in MSC.Natran in order to change the vector λ of eigenvalues based on the vector θ of structure updating parameters as stated in Eq. (4).

$$\lambda_{i+1} = \lambda_i + [S_i](\delta\theta) \tag{4}$$

where $[S_i]$ is a sensitivity matrix of i-th iteration, which signify the rate of change of the structural eigenvalues λ_i with respect to changes in $\delta\theta$.
The expression for the eigenvalue sensitivity S_i can be stated as Eq. (5).

$$S_i = \frac{\delta\lambda_i}{\delta\theta} = \phi_i^T \left[\frac{\delta K}{\delta\theta} - \lambda_i \frac{\delta M}{\delta\theta}\right] \phi_i \tag{5}$$

The objective function for the prediction error is defined as Eq. (6).

$$g(x) = \sum_{i=1}^{n} W \left(\frac{\omega_i^a}{\omega_i^e} - 1\right)^2 \tag{6}$$

where ω_i^e and ω_i^a are the experimental and finite element natural frequencies respectively; W is the weighting factor for each mode.

The number of updating parameters was kept to be less than number of modes to be updated in order to avoid ill-conditioning problem in updating procedure. The sensitivity analysis was carried out in the first place in the interest of ensuring that the selected updating parameters were meaningful. The prediction of objective function $g(x)$ in updating procedure was set to obtain the minimized value.

Therefore, the procedure was kept going until convergence was accomplished when the values of $g(x)$ is sufficiently small. Sensitivity analysis is to determine the sensitive parameters to be used for updating the structure with bolted joints. Finite element model updating only works to adjust the uncertainties and assumptions of modelling parameters [21-22]. Table 3 shows the sensitivity matrix for updating.

Table 3. Sensitivity matrix for updating

Output Type	Young Modulus (structure)	Young Modulus (Joint)	Density (structure)	Poison Ratio (structure)
NF1	11.55	2E-08	0.52	39.41
NF2	17.36	1.6E-08	2.49	59.27
NF3	18.23	1.6E-09	0.00046	62.21
NF4	18.33	9.2E-09	0.023	62.57

Based on the facts tabulated by the sensitivity matrix, Young modulus and position ratio of the structure are the most sensitive among all 4 parameters, but not to ignore the density parameter which contributes to reduce the discrepancies between FEA and EMA. Joint young modulus gives an extremely low value; this will not give any significant change to the natural frequency. Table 4 display the error between initial and updated results of FEA.

Table 4. Comparison of discrepancies between initial and updated results.

Mode	EMA (Hz)	Initial FEA		Updated FEA	
		Nat Freq (Hz)	Error (%)	Nat Freq, (Hz)	Error (%)
1	39.6	39.45	0.38	39.92	0.81
2	71.7	70.65	1.47	71.51	0.26
3	84.4	80.47	4.66	81.42	3.53
4	90.4	91.84	1.59	92.91	2.78
Average Error (%)			2.03		1.85

The results above show 3 out of 4 natural frequency improved except for mode 1 natural frequency. Average percentage error reduced from 2.03% to 1.85% and this proved model updating with SOL 200 able to reduce the discrepancies between FEA and EMA. However, Table 5 explain the changes in value of parameter prior to model updating

Table 5. Updated value of parameters.

Parameter	Initial value (i)	Updated value (ii)	Deviation |(ii-i)/i|
Young Modulus of structure (GPa)	190	192.1	0.01
Density of structure (kg/m^3)	8000	7600	0.05
Poison Ratio of structure	0.29	0.287	0.01

7. Conclusion

This study was undertaken to correlate the natural frequency between FEA and EMA on frame structure with bolted joints. Few joint strategies have been introduced to this structure with bolted joints are CBAR, RBE2 and CELAS. The average percentage of error being tabulated, CELAS element is selected as a reliable model for bolted joint and the sensitivity analysis with SOL 200 was executed to improve the discrepancies. Based on the sensitivity matrix, it is found that Young Modulus and poison ratio gives a significant number impacting the natural frequency and as well as the density. The discrepancy between EMA and FEA has been brought down prior to model updating.

Acknowledgments

The authors of this paper would like to acknowledge a great support and encouragement by focus group of Advanced Structural Integrity of Vibration Research (ASIVR), Universiti Malaysia Pahang (UMP) for providing all the equipment used for this work Fundamental Research Grant Scheme (FRGS/1/2017/TK03/UMP/02-19) – RDU 170123.

References

[1] Tien T L 1999 Space Frame Structures in Structural Engineering Handbook.
[2] Marwala T, Mdlazi L and Sibisi S 2007 Finite Element Model Updating Using Bayesian Approach *22th Int. Modal Anal. Conf.*, vol. 42 (1) 1-7
[3] He K and Zhu W D 2009 Finite Element Modeling of Structures with L-shaped Beams and Bolted Joints: Model Updating and Predictive Modeling Approaches ASME 133 1–18.

[4] Mottershead J E and Friswell M 1993 Model Updating In Structural Dynamics: A Survey *Journal of Sound and Vibration* 167(2) 347–375.
[5] Kim J, Yoon J C and Kang B S 2007 Finite element analysis and modelling of structure with bolted joints *Appl. Math. Model* 31(5) 895–911.
[6] Montgomery J 2002 Methods for Modeling Bolts in the Bolted Joint *ANSYS User's Conf.* 15.
[7] Yunus M A 2011 Finite Element Modelling and Updating of Structure of Sheet Metal with Bolted and Welded Joints 227.
[8] Zahari S N, Zakaria A A R, Sani M S M and Bujang I 2016 A review on Model updating of joint structure for dynamic analysis purpose MATEC Web of Conferences 74 (00023)
[9] Abdullah N A Z, Sani M S M and I Zaman 2017 Correlation of Structural Modal Properties of Go-kart Frame Structure using Different Type of Joint in Finite Element Modeling *Adv. Sci. Letter* 23(11) 11315–113193.
[10] Nazri N A and Sani M S M 2017 Finite element normal mode analysis of resistance welding jointed of dissimilar plate hat structure *IOP Conf. Ser. Mater. Sci. Eng* 257(1) 012059
[11] Schwarz B J and Richardson M H 1999 Experimental Modal Analysis *J. Sound Vib.*
[12] Avitabile P 2001 Experimental Modal Analysis *J Sound Vib.* 35(1) 1–15.
[13] Larbi N and Lardies J 2000 Experimental Modal Analysis of a Structure Excited By a Random Force *Mech. Syst. Signal Process.* 14(2) 181–192.
[14] Bakir P G, Reynders E, and De Roeck G 2007 Sensitivity-based finite element model updating using constrained optimization with a trust region algorithm *J. Sound Vib.* 305(1-2) 211–225.
[15] Abu Husain N, Khodaparast H H and Ouyang H 2012 Parameter selection and stochastic model updating using perturbation methods with parameter weighting matrix assignment *Mech. Syst. Signal Process.* 32 135–152.
[16] Li W L 2002 A New Method for Structural Model Updating and Joint Stiffness Identification *Mech. Syst. Signal Process* 16(1) 155–167.
[17] Fouzi M S M, Jelani K M, Nazri N A and Sani M S M 2018 Finite Element Modelling and Updating of Welded Thin-Walled Beam International Journal of Automotive and Mechanical Engineering (IJAME) 15(4) 5874-89
[18] Zahari S N, Sani M S M, Husain N A, Ishak M and Zaman I 2016 Dynamic analysis of friction stir welding joints in dissimilar material plate structure J. Teknol. 78
[19] Abdullah N A Z, Sani M S M, Hussain N A, Rahman M M and Zaman I 2017 Dynamics properties of a Go-kart chassis structure and its prediction improvement using model updating approach International Journal of Automotive and Mechanical Engineering (IJAME) 14 3887-97
[20] Mottershead J E, Link M and Friswell M I 2011 The sensitivity method in finite element model updating: A tutorial Mech. Syst. Signal Process. 25 2275–96
[21] Zahari S N, Sani M S M and Ishak M 2017 Finite element modelling and updating of friction stir welding (FSW) joint for vibration analysis MATEC Web of Conferences 90 (01021)
[22] Sani M S M and Ouyang H 2016 Identification of Nonlinearities in Joints of a Wing Structure MATEC Web of Conference 83 (03006)

IOP Conf. Series: Journal of Physics: Conf. Series **1262** (2019) 012024 doi:10.1088/1742-6596/1262/1/012024

Tremor Quantification and its Measurements Using Shimmer

Reem Musab[1]*, Azizan As'arry[1], Khairil Anas Md. Rezali[1], Nawal Aswan Abdul Jalil[1], Raja Mohd Kamil Raja Ahmad[2] and Mohd Zarhamdy Md Zain[3]

[1]Department of Mechanical and Manufacturing Engineering, Faculty of Engineering, Universiti Putra Malaysia (UPM), 43400 UPM Serdang, Selangor, Malaysia
[2]Department of Electrical and Electronic Engineering, Faculty of Engineering, Universiti Putra Malaysia (UPM), 43400 UPM Serdang, Selangor, Malaysia
[3]Department of Applied Mechanics & Design, School of Mechanical Engineering, Faculty of Engineering, Universiti Teknologi Malaysia (UTM), 81310 UTM Skudai, Johor, Malaysia

*Corresponding author: Reemusab244@gmail.com

Abstract. This study aimed to provide a quantitative assessment of parkinsonian tremor through the precise measurement of tremor among Parkinson patients while taking into consideration specific parameters including acceleration, velocity, and displacement. The parameters were obtained by using transducers. For this assessment, three-axis accelerometer and three axis gyroscopic transducers embed in one device called shimmer. In this present paper, sensitivity assessment was used to measure the severity of hand tremors in two positions; resting and postural. Besides, the researchers obtained rotational movement along with tremors' acceleration movements. The amplitude and frequency was taken from Parkinson patients' hand. Shimmer sensor showed excellent correspondence of amplitude and frequency measurement with rotational transducer. The frequency of the patient's hand tremor occurred between 1.594 Hz and 4.813 Hz. However, the accelerometer showed poorer correspondence compared to gyroscope measurement. At the end, gyroscope responses are more remarkable because of the low level of muscle activity required in the task for precise tremor measurements to appear in high sensitivity to this impairment tremor.

1. Introduction

Parkinson's disease (PD) is the second common neurodegenerative disease after Alzheimer's. One of the most visible signs of this disease is tremor. Other diseases have similar oscillation movements, but each disease has a different recognition and specification. Parkinson's patients face challenges in terms of managing active life due to the oscillation movement of the hand or any body part affected by PD. Significantly, the involuntary movement brings about frustration and depression among patients [1-2]. Patients should undergo medical diagnosis as soon as they develop early signs of Parkinson's disease to monitor the disease progression and find optimal treatment regimens.

Medical scientists and drug developers are searching for alternative ways to minimize the cost and efforts for tremor validation. In return, patients wish to increase their life quality at the same time reduce clinic visits. Surgeries may involve some risks, and patients have different responses to treatment. For this reason, alternative solutions caught researchers' interest to suppress the tremor with the least adverse effects.

Measuring tremor assists researchers to save their time and efforts during conducting experiments and use collected data as a reference. [3]. Parkinson's disease is recorded from all limb segments under resting, postural or movement conditions [4]. Kinetic tremor happens when a voluntary contraction of a muscle follows a specific action like holding a cup [5]. Rest tremor occurs when the body part is wholly sported against the gravity just like a sported hand on a table [6] while postural tremor occurs when the body part maintains its position against gravity like extending the arms in front of the body [7].

Numerous studies have been conducted to obtain the most important parameters in tremor assessment including acceleration, velocity and displacement [8-9]. Researchers agree that it is possible to obtain those parameters by using differentiation for the velocity then finding displacement [9]. Realistically, the velocity and displacement can be found by using transducers to eliminate the noise amplification resulted from differentiated signal [10-11]. For displacement signal, it can be obtained by the double integration of the acceleration signal. However, the result is not satisfying because of the amplified noise produced by the first and second level of integration process [12]. Besides, whenever an integration part is needed, filter application with time and efforts is required to tune the filter's parameter for getting realistic output.

Generally, there are three approaches for measuring the hand tremor; acceleration transducer, velocity transducer and displacement transducer. Digital technology, such as cameras, graphical tablets and laser, is essential for the displacement sensor to present the coordinates for hand motion [13-15]. However, using the accelerometer for hand tremor measurement is more convenient for its lightweight and low cost [16].

Combining medicine and engineering fields promises great invention in technologies and enhances patient's comfort. This study demonstrated how acceleration and angular velocity, using Shimmer, could record arm tremor. The tremor behaviour was displayed in the time and frequency domain. Identified tremor parameters, in terms of amplitude and frequency, were essential in developing the test rig that can emulate human hand tremor.

2. Experiment Method

For this raw data assessment there are few steps needed to be followed which are listed as:

2.1. Subjects and experimental protocol

2.1.1 Test Tasks and Relevant Parameters

The present study collected data from a sample of ten human subjects; male and female. They ranged between 58 to 66 years old (mean age 26 years). All the subjects were physically active without any impairments whether in moving, hearing or seeing. They consumed their medication about four hours before the experiment. Written questionnaire form was collected from each patient to keep their personal record, followed by the collection of essential tremor record within one session for two trials including two tasks; one for the resting tremor while the other for the postural tremor as shown in Figure 1 and 2. Each task lasted for 10 seconds.

Figure 1. Test tasks for parkinsonian tremor assessment for rest tremor

Figure 2. Test tasks for parkinsonian tremor assessment for postural tremor

2.1.2 System Concept

Data was collected by applying ConsensysPRO Software, which worked alongside Shimmer sensors via Bluetooth and provided accelerometer data in unit of meters per Second Square and gyroscope data in unit of a degree per second. Transferred data from Shimmer to computer can be though two ways: by wires and wireless communication, while this data was recorded in real time. The sampling rate was 512 Hz [17]. To validate the tremor oscillation data in terms of the frequency domain, the acceleration versus time graph was converted to the frequency graph using MATLAB software. As shown in Figure 3, the system consisted of two parts; Shimmer part (left) and computer part (right). The lightweight Shimmer sensor is placed on the top of the hand for tremor assessment [14] [17]. The command module sent the sensor data to the computer via the wireless transmission of Bluetooth.

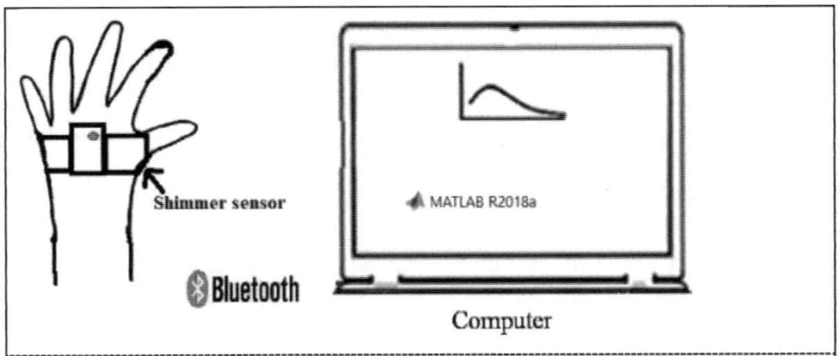

Figure 3. System diagram of the tremor assessment system.

2.1.3 Procedure for Postural Tremor

Before measurement assessment, Shimmer needed to be calibrated. Calibration was done earlier in the lab by using calibrator then using calibration procedure on each patient to determine the sensor-to-segment coordinate systems. Patients were comfortably seated on a chair and asked to hold their hands at 90 degrees perpendicular to their body as shown in Figure 4. Then Shimmer sensor was placed on the hand to calibrate it and start recording the data.

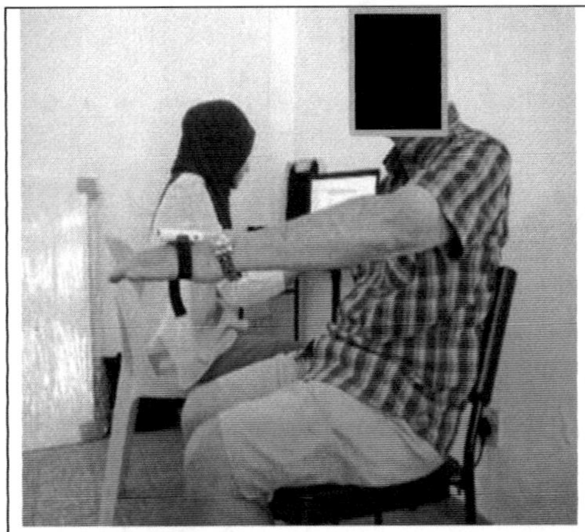

Figure 4. Data collection from a PD patient

2.2 Data Collection and Analysis

Quantification of tremor oscillation severity is essential. It has been always done by using time-domain analysis, time-frequency analysis, spectral analysis, nonlinear analysis or spectral analysis [17-20]. After that, the signal goes through some processes with PSD (auto power spectrum) calculation to detect tremor within time interval (10 seconds). However, gyroscope in Shimmer sensor measures the angular velocity for tremor that can be integrated over time for once to get displacement while for acceleration is required for double integration. Data collected was used for tremor quantification. The accelerometer and gyroscope outputs for three axis in addition to the gravitational acceleration, which is equal to 9.81 m/s2 in vector product and linear acceleration. Using the calibration removed gravitational acceleration from Shimmer accelerometer output signal. The range of Parkinson tremor was 3.5-9 Hz which was it based on the dominant frequency of the tremor.

2.3 Parkinsonian Tremor Assessment Methods
MATLAB software was used for data processing and analysis, providing displacement along the three axes X, Y, and Z, respectively. While, rotational data for angular movement for the hand was also inserted for gyroscopic movement directions were obtained for each marker. This assessment data presented in acceleration, velocity and displacement chart either in time domain series or in Fast-Fourier transformation (FFT) analysis [8-9], where raw data captured by Shimmer to analysis it using MATLAB.

3. Experiment Method

Tremor recording system consisted of lightweight Shimmer device, which was placed on the top of the patient's hand, and it had acceleration sensor and gyroscope sensor. Both sensors were used to record the six axis data while the gyroscope recorded the rotational movement. The current study focused on the resting and postural tremor along with Z axis for ten PD patients with different rhythmic movements and amplitude ranged within 10 seconds. Firstly, data was collected from PD patients while they were holding their hands in rest position. Secondly, the PD patients held their hands at 90 degree upright the body. The sensitivity of Shimmer sensor was 5.5389 Hz as calculated from the graph in figure 5. The researchers discussed the data related to the postural tremor only.

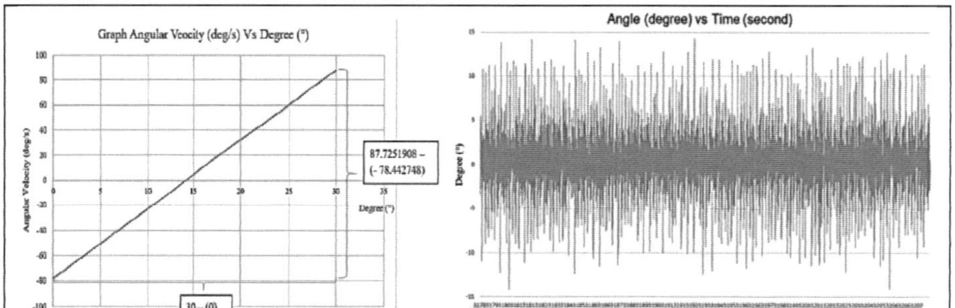

Figure 5. Graphs of measuring the sensitivity of Shimmer device.

Based on the graph, the sensitivity was calculated by,
X-values = 0 degree to 30 degree
Y-values = -78.442 degree to 87.725
By applying these values into the equation, we got:

$$\frac{87.725 - (-78.442) deg/sec}{30 - (0) deg} = 5.538\, Hz$$

Table 1. Results of Parkinson hand tremor

Patient	Time Domain		Frequency Domain
	Acceleration (m/s^2)	Angular Velocity (deg/s)	Acceleration (Hz)
1	-0.81 to 0.82	-7.5 to 11.1	2.031
2	-2.65 to 2.51	-64.8 to 44.2	4.813
3	-1.49 to 1.49	-28.0 to 46.1	2.125
4	-1.96 to 1.95	-48 to 36	4.063
5	-0.84 to 0.86	-33.9 to 61.9	3.188
6	-0.59 to 0.59	-8.8 to 8.8	1.844
7	-1.52 to 1.52	-74.5 to 105.3	4.813
8	-1.49 to 1.50	-19.9 to 20.1	1.75
9	-1.35 to 1.32	-15.1 to 26.1	4.188
10	-1.30 to 1.32	-10.8 to 10.8	1.594

Table 1 above presents the summary results of postural tremor for the ten patients in the Z-axis and rotational in the Y-axis. It shows the maximum and minimum of acceleration and rotational amplitude in time domain, vibration frequency in frequency domain. Every person had a different rhythmic movement. Within a time, interval of 10 seconds, some tremors vibrated extremely while some tremors vibrated a little. This was because some patients were under medication during the period of this assessment was conducted. Those obtained results are unpredictable. From figure 5, which shows the highest peak 121.4 Hz in frequency, domain graph represented the vibration frequency. Based on the present research, the range of tremor frequency occurred in Parkinson's disease was up to 9 Hz [5] [18]. The acceleration in time domain was converted to the frequency domain by using Fast Fourier Transform.

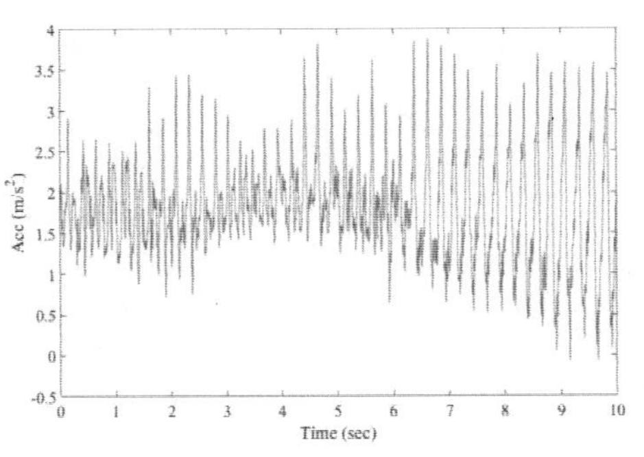

(a) Acceleration versus time graph.

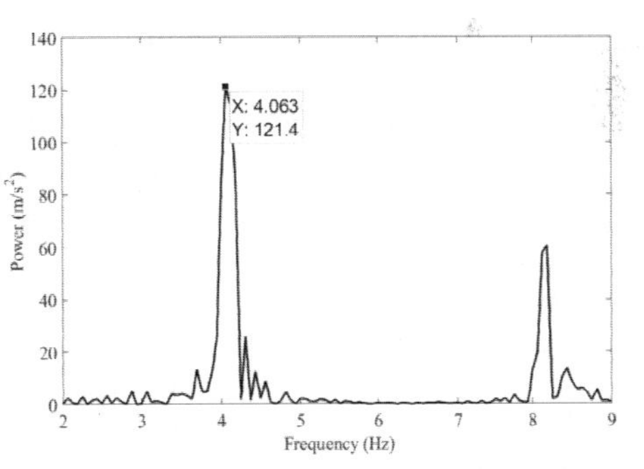

(b) Frequency graph

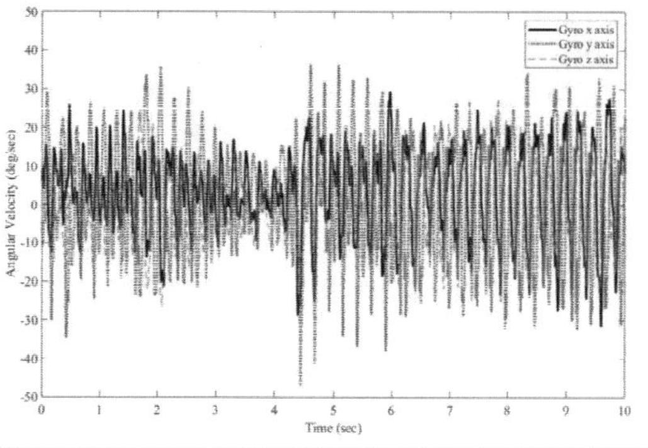

(c)Angular velocity versus time graph

Figure 6. The graphs for fourth patient

Figure 6 above shows the tremor behaviour in time and frequency domain for patient 4. It was noticed to be different from the other patients' because there was no effect of medication on this patient before the test was conducted. By comparing angular velocity, the Y-axis shows the highest amplitude compare to the X and Z-axes. Finally, the frequency occurred for this patient was at 4.063 Hz.

4. Conclusion

In summary, for this measurement for the features of Parkinson patients hand tremor frequency and its quantitative assessment method have been presented. A time-frequency and time-Angular velocity for tremor state detection was adopted. The task was conducted within 10 seconds for postural and another for rest positions, for both frequency and time domains. The peak power of measured frequency domain determined the quantitative the natural frequency of PD patients. However, this result is different from different tremor types.

An easy way to get precise measurement quantification for data collection which is collected by shimmer device. Based on the results of the present study, each patient have own natural identical frequency and rotational values. That can be attributed to the un-avoided noise during data collection process. Besides, some patients were tired, while most of them were under medication affect. Shimmer device is a very sensitive device that a slight movement affects the reading. As a result, it was challenging to make the Shimmer record the pure vibration of the postural tremor.

Then future works should involve following points:

1. Comparison among other latest sensor for tremor assessment as for this study where between Gyroscope and accelerometer placed in one Shimmer device.
2. The repeatability of tremor amplitude with the same patient at different times should be investigated further.
3. More verity tasks for the patients to declare the difference among tremor types where more experiments that are clinical are needed to modify the scale factors to be quantitatively assessed.
4. With medication, clinical history and more parameters should be evaluated for Patients to be monitored before conducting the assessment to have accurate overall assessment of Parkinsonian symptoms.

The results showed different between gyroscope and accelerometer reading for time and frequency domain where the response of gyroscope for tremor osculation is more noticeable. For making comparison to the previous methods, the current one is more affordable, easier and higher in quality. It measured the signal main parameters to evaluate the tremor severity stage of Parkinson tremor. In this study, it is show that the influence of patients' medication should not be ignored in data assessment.

Acknowledgments
The authors would like to thank Universiti Putra Malaysia (UPM) and the Malaysian Ministry of Education (MOE) for their continuous support in their research work. This research work was supported and financed by The Geran Putra IPS Fund, under Grant Vot. Number 9540600.

References

[1] C. G. Goetz et al., "Movement Disorder Society-Sponsored Revision of the Unified Parkinson's Disease Rating Scale (MDS-UPDRS): Scale presentation and clinimetric testing results," Mov. Disord., vol. 23, no. 15, pp. 2129–2170, 2008.
[2] K. D. Sethi et al., "Benefits and Risks of Pharmacological Treatments for Essential Tremor," Drug Saf., vol. 26, no. 7, pp. 461–481, 2006.
[3] S. Smaga, "Tremor," Am. Fam. Physician, vol. 68, no. 8, pp. 1545–1553, 2003.
[4] J. H. Mcauley and C. D. Marsden, "Physiological and pathological tremors and rhythmic central motor control," Brain, vol. 123, pp. 1545–1567, 2000.
[5] G. Deuschl, P. Bain, and M. Brin, "Consensus Statement of the Movement Disorder Society on Tremor," Mov. Disord., vol. 13, no. S3, pp. 2–23, 2008.
[6] S. Morrison, G. Kerr, and P. Silburn, "Bilateral tremor relations in Parkinson's disease: Effects of mechanical coupling and medication," Park. Relat. Disord., vol. 14, no. 4, pp. 298–308, 2008.
[7] S. Morrison and K. M. Newell, "Postural and resting tremor in the upper limb," Clin. Neurophysiol., vol. 111, no. 4, pp. 651–663, 2000.
[8] E. Randall, "spectral analysis of finger acceleration," 2018.
[9] R. H. Wyatt, "A Study of Power Spectra Analysis of Normal Finger Tremors," IEEE Trans. Biomed. Eng., vol. BME-15, no. 1, pp. 33–45, 1968.
[10] S. Gebai and M. Hammoud, "Parkinson's Disease Treatment as Seen from a Mechanical Point of View," Adv. Park. Dis., vol. 05, no. 04, pp. 97–106, 2016.
[11] Z. Ladin, W. C. Flowers, and W. Messner, "A quantitative comparison of a position measurement system and accelerometry," J. Biomech., vol. 22, no. 4, 1989.
[12] K. E. Norman, R. Edwards, and A. Beuter, "The measurement of tremor using a velocity transducer: Comparison to simultaneous recordings using transducers of displacement, acceleration and muscle activity," J. Neurosci. Methods, vol. 92, no. 1–2, pp. 41–54, 1999.
[13] A. Izworski, M. Michałek, R. Tadeusiewicz, J. Bulka, and I. Wochlik, "Acquisition and Interpretation of Upper Limbs Tremor Signal in Parkinsonian Disease Department of Automatics," no. January, pp. 2–6, 2005.
[14] Z. Szabo, B. Storkova, J. Hozman, and V. Zanchi, "Objective assessment of patients with Parkinson's disease," Challenges Remote Sens. Proc. 3Rd Wseas Int. Conf. Remote Sens. (Remote `07), pp. 39–42, 2007.
[15] R. P. Meshack, "The effects of weights on the amplitude and frequency of postural hand tremor in people with parkinson's disease," p. 86 p, 2001.
[16] M. Z. Md. Zain, A. As'arry, S. Kazi, M. Mailah, M. Hussein and M. Y. Abdullah "Development of experimental-rig for human postural tremor behaviour" , International Journal of Human Factors Modelling and Simulation, vol. 1, no. 3, pp. 339–351, 2010.
[17] R. M. A. de Bie et al., "Quantification of Hand Motor Symptoms in Parkinson's Disease: A Proof-of-Principle Study Using Inertial and Force Sensors," Ann. Biomed. Eng., vol. 45, no. 10, pp. 2423–2436, 2017.
[18] F. W. Wittich, "Diagnosis and treatment of common forms of respiratory allergy," Minn. Med., vol. 32, no. 7, pp. 702–708, 1949.

IOP Conf. Series: Journal of Physics: Conf. Series **1262** (2019) 012025 doi:10.1088/1742-6596/1262/1/012025

Application of Micro Electro-Mechanical Sensors (MEMS) Devices with Wifi Connectivity and Cloud Data Solution for Industrial Noise and Vibration Measurements

A. S. Sahdom
Geonoise Malaysia Sdn. Bhd., D-20-06, Sunway Nexis, No.1, Jalan PJU 5/1, Kota Damansara,47810 Petaling Jaya, Selangor, Malaysia

*Corresponding author: malaysia@geonoise.com

Abstract. This paper introduces the application of Microelectromechanical Systems (MEMS) for industrial noise and vibration measurements. Sound measurement was done using a compact sound level meter with type 1 Micro-Electro-Mechanical Systems (MEMS) microphone while the vibration measurement used a compact vibration measuring device with 3-axial accelerometer, which is also a MEMS type sensor. The MEMS Sensors being used in both measuring devices provide advantages in terms of features and specifications such as weight reduction, compact in size and economically affordable for various applications. Besides MEMS Sensors, the devices are equipped with wireless connectivity to a local server via a WIFI router. This eliminates the usage of wired connection in data handling. Furthermore, Cloud connectivity is an add-on feature to the wireless capabilities of both devices. Using the Cloud features enable threshold alert alarm, data monitoring and collection to be done regardless of time and location of devices, provided that the internet connectivity is available to the user and the devices.

1. Introduction

A typical sound level meter is a hand-held instrument with a microphone used for acoustic measurement (in decibels dB SPL) especially in industrial and environmental applications. The microphone of this sound level meter consists of a diaphragm which moves corresponding to changes in air pressure caused by sound waves and the movement is converted into electrical signal [1-4]. The problem with this type of sound level meter is that its size, weight and design is not suitable to leave it stand alone for measurement and required to be hand-held. Furthermore, these sound level meters usually have high power consumption as such they are not suitable for long term continuous measurement.

In present work, we used a sound level meter that consists of Microelectromechanical systems (MEMS) microphone as in Figure 1:

Figure 1. Sound level measuring device used in this project.

As could be seen in Figure 1, physical appearance and design of MEMS sound level meter is totally distinct from typical sound level meters. It is compact in terms of size, weight and design due to the adoption of MEMS microphone instead of typical diaphragm microphone. MEMS are made up of components that are microscopic in size, thus it is known as a technology of microscopic devices. These components include a microprocessor and microsensors that interact with surroundings. MEMS has a lot of application ranging from our daily devices such as smartphones to autopilot and navigation of an airplane [5-7].

MEMS can be used as an accelerometer [7], which is why it can be used for navigation as mentioned above. Therefore, in addition to measure sound using MEMS sound level meter, we also conducted vibration measurement using a vibration meter that is made up of MEMS accelerometer which was as Figure 2:

Figure 2. Vibration level measuring device used in this project.

Typical vibration measurement set up and configuration using external data acquisition system connected to sensors and computer are time consuming and inconvenient. MEMS vibration meter being used in this work does not require external data acquisition system and comes with standalone built-in storage for continuous measurement. Like the MEMS sound level meter, the vibration meter also has a lower power consumption compared to typical vibration meter, thus can also be used for long term measurement and monitoring (number of hours). Both sound and vibration devices being used in this work are also equipped with WIFI connectivity and Clouds storage capability using the Clouds webpage as illustrated in Figure 3.

IOP Conf. Series: Journal of Physics: Conf. Series **1262** (2019) 012025 doi:10.1088/1742-6596/1262/1/012025

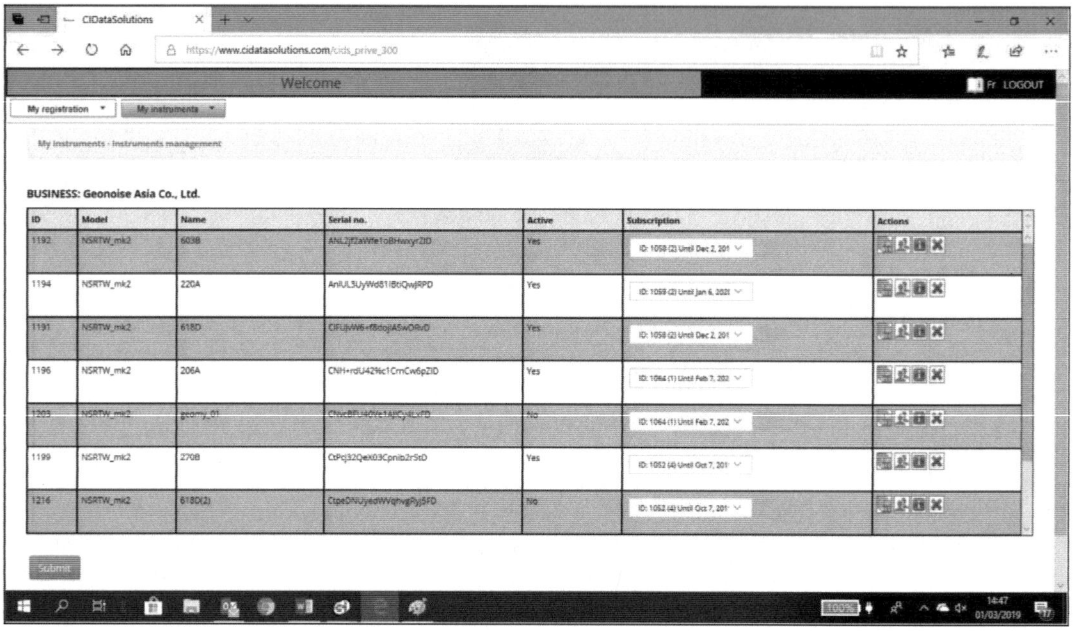

Figure 3 Clouds page of one of MEMS-equipped sound and vibration measuring devices.

It eliminates wired connection and allows data handling to be done anytime and anywhere. These features ease the measurement and data handling works, thus is convenient for industrial applications.

2. Measurement Set Up

The device used for sound and measurement was known as Convergence Instruments NSRTW_mk2 and Convergence Instruments VSEW_mk2 respectively. The measurements were carried out at a Mass Rapid Transit (MRT) station for 1 hour and the train platform that the devices were placed was the platform where the train bound for Sungai Buloh. The level of sound and vibration at the station every time a train arrived were observed in this project. The MEMS sound level meter was connected to a PC and configured a computer software named Instrument_Manager which was installed on the PC. The instrument set up instructions can be obtained from the user manual which is available at (https://convergenceinstruments.com/pdf/NSRTW_mk2_UserManual.pdf). The following figure shows sound measurement parameters set-up using the software for this project:

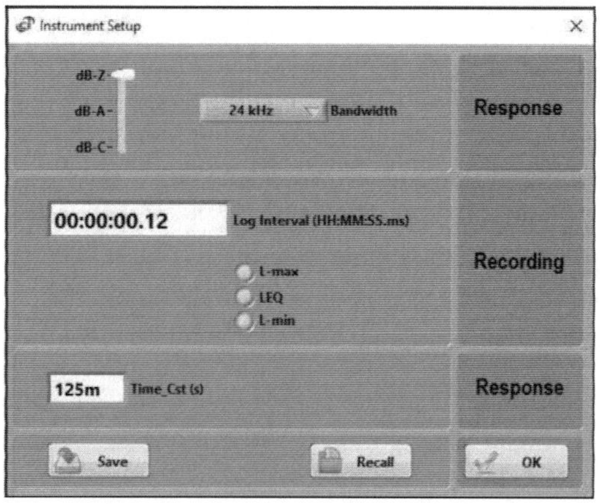

Figure 4. Sound level measurement parameter set up using Instrument_Manager software.

Meanwhile, a MEMS vibration sensor was configured using similar method and the measurement parameters for vibration were as following figure:

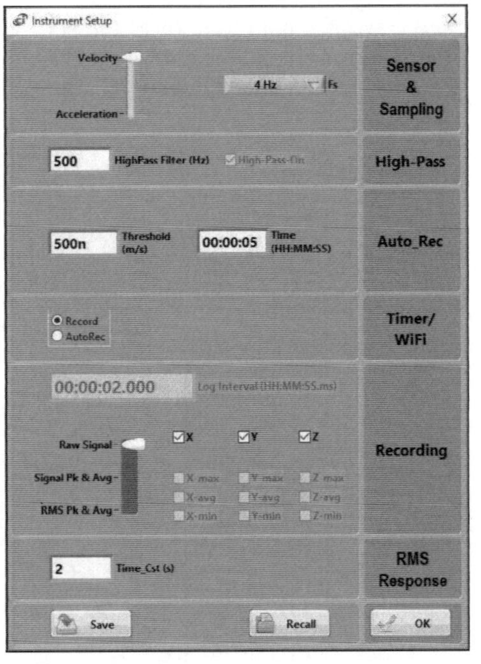

Figure 5. Vibration measurement parameter set up using Instrument_Manager software.

Both sensors were configured to automatically begin measurement at 11.25AM. Then, the sensors were placed on the glass barrier that separate the platform with the train track as can be seen in the following figure:

(a) (b) (c)

Figure 6. Sound and vibration measurement devices placement on the glass barrier. (a) Device in top circle is the sound measurement device while device in the bottom red circle is the vibration measurement device; (b) Sound measurement device in closer look; (c) vibration measurement device in closer look.

IOP Conf. Series: Journal of Physics: Conf. Series **1262** (2019) 012025 doi:10.1088/1742-6596/1262/1/012025

From Figure 6, we could see that the compact size eased the placement and set up of both devices such that no additional mounting support accessories such as tripod stand was required as typical sound level meter. As the measurement automatically started at 11.25AM indicated by flashing red LED light on the sensors, time for each train arrival for both train direction was recorded throughout the 1-hour measurement duration. After 1-hour measurement, the measurement by the sensors were stopped and the data was collected and saved to the PC using the software mentioned above.

3. Results and Discussion
The following table shows recorded train movement time at the for both directions.

Table 1. Train movement and time.

	Bound	
	Kajang	Sungai Buloh
Time	11.29AM	11.31AM
	11.36AM	11.38AM
	11.43AM	11.45AM
	11.50AM	11.51AM
	11.57AM	11.59AM
	12.04PM	12.06PM
	12.11PM	12.13PM
	12.18PM	12.20PM
	12.25PM	

From the table, we could observe that the train movement direction alternates with another at every single time.The measurement was carried out at the platform where the train bound for Sungai Buloh which was a factor that contribute to sound and vibration level of each train movement.

3.1. Sound Level
Measurement data from both devices were collected after 1 hour measuring duration. Sound level measurement result was as Figure 6 (a), (b) and (c).

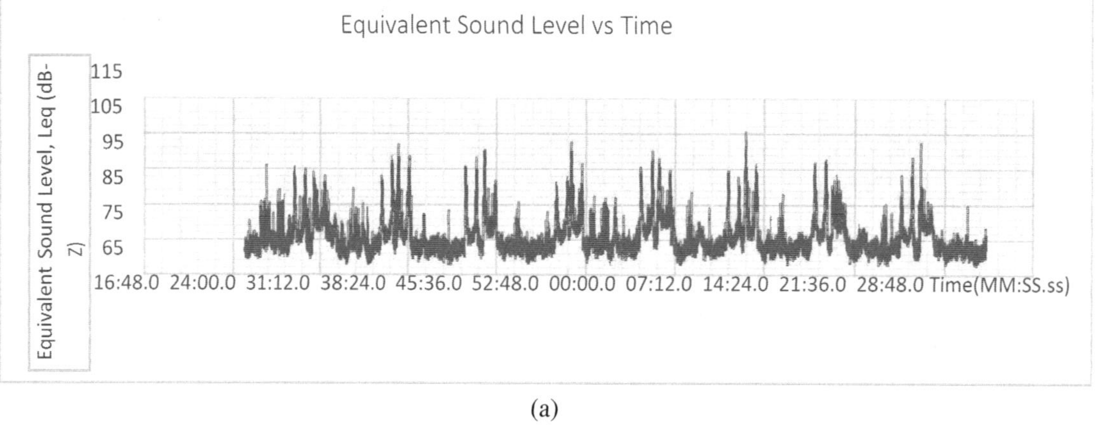

(a)

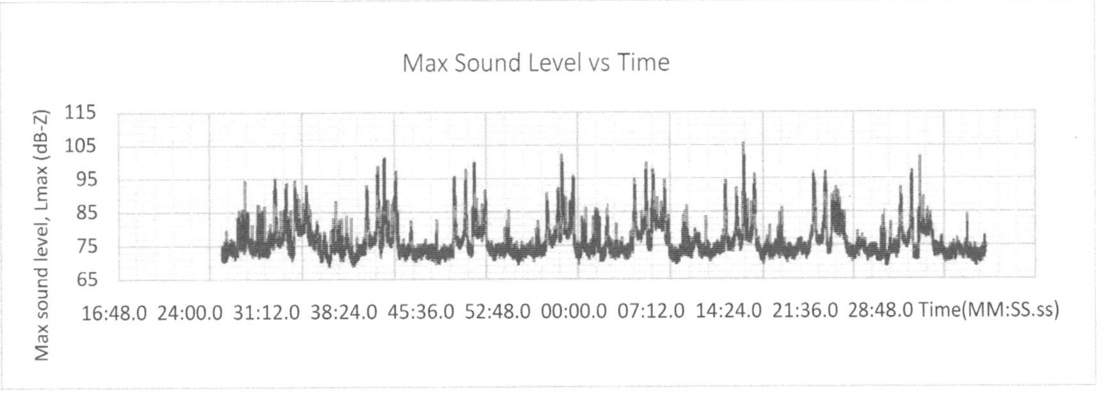

(b)

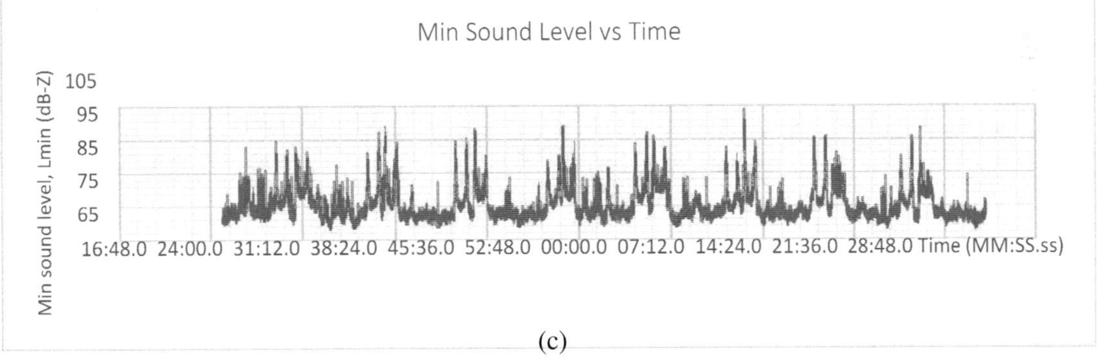

(c)

Figure 7 Sound level data collected. (a) L_{EQ}; (b) L_{MAX}; (c) L_{MIN}.

Data presented in Figure 6 (a), (b) and (c) are Z-weighted Sound Levels in L_{MAX}, L_{MIN}, and L_{EQ} in decibels (dB-Z) against measurement time respectively. From the data, one could see that the sound level will be highest at times when there was train movement close to the platform that the device is placed, in other words train that was bound for Sungai Buloh. This was followed by times when there was train movement in the opposite direction where the track is further from the platform.

3.2 Vibration Level

Velocity data collected from vibration measuring device were presented in the following Figure.

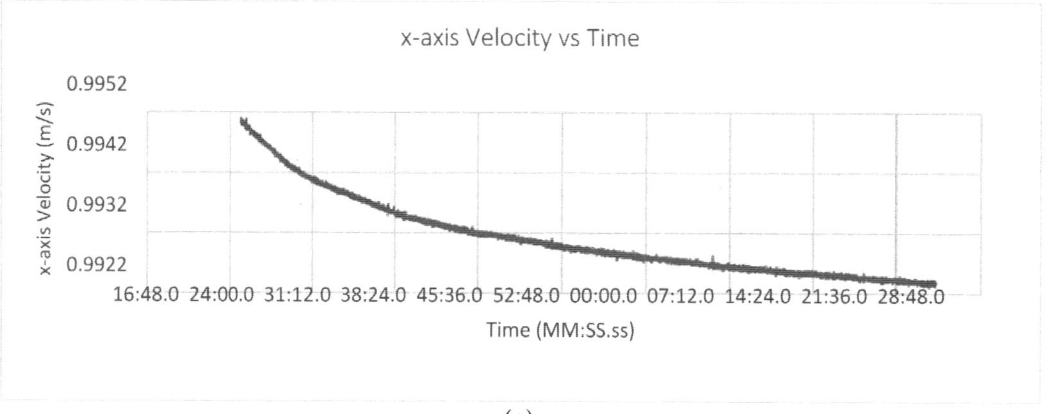

(a)

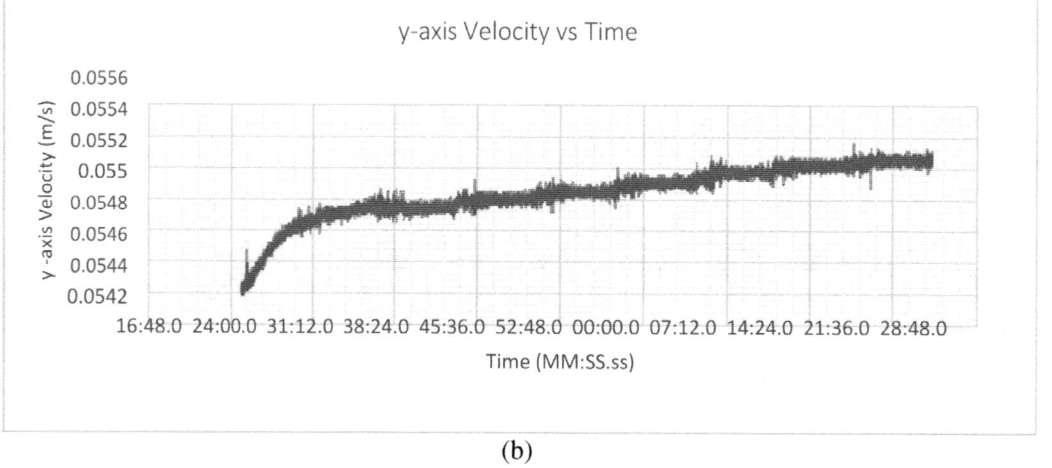

(b)

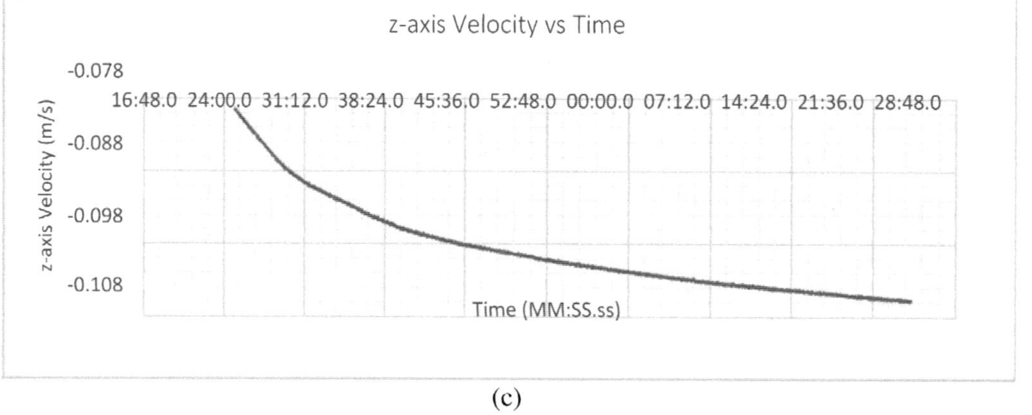

(c)

Figure 8 Vibration data collected in velocity. (a) x-axis vibration; (b) y-axis vibration; (c) z-axis vibration.

The vibration measuring device consists of 3-axial sensors. Therefore, three measurement data, one for each axis were collected and presented where Figure 7 (a), (b) and (c) represent vibration data in for x-, y- and z- axis respectively.

4. Conclusion

In this paper, the technology of MEMS was introduced and applied in sound and vibration measurement. Sound and vibration measuring devices used in this project were compact in size, lightweight, wireless and easy to use without the need of other extra support equipment or accessories.

Acknowledgements

Author thanks CNVC 2019 secretariat for organizing the event which is an opportunity to introduce MEMS technology in sound and vibration measurement; Geonoise Malaysia management for financial and equipment support; and colleagues for useful ideas, assistance and co-operation.

References

[1] Sani M S M, Rahman M M , Baharom M Z and Zaman I 2015 Sound Intensity Mapping of an Engine Dynamometer *International Journal of Automotive and Mechanical Engineering (IJAME)* **11** 2820-9

[2] Sani M S M, Rahimi M A F A and Mohamed D 2019 Sound Intensity Mapping of Two Stroke Engine by Using Hemispherical Surface Coordinate Arrangement *IOP Conf. Ser. Mater. Sci. Eng.* 506 012060

[3] Razak N F D, Sani M S M, Azmi W H and Zhang B 2017 Noise and vibration analysis for automotive radiator cooling fan *IOP Conf. Ser. Mater. Sci. Eng.* 257 012083

[4] Sani M S M, Zaman I and Rahman M M 2015 Analysis of Split Air Conditioner Noise using Sound Intensity Mapping *Journal of Advanced Research in Fluid Mechanics and Thermal Sciences* 9(1) 28-33.

[5] Celestina M, Hrovat J and Kardous C A 2018 Smartphone-based sound level measurement apps: Evaluation of compliance with international sound level meter standards *Applied Acoustics* 139 119-128.

[6] Garg S, Lim K M and Lee H P 2019 An averaging method for accurately calibrating smartphone microphones for environmental noise measurement *Applied Acoustics* 143 222-228.

[7] Murphy E and King E A 2016 Testing the accuracy of smartphones and sound level meter applications for measuring environmental noise *Applied Acoustics* 106 16-22.

IOP Conf. Series: Journal of Physics: Conf. Series **1262** (2019) 012026 doi:10.1088/1742-6596/1262/1/012026

The Effects of Posture on Suspension Seat Transmissibility during Exposure to Vertical Whole-Body Vibration

S. Aisyah Adam[1], N. A. Abdul Jalil[1*], K. A. Md Razali[1], Y. G. Ng[1]
[1]Sound and Vibration Research Group, Department of Mechanical and Manufacturing Engineering, Faculty of Engineering, Universiti Putra Malaysia (UPM), 43400 UPM Serdang, Selangor, Malaysia

*Corresponding author: nawalaswan@upm.edu.my

Abstract. Suspension seat is used in the off-road condition to attenuate excessive vibration exposed to the human body. The efficiency of a seat reducing vibration not only depends on the dynamic characteristics of the seat, but the dynamic characteristics of human body and the characteristics of the input vibration as well. Tractor drivers adopted different postures during their farm work activities, which may influence the dynamic characteristics of the human body. However, the influenced of the driver's posture on suspension seat transmissibility has received less systematic attention. Thus, this study is carried out with the objective to investigate the effect of different postures on seat transmissibility when seated on a suspension seat. Three male subjects were exposed to random vibration at 2.0 m/s^2 r.m.s with frequency ranging from 1-20 Hz, while seated on a vibration simulator for 60 seconds. The subjects adopted four seating postures: (i) relaxed, (ii) slouched, (iii) tensed and (iv) with backrest support. The study found that relaxed and slouched postures have a resonance frequency at 2.0 Hz. However, as the posture changed to backrest support, the resonance frequency of the seat transmissibility slightly increased by 0.25 Hz. This study suggested that changing the postures caused changes in the dynamics of human body, and thus affected the suspension seat transmissibility. It is concluded that, non-linearity in suspension seat transmissibility is influenced by the changes of body postures.

1. Introduction

The characteristics of the vibrations to which the human body is exposed differ widely between environments. Off-road vehicles including agricultural tractors, construction trucks and heavy machineries operate on the uneven surface, and thus exposed to higher levels of whole-body vibration (WBV). When agricultural vehicles moving on the uneven ground's surface, the whole of its framework is exposed to complex oscillatory motion, which is caused by the vibration from the engine, and the interaction with the uneven terrain on the ground's surface [1]. Intense low frequency of vibration (0.5 to 80 Hz) mainly induced by the unevenness of the ground surface may reduce levels of comfort and increase the potential of musculoskeletal disorders (MSDs), particularly low back pain (LBP) [2-4]. It is evident that higher vibration magnitude is transmitted to the human body on off-road condition as compared to the on-road condition [5-7].

Primary contact of the vertical WBV transmitted to the human body is through the seat. Hence, suspension seat is introduced to attenuate vibration transmitted to the human body. The efficiency of a seat in attenuating the vibration depends on: (i) seat transmissibility, (ii) the sensitivity of the human body to the vibration, and (iii) the input vibration at the seat base. Seat transmissibility can be defined as the ratio of output vibration (on the seat surface) to the input vibration (on the seat base), and often

IOP Conf. Series: Journal of Physics: Conf. Series **1262** (2019) 012026 doi:10.1088/1742-6596/1262/1/012026

be used to quantify the seat performance. There are three different types of suspension seat commonly use in the field, which are; passive, semi-active and active. A recent study found that active suspension system with fuzzy logic controller could improve the suspension seat performance [8]. Even though various studies demonstrated the reduction of the vibration transmitted from optimising the suspension seat, the influence of human body should not be ignored.

Human body formed a coupled system with suspension seat and affect the efficiency of a seat [9]. Drivers adopted different postures during farm work activities, and it is highly dependent on the task they are doing. When the body moves from its original posture, the weight is shifted, and thus may prevented from the suspension seat to work properly. Nevertheless, less is known about the effect of posture on suspension seat transmissibility. Thus, this study was carried out with the objective to study the effect of different postures on vertical suspension seat transmissibility. It is hypothesised that the suspension seat transmissibility is influenced by the posture of human body.

2. Methods and Procedures

2.1. Experimental Setup

Suspension seat use in this study is from agricultural tractor that is commonly used in Malaysia. The cushion is removed, and the suspension system is attached to a rigid seat. Then, the seat is secured on top of the shaker. Vibration analysis of the structure was done prior to the experiment with SOLIDWORKS Simulation 2017.

Shaker was used to generate vibration in the vertical direction (z-axis). Random vibration was generated with the frequency range of 1 to 20 Hz at 2.0 m/s^2 of vibration magnitude. Vibration in the vertical direction was measured using single-axial integrated circuit piezo-electric (IEPE) accelerometers (B&K type 4514) with 100 mV/g sensitivity. The accelerometers were attached on top of the shaker (seat base) and on the suspension seat. Both of the accelerometers were calibrated prior to the experiment. The signals were acquired using data acquisition (NI-9234) and the signal processing was analysed in the MATLAB 2018. The waveform were low-pass filtered at 50 Hz via anti-aliasing filter.

The schematic diagram of the experimental setup is shown in Figure 1.

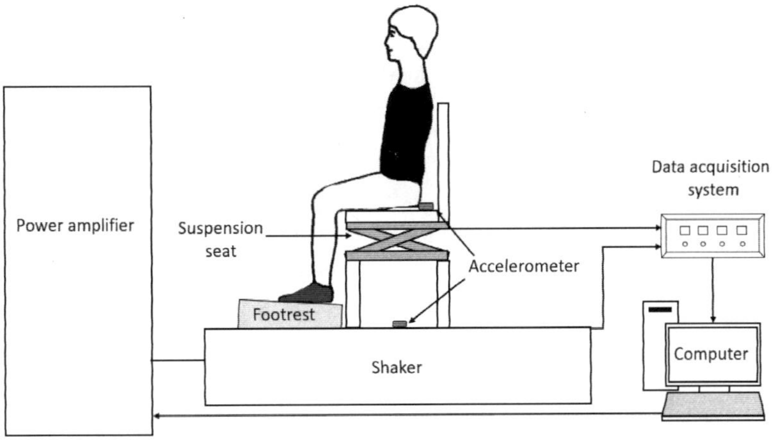

Figure 1. Schematic diagram of the experimental setup

2.2. *Subjects and Postures*

Three healthy male subjects participated in the study. Healthy here means the subject did not have background of accident, incident or disease that cause low back pain and other musculoskeletal disorders. Written consent form was collected prior to the experiment. Ethical approval was obtained from the Ethics Committee for Research involving Human Subjects of Universiti Putra Malaysia (JKEUPM). The physical characteristics of the subjects are summarised in Table 1.

Table 1. Subjects' physical characteristics.

	Mean	Standard deviation	Minimum	Maximum
Age (years)	27.3	5.43	23	35
Weight (kg)	62.0	5.9	55.8	70.0
Height (m)	1.68	0.02	1.65	1.72

The subjects adopted four seating postures: (i) relaxed, (ii) slouched, (iii) tensed and (iv) with backrest support. They are instructed to put their hands on their laps, and rested their feet on a footrest that was attached to the shaker platform. Subjects wore a loose fitting lap belt for safety. The subjects were asked to maintain the desired posture for 60 seconds during exposure to vertical vibration. The sitting conditions were shown in Figure 2.

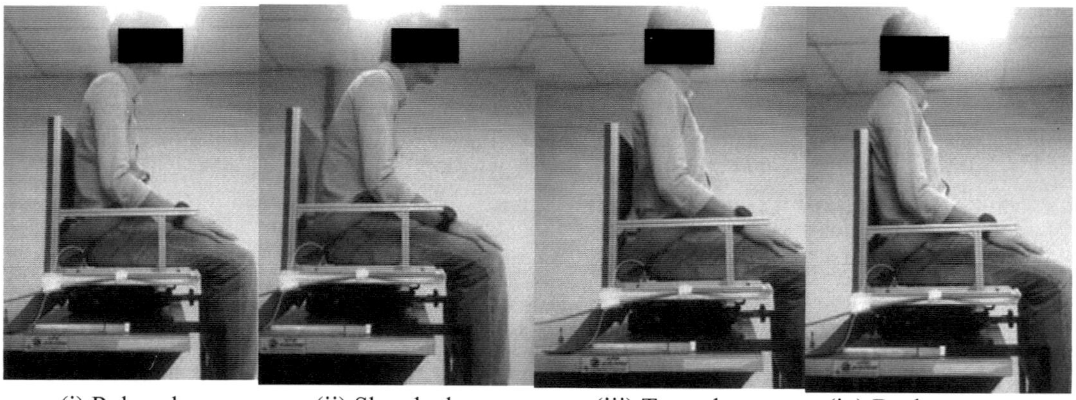

(i) Relaxed (ii) Slouched (iii) Tensed (iv) Backrest support

Figure 2. Postures adopted by subjects.

2.3. *Analysis*

Transfer functions between the acceleration at the base of the seat and the acceleration on the suspension seat were calculated using cross spectral density (CSD) method with a 0.25 Hz of frequency resolution. The seat transmissibility is determined from the ratio of the input and output acceleration of the CSD to the power spectral density (PSD) of input acceleration. The CSD method is used to minimise the effect of noise, and can be expressed as:

$$H(f) = \frac{G_{io}(f)}{G_{ii}(f)} \quad (1)$$

where $H(f)$ is the transfer function by CSD method, $G_{io}(f)$ is the cross spectrum of the input (seat base) and output (suspension seat) and $G_{ii}(f)$ is the power spectrum of input acceleration.

Using CSD method gives advantage in a way that it is a complex function, and provides both modulus and phase of the transfer function, as shown in Eq. 2 and Eq.3:

$$Modulus = \sqrt{(R(H(f)))^2 + (I(H(f)))^2} \tag{2}$$

$$Phase = tan^{-1}\left[\frac{I(H(f))}{R(H(f))}\right] \tag{3}$$

where $R(H(f))$ is the real part of the transfer function, and $I(H(f))$ is the imaginary part of transfer function.

In order to estimate how the output acceleration are related to input functions, the coherency may be determined such as:

$$Coherency\,\gamma_{io}{}^{2}(f) = \frac{|G_{io}(f)|^2}{G_{ii}(f)G_{oo}(f)} \tag{4}$$

where $G_{oo}(f)$ is the PSD of output acceleration.

The principal resonance frequency was defined as the frequency at which resulted to the highest peak of the suspension seat transmissibility. It is expected that the principal resonance frequency is at the range of 1-20 Hz.

3. Results and Discussion

3.1. Effect of subject posture on suspension seat transmissibility

Seat transmissibility of seated subjects for all postures are shown in Figure 3-6. All the subjects showed three peak with the highest peak is around 1-3 Hz for all postures. This is consistent with previous study done by Qiu where the author found three peaks with a primary peak in the frequency range below 6 Hz [10]. All the data showed high coherency (above 0.8), which suggested the system is linearly correlated. The median data for all the postures adopted in the study is shown in Figure 7.

Similar resonance frequency is observed during relaxed and slouched posture, which is at 2.0 Hz. However, slightly higher peak of seat transmissibility was found for relax posture as compared to slouched posture (from 1.245 to 1.324). When the posture is changing from slouched to tensed, the resonance frequency increased by 0.25 Hz (from 2.0 Hz to 2.25 Hz).

Previous studies reported the resonance frequency of apparent mass of a seated human body is higher in tensed posture as compared to relaxed posture [11]. Changes in muscle tension resulting from the changes of upper-body posture would influence the magnitude of the muscular activity in response to vibration [12]. Tensed posture caused the pressure under the ischial tuberosity to the seat increased, and thus, resulted to slightly increases in the resonance frequency of the seat transmissibility.

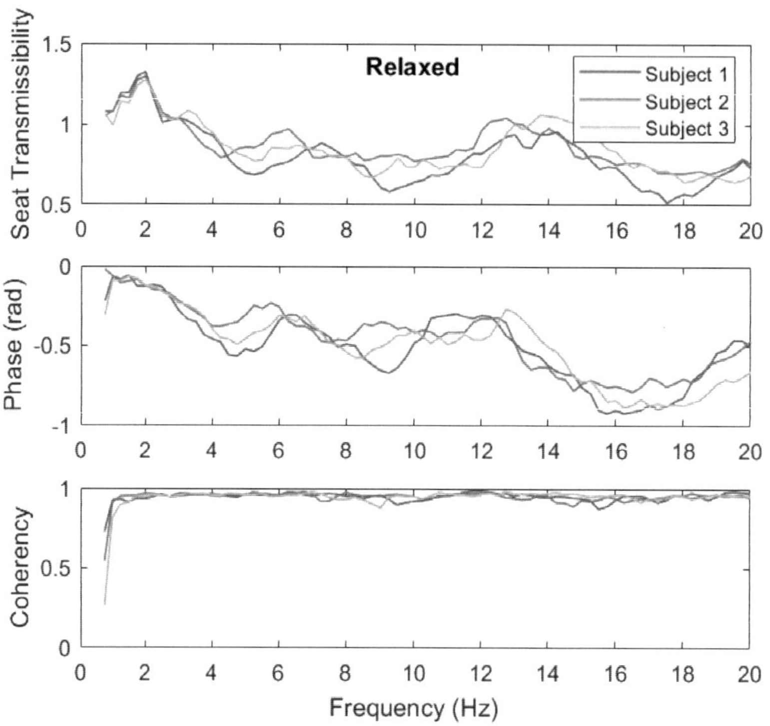

Figure 3: Seat transmissibility, phase and coherency for relaxed posture.

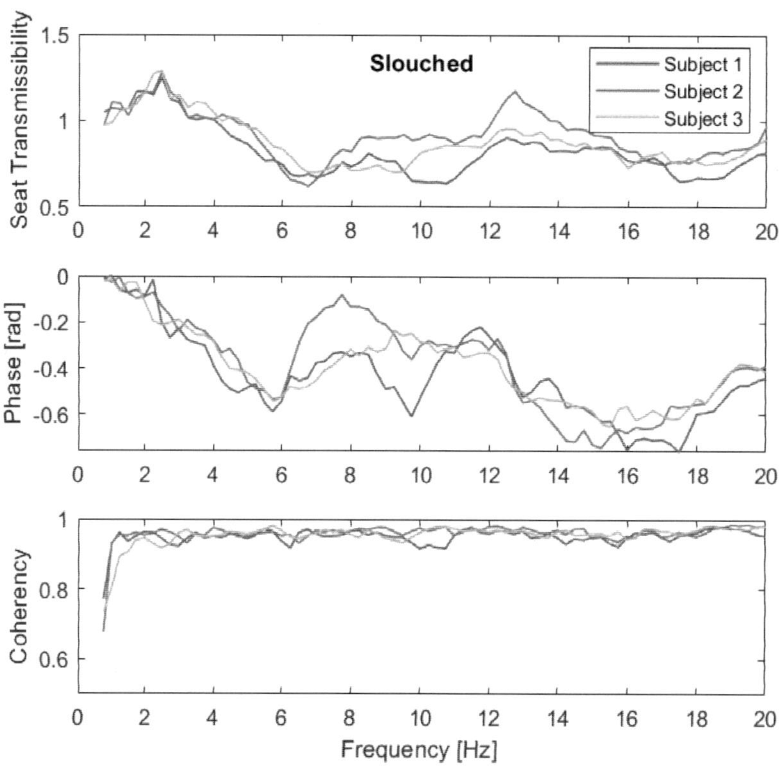

Figure 4: Seat transmissibility, phase and coherency for slouched posture.

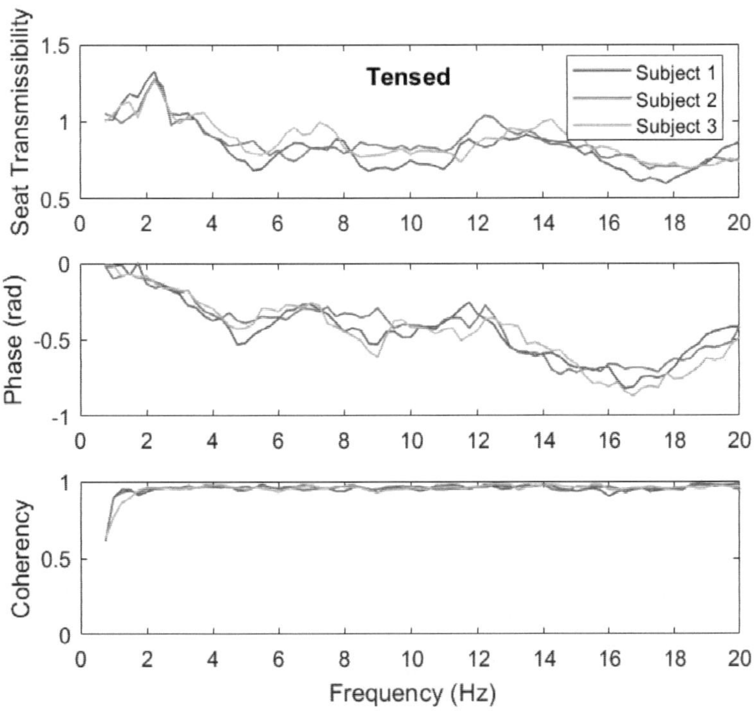

Figure 5: Seat transmissibility, phase and coherency for tensed posture.

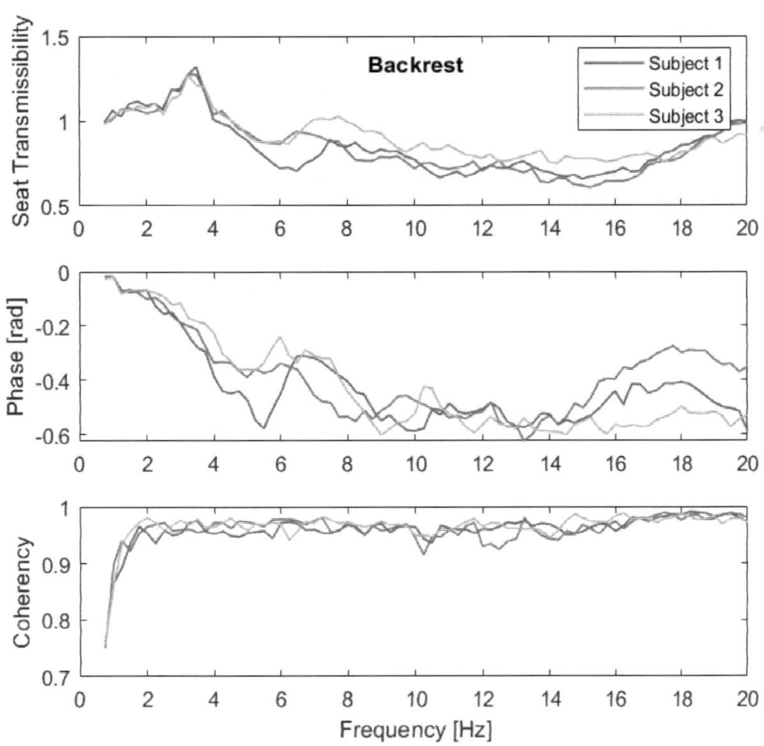

Figure 6: Seat transmissibility, phase and coherency for backrest posture.

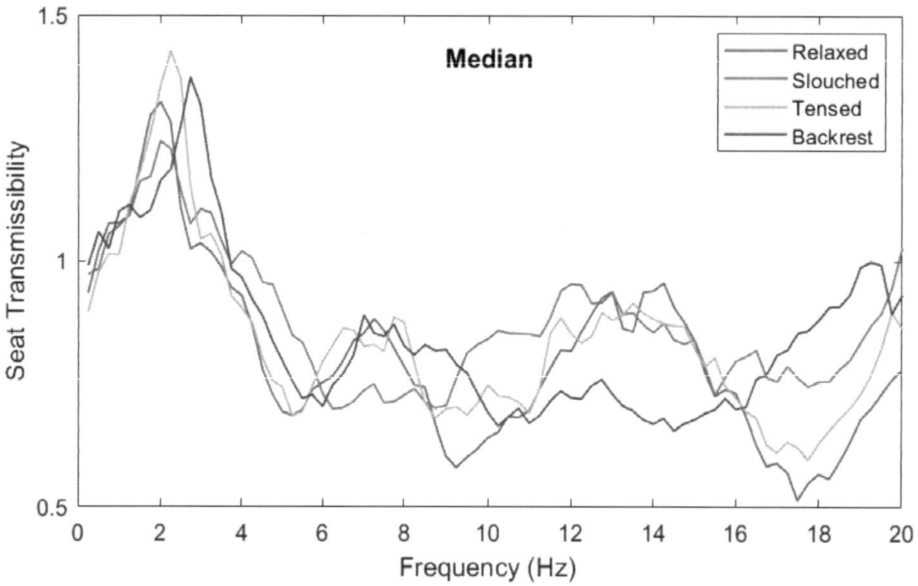

Figure 7: Median data for all the postures.

3.2. Effect of backrest

Adding a rigid backrest support slightly increased the resonance frequency of seat transmissibility, as compared to different postures without backrest support (relaxed, slouched and tensed). Backrest may introduce additional vibration to the seated body. Note that in this study, a rigid backrest support is used, hence there is no attenuation of vibration from the backrest support to the seat. However, recent study found that reclined backrest support may increase transmission to body resonance as compared to upright backrest support [13]. Contact with the backrest support could constraint the movement of upper body parts, and thus alters the dynamic response of human body [14]. Consequently, changes in the dynamic response influenced the suspension seat transmissibility.

4. Conclusions

The results showed different postures slightly affect the resonance frequencies of suspension seat transmissibility. Changes in the resonance frequency is apparent when the body postures changes from relaxed and slouched to tensed. This is due to changes of stiffness and damping exhibits by the body, which alters the pressure under the ischial tuberosities to the seat. In this study, it is proved that the influence of human body should not be ignored in optimising the performance of suspension seat.

Acknowledgements

The authors would like to acknowledge MyBrain15 and Putra Grant GP-IPS/2018/9605700 for providing financial support for the study.

References

[1] Chvetanovic B and Zlatkovic 2013 Evaluation of whole-body vibration risk in agricultural tractor drivers. *Bulgarian Journal of Agricultural Science* **19** 1155 – 60
[2] Bovenzi M, Rui F, Negro C, D'Agostin F, Angotzi G, Bianchi S, Bramanti L, Festa G, Gatti S, Pinto I, Rondina L, Stacchin N 2006 An epidemiological study of low back pain in professional drivers. *Journal of Sound and Vibration* **298** 514 – 39
[3] Raffler N, Ellegast R, Kraus T and Ochsmann E 2016 Factors affecting the perception of whole-

body vibration of occupational drivers: an analysis of posture and manual materials handling and musculoskeletal disorders. *Ergonomics* **59** No. 1 48 – 60

[4] Burström L, Nilsson T and Wahlström J 2014 Whole-body vibration and the risk of low back pain and sciatica: A systematic review and meta-analysis. *International Archives of Occupational and Environmental Health* **88** 10.1007/s00420-014-0971-4

[5] Aisyah Adam S and Abdul Jalil NA 2017 Vertical suspension seat transmissibility and SEAT values for seated person exposed to whole-body vibration in agricultural tractor preliminary study. *Procedia Engineering* **170** 435–42

[6] Ho J, Dennerlein JT, Johnson PW 2018 The effect of a multi-axis suspension on whole body vibration exposures and physical stress in the neck and low back in agricultural tractor applications. *Applied Ergonomics* **68** 80-9

[7] Smets MPH, Eger TR and Grenier SG 2010 Whole-body vibration experienced by haulage truck operators in surface mining operations : A comparison of various analysis methods utilized in the prediction of health risks. *Applied Ergonomics* **41** 763–70

[8] Alfadhli A, Darling J and Hillis AJ 2018 The Control of an Active Seat Suspension Using an Optimised Fuzzy Logic Controller, Based on Preview Information from a Full Vehicle Model. *Vibration* **1** 20-40

[9] Toward MGR and Griffin MJ 2011 The transmission of vertical vibration through seats: Influence of the characteristics of the human body. *Journal of Sound and Vibration* **330**(26) 6526-43

[10] Qiu Y 2017 Dynamic Characteristics of a suspension seat determined in laboratory study. *Journal of Ergonomics* **7(6)**: 220

[11] Kitazaki S and Griffin MJ 2008 Resonance behaviour of the seated human body and effects of posture. *Journal of Biomechanics* **31** 143-9

[12] Yang M, Qiu Y and Griffin MJ 2016 Effect of muscle activity on the apparent masses of the human body when exposed to vertical whole-body vibration, *51st UK Conference on Human Responses to Vibration*

[13] Zhang X, Qiu Y and Griffin MJ 2015 Transmission of vertical vibration through a seat: Effect of thickness of foam cushions at the seat pan and the backrest. *International Journal of Industrial Ergonomics* **48** 36-45

[14] Mansfield NJ and Maeda S 2005 Effect of backrest and torso twist on the apparent mass of the seated body exposed to vertical vibration. *Industrial Health* **43**, 413–20

IOP Conf. Series: Journal of Physics: Conf. Series **1262** (2019) 012027 doi:10.1088/1742-6596/1262/1/012027

Development of Vibration Test System to Measure Human Biodynamic Responses to Hand-Transmitted Vibration (HTV)

A M Aizuddin[1,2], N A Abdul Jalil[1*], K A Md Rezali[1] and Y G Ng[1]
[1] Sound and Vibration Research Group, Faculty of Mechanical and Manufacturing Engineering, Universiti Putra Malaysia (UPM), 43400 UPM Serdang, Selangor, Malaysia
[2] Department of Mechanical and Manufacturing Engineering, Faculty of Engineering, Universiti Malaysia Sarawak (UNIMAS), 94300 Kota Samarahan, Sarawak, Malaysia

*Corresponding author: nawalaswan@upm.edu.my

Abstract. Workers commonly experience hand-transmitted vibrations (HTV) especially in industries involving power tools. Vibration exposure will differ by individuals where prolonged and unmonitored exposure can lead to the development of symptoms known as hand arm vibration syndrome (HAVS). A deep understanding of human response to HTV is required to provide feasible solutions to lessen its adverse effects. Research has been extensively conducted on exploring the vibration transmission to the hand-arm system. However, there is limited study on the effects of handles mounted on a damping material or suspension systems. The objective of this study is to develop a vibration test system to measure the human biodynamic response to hand-transmitted vibration. Engineering design theory were used to formulate the essential design requirements of the system. The evaluation and selection of concepts were done. The design was modelled, and frequency analysis was performed using Computer-Aided Engineering (CAE) software. The frequency analysis results showed the first 10 modes in the frequency range of 1320 – 5088 Hz and therefore will not have resonance issues in the 25 – 1250 Hz frequency excitation range as per requirement. The prototype design was shown to comply with the product design specification and final detail design drawings were approved for fabrication. In conclusion, an instrumented handle with a modular expansion slot had been appropriately designed to measure human biodynamic responses due to hand-transmitted vibration.

1. Introduction

Hand-transmitted vibration (HTV) are often experienced daily depending on the type of working environment that the individuals are exposed to. Industrial workers in various fields such as in agriculture and manufacturing operate hand power tools and machineries and are more susceptible to severe vibration exposure. The exposure if left unsupervised can adversely affect the health of these workers. Hand-arm-vibration syndrome (HAVS) are commonly reported amongst workers as a result of prolonged exposure especially in cold climates [1].

Ongoing research is conducted to investigate the four vibration components that can affect the amount of vibration exposure which are the vibration's frequency, magnitude, duration and frequency [2]. A good understanding of human biodynamic response to HTV with the variation of these components can provide future solutions and create a better work environment.

Current vibration test systems are mostly developed in the area of research. There are various configurations such as the split handle design [3] and the suspended handle [4]. The types of sensors used to measure force and acceleration also differs among experimenters depending on the purpose of

IOP Conf. Series: Journal of Physics: Conf. Series **1262** (2019) 012027 doi:10.1088/1742-6596/1262/1/012027

their research. Besides that, there are instrumented handle design as suggested by international standards and guidelines to evaluate the vibration transmissibility of gloves at the palm of the hand[5]. Some researchers utilise indenter rigs to measure dynamic stiffness of materials to predict vibration transmissibility based on human biomechanical models [6].

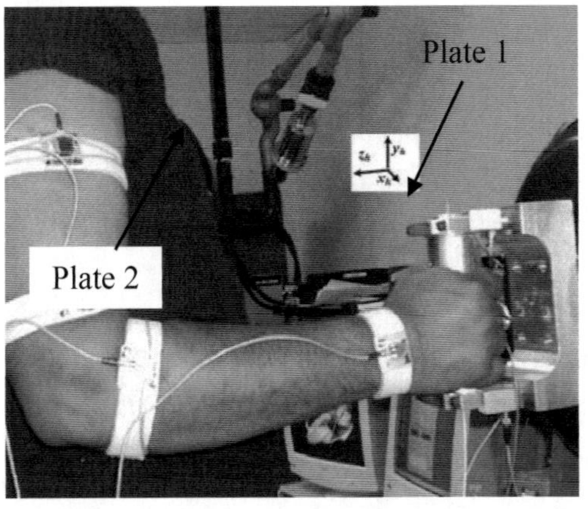

Figure 1. Instrumented handle setup used by Adewusi et al [3].

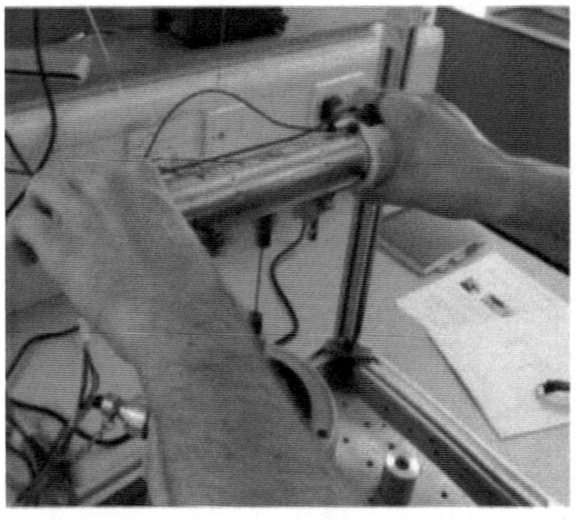

Figure 2. Almagirby et al [4] experimental setup of the instrumented handle.

As of now, there are no instrumented handle designs which can be mounted on materials and suspension systems. This would be useful to further investigate the effects of materials or suspension systems on the human biodynamic response.

This paper addresses the design process of the development of a vibration test system to measure human biodynamic response to HTV which also have a modular expansion feature to accommodate for future experiments with materials and suspension systems.

2. Methodology

The basic methodology follows a standard engineering design process of conceptual design followed by the embodiment design phase and finally the detail design for fabrication [7].

2.1. Conceptual Design

This phase starts with defining the problem and generating the product design specification (PDS) as shown in Table 1. Table 2 shows the functional decomposition of the instrumented handle. Various concepts are then brainstormed, and conceptual drawings were produced. The evaluation of the concepts was done using a weighted rating method based on the selection criteria.

Table 1. Product design specifications for the instrumented handle.

Aspect	Objective	Criteria
Materials	Appropriate for design	High stiffness to ensure structure rigidness
Weight	Low	Below 3 kg
Safety	High	No sharp edges that can injure human subjects
Handle dimensions	Convenient for gripping purposes	Diameter (40 – 50 mm) & minimum length 110 mm
Handle resonance characteristics	Appropriate for design	No modes in the range of 25 – 1250 Hz

Table 2. Functional decomposition of instrumented handle.

Part	Function
Base	Acts as the foundation for the instrumented handle. Provide extension for modular feature.
Handle	The core part of the assembly. Provide the housing for the sensors.
Grip handle	Allow for human subjects to grip the handle. Attached to the force sensor.
Front bracket	Secure alignment of the handle's position.
Rear bracket	Secure alignment of the handle's position. Provide extension for modular feature.
Force Sensors	Measure the dynamic force.
Accelerometers	Measure the acceleration of the handle.
Plate	Connect the instrumented handle to the shaker.

2.2. Embodiment Design

The next phase carried out in the arrangement and sizing of the physical parts of the chosen concept. The concept was then modelled in a computer-aided engineering (CAE) software to check out for function and spatial constraints. Selection of materials were also investigated to comply with the PDS. Finite element analysis (FEA) was conducted to simulate and predict the first 10 modes of the design.

2.3. Detail Design

This design phase involved the complete product engineering description of the instrumented handle. The detailed design drawings for each parts and overall assembly were prepared before it was sent for fabrication.

3. Results

3.1. Prototype Design

Figure 3 shows the various parts in the instrumented handle assembly design. All the main parts are fastened together via M8 socket button head cap screws while the grip handles are joined to the force sensors via ANSI ¼ - 28 screws. The sensors have their own stud mounting screws connected to the tapped holes in the handle part. The base part will be fixed to a plate using M8 screws. Materials for all parts excluding the accelerometer, force sensors, and screws are made of Aluminium Alloy 7075. It is chosen based on its high stiffness, relatively lightweight and readily available in the market. The overall weight of the design is 2.62 kg. The grip diameter is 50 mm.

All sensors are placed inside the instrumented handle to minimise errors in the vibration path transmission. However, there is an extra accelerometer hole placement for future investigation of the vibration transmissibility of the instrumented handle structure. Cable holes are available to allow for best measurement practice in terms of safety, quality of sensors data acquisition and convenient for human subjects.

The instrumented handle had been designed with a modular feature. This is achieved by the split design of the rear bracket and the base parts. Any module can be attached here depending on the objectives of the experiment.

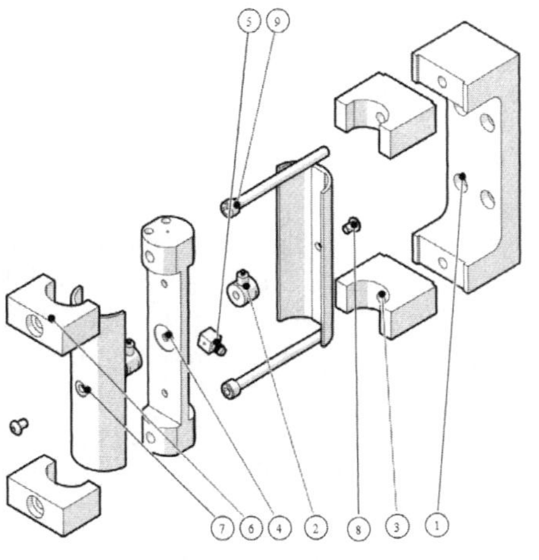

Description	BOM ID	Qty
Accelerometer	5	1
Base	1	1
Force Sensor	2	2
Front bracket	6	2
Grip handle	7	2

Description	BOM ID	Qty
Handle	4	1
Rear bracket	3	2
socket button head cap screw 1/4-28	8	2
socket head cap screw M8	9	2

Figure 3. Exploded view of the instrumented handle assembly.

For the experimental setup of the vibration test system, the instrumented handle will be positioned on top of a plate component that are rigidly attached to the vibration shaker as shown in Figure 4. A vibration measurement and control system usually in the form of a personal computer, will control the vibration input of the shaker to the handle as well as measuring the data output from the accelerometers and force sensors measured from the instrumented handle. A data acquisition system converts the analog output signals into digital signals for further post processing. Human subjects will need to grip the handles at a constant grip and push force as required by the experimenter. Thus, the forces will be displayed on a monitor for the subjects.

IOP Conf. Series: Journal of Physics: Conf. Series **1262** (2019) 012027 doi:10.1088/1742-6596/1262/1/012027

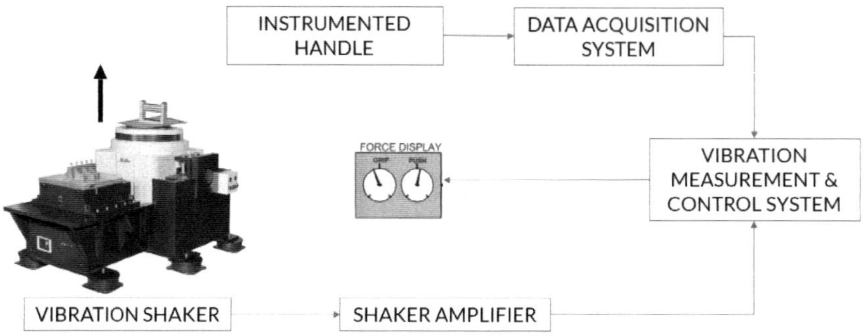

Figure 4. Overall setup of the vibration test system to measure human biodynamic response to hand-transmitted vibration.

3.2. Finite Element Analysis (FEA) for Frequency Analysis of the Handle

The frequency analysis of the handle was conducted to predict the handle resonance characteristics based on the PDS in Table 1. The criteria specified that the design shall not have any modes or natural frequencies in the range between 25 to 1250 Hz. The parameters that can affect the frequency analysis were the boundary conditions and the quality of meshing. The instrumented handle design was fixed at the bottom of the base part as that will be the configuration of the handle in the experiment. The quality of meshing was determined by doing a convergence plot and the final meshing parameters were shown in Table 3. The results of FEA from the CAE software shows the first 10 natural frequencies and their corresponding mode shapes as shown in Figure 5. It is calculated that the natural frequencies are in the range of 1320 Hz till 5088 Hz. The first mode shape is a translational mode shape to the sides of the handle. The second and the third mode shape represent a twisting pattern of the base and grip handles. Most of the subsequent mode shapes predict resonances occurring at the grip handles. This is coherent with the fact that the grip handles have small thickness dimensions and therefore are proved to be the weakest link of the instrumented handle design.

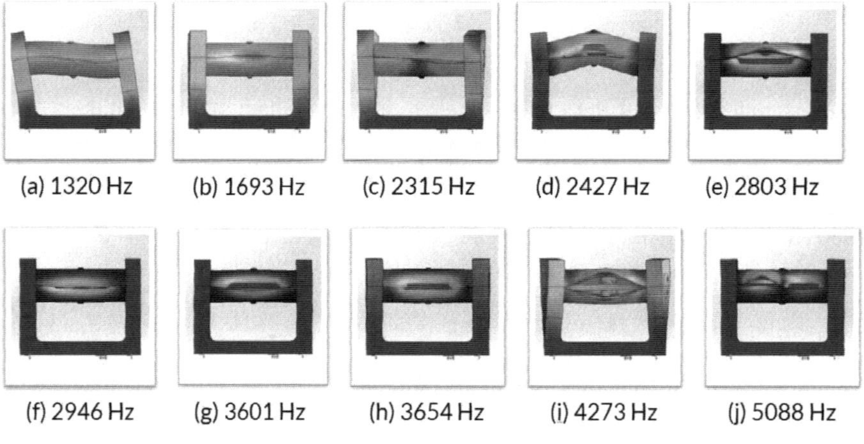

Figure 5. The first 10 modes of the instrumented handle calculated using finite element method.

Figure 6 shows the convergence of the value of the first mode of the design. A convergence of FEA results is important to validate the accuracy of the study. It is shown that there is a convergence trend as the number of elements increases. A difference of 16 Hz was observed between the lowest quality mesh at 1336 Hz and 1320 Hz for the final mesh. The computation time was quick at 1 minutes 46 seconds. The same convergence trend is also evident for the other nine modes. Therefore, the FEA results are accepted with confidence. Table 3 shows the parameters of the frequency analysis used for the simulation.

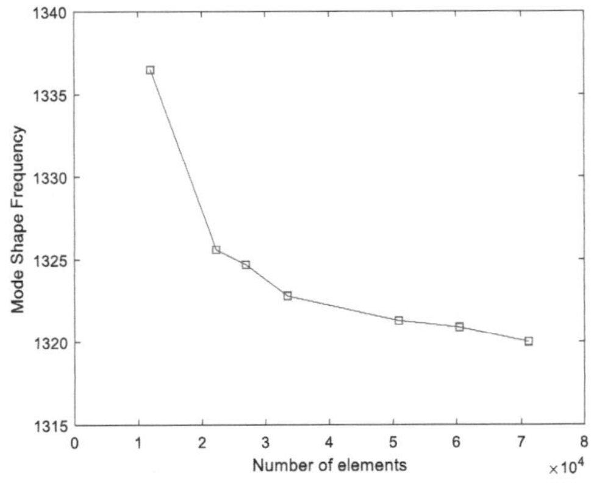

Figure 6. Convergence plot for the first mode of the instrumented handle.

Table 3. Meshing parameters used for frequency analysis.

Parameters	Value
Number of elements	71,105
Number of nodes	111,887
Degree of freedoms	328,212
Total solution time	1:46 minutes

4. Discussion

From the results, it is evident that the design achieved the PDS and functional requirements. The first mode occurs outside the maximum 1250 Hz. The detailed design drawings had been sent for fabrication. Once the handle is fabricated, experimental modal analysis (EMA) will be performed to validate the FEA results. This step is important to ensure there would be no resonances occurring during experimentation in the frequency range of interest [4]. Resonances will negatively affect outcomes and reduce the acceptable frequency range of measured vibration spectrums.

Figure 7 show the general configuration of modules containing a certain thickness of a variety of materials can be slotted between the rear bracket part and the base part of the instrumented handle. Likewise, suspension modules consisting of springs and damper can be designed and slotted in the same configuration. The purpose is to measure the human biodynamic response with the inclusion of these material or suspension variables and investigate the effects. Previous research employed several methods to understand the effects of materials in reducing HTV. Hamouda *et al* used the obvious method of human subjects wearing actual vibration reducing gloves whilst gripping an instrumented handle rig [8], [9]. Other researchers developed a more systematic approach by investigating the effects of numerous factors such as thickness of materials and effective contact area towards the human biodynamic response using an indenter rig [10]–[12]. The data can then be fitted to a biomechanical model to better predict vibration experienced by humans [13], [14].

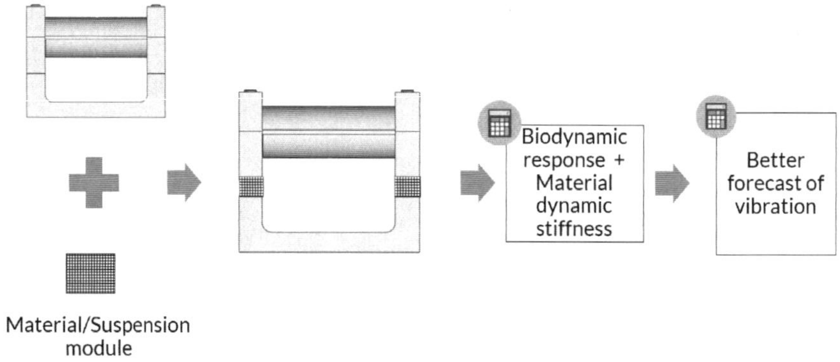

Figure 7. Instrumented handle with material or suspension module configuration for future biodynamic response experimentation.

5. Conclusions

The instrumented handle with a modular expansion feature was designed to be compatible with a vibration test system that can measure human biodynamic responses to hand-transmitted vibration. The design was modelled in a CAE software and a frequency analysis study was conducted using finite element method. The design was shown to comply with the product design specification and final detail design drawings were approved for fabrication.

Acknowledgments

The authors would like to acknowledge the contributions from Universiti Putra Malaysia (UPM) for providing financial support.

References

[1] M. J. Griffin, "Negligent exposures to hand-transmitted vibration.," *Int. Arch. Occup. Environ. Health*, vol. 81, no. 5, pp. 645–659, Apr. 2008.
[2] M. J. Griffin, "Handbook of human vibration," *London Acad.*, 1990.
[3] S. A. Adewusi, S. Rakheja, P. Marcotte, and J. Boutin, "Vibration transmissibility characteristics of the human hand-arm system under different postures, hand forces and excitation levels," *J. Sound Vib.*, vol. 329, no. 14, pp. 2953–2971, 2010.
[4] A. Almagirby, M. J. Carré, and J. A. Rongong, "A new methodology for measuring the vibration transmission from handle to finger whilst gripping," *Int. J. Ind. Ergon.*, vol. 58, pp. 55–61, 2017.
[5] International Organization for Standardization (ISO), *ISO 10819:2013 Mechanical vibration and shock -- Hand-arm vibration -- Measurement and evaluation of the vibration transmissibility of gloves at the palm of the hand*, 2nd ed., no. July. 2013.
[6] K. A. M. Rezali, A. As'Arry, Z. A. Zulkefli, R. Samin, and N. A. A. Jalil, "Predicting the transmissibility of a glove material to the palm using a simple lumped parameter model of the hand and the glove," *J. Teknol.*, vol. 78, no. 6–10, pp. 17–22, 2016.
[7] G. Dieter and L. C. Schmidt, *Engineering Design*, 5th ed. McGraw-Hill Education, 2012.
[8] K. Hamouda, S. Rakheja, P. Marcotte, and K. N. Dewangan, "Fingers vibration transmission performance of vibration reducing gloves," *Int. J. Ind. Ergon.*, vol. 62, pp. 55–69, 2017.
[9] K. Hamouda, S. Rakheja, K. N. Dewangan, and P. Marcotte, "Fingers' vibration transmission and grip strength preservation performance of vibration reducing gloves," *Appl. Ergon.*, vol. 66, pp. 121–138, 2018.
[10] K. A. Md Rezali and M. J. Griffin, "The Transmission of Vibration through Gloves to the Hand and to the Fingers: Effects of Material Dynamic Stiffness," *Appl. Mech. Mater.*, vol. 564, pp. 149–154, 2014.

[11] K. A. Md Rezali and M. J. Griffin, "Transmission of vibration through gloves: effects of contact area," *Ergonomics*, vol. 60, no. 1, pp. 69–81, Jan. 2017.

[12] K. A. Md Rezali and M. J. Griffin, "Transmission of vibration through gloves: effects of material thickness," *Ergonomics*, pp. 1–12, Dec. 2015.

[13] S. Adewusi, S. Rakheja, and P. Marcotte, "Biomechanical models of the human hand-arm to simulate distributed biodynamic responses for different postures," *Int. J. Ind. Ergon.*, vol. 42, no. 2, pp. 249–260, 2012.

[14] A. As'arry, H. Jamaluddin, T. A. Z. Rahman, K. A. Md Rezali, and M. Z. Md Zain, "Suppression of Parkinson's hand-like tremor using fuzzy-PID controller," in *Proceedings - 2016 IEEE International Conference on Automatic Control and Intelligent Systems, I2CACIS 2016*, 2017, vol. 2017–March.

IOP Conf. Series: Journal of Physics: Conf. Series **1262** (2019) 012028 doi:10.1088/1742-6596/1262/1/012028

Predicting the Dynamic Stiffness of a Glove Material using Mechanical Impedance Model

K. A. Md Razali[1*], R. Samin[1], A. As'arry[1], N. A. A. Jalil[1]

[1]Sound and Vibration Research Group, Department of Mechanical and Manufacturing Engineering, Faculty of Engineering, Universiti Putra Malaysia (UPM), 43400 UPM Serdang, Selangor, Malaysia

[*]Corresponding author: khairilanas@upm.edu.my

Abstract. Anti-vibration glove has been developed to be able to attenuate vibration transmitted to the hand. However, the performance depends on lots of factors such as the contact area and contact force. In addition, the materials used in gloves are commonly made of foam and gel, which produce complex response when these factors are changed. This study aimed to predict the dynamic stiffness of a glove based on the measured transmissibility and apparent mass, using mechanical impedance model. This is used so that the dynamic response of the material can be well understood when parameters such as contact area and contact force are changed. The mechanical impedance model has previously been used to predict transmissibility and showed prediction that is comparable to the measured transmissibility. Measured transmissibilities and apparent masses of the hand of five subjects were used in the model to predict the dynamic stiffness of the glove material. The predicted dynamic stiffness was then compared to the dynamic stiffness measured experimentally. The results showed that the predicted dynamic stiffnesses were similar to the measured dynamic stiffness. In conclusion, the mechanical impedance model is able to predict dynamic stiffness of the glove material.

1. Introduction

A glove shall have transmissibility of less than 0.9 at frequencies less than 200 Hz and less than 0.6 at frequencies greater than 200 Hz, before it can be considered as an anti-vibration glove, as required in the International Standard ISO10819:2013 [1]. There are many gloves passed the requirement in the standard but the performance of the glove in attenuating vibration transmitted to the hand is still unclear. A glove that can attenuate vibration, can also amplify vibration (i.e., at resonance). Coupling with the dynamic response of the hand and the dynamic response of the glove material, the performance of the glove could not well be understood. Both parameters (i.e., dynamic response of the hand and the dynamic response of the glove material) depends on factors such as the contact area [2], contact force [3, 8], location of measurement [4], vibration magnitude [5], posture of the hand and the arm [6], as well as the frequency of vibration [2-6]. It has been suggested that the contact area and the contact force are two prominent factors influencing the transmissibility of vibration through glove to the hand [7]. Changing the contact area, or the contact force, will change the dynamic characteristics of the material in attenuating vibration [2-3]. Hence, in order to understand the glove performance in various conditions, it is important to be able to quantify the material dynamic stiffness so that how the material response during vibration exposure in different situation can be well understood.

The dynamic stiffness of the glove material can be measured using an indenter rig [8]. An indenter rig contains a force cell fixed at one direction measuring the output force, a shaker placed opposite to the force cell that generates vibration exposure to the material, and an accelerometer attached to the table of the vibrator, measuring the input acceleration. The dynamic stiffness is the ratio of the cross spectral densities of the input acceleration and output force, to the power spectral density of the input acceleration. An indenter rig was used to measure the effects of contact area, and contact force on glove material dynamic stiffness, and found that the dynamic stiffness increases with increasing the contact force or contact area [2-3, 8].

Glove transmissibility can be predicted based on the method specified in the International Standard ISO13753:2008 [9]. The material proposed to be used in an anti-vibration glove, is placed on a shaker, with mass placed on top of the material, before vibration is exposed to the material and mass system. The input acceleration and output acceleration are used to predict glove transmissibility. The glove transmissibility is used to analyse whether the proposed material is able to be used in an anti-vibration glove. A study conducted previously based on the ISO13753 and ISO10819 has found discrepancies in terms of glove effectiveness in reducing vibration [10].

Mechanical impedance model was first used in 1986 by Fairley and Griffin, to predict the transmissibility of seat to the body [11]. In 2004, the mechanical impedance model was able to predict the effects of material dynamic stiffness on the glove transmissibility [8]. Later, Rezali and Griffin in 2016, 2017, and 2018 showed a good comparison between predicted transmissibility and measured transmissibility of glove material to the hand when material thickness, contact area and contact force were changed [2-4]. In all experiments conducted previously, measured dynamic stiffnesses and measured apparent masses of the hand were used in the model to predict glove transmissibility.

In this study, the dynamic stiffness of the glove material, apparent mass of the hand, and glove transmissibility were previously measured [2] and the glove dynamic stiffness was predicted using the measured apparent mass of the hand, and the measured glove transmissibility. It is expected that the predicted glove dynamic stiffness was similar to the dynamic stiffness measured experimentally.

2. Methods and Procedures

2.1. Measured glove transmissibility, apparent mass of the hand, and the dynamic stiffness of the glove material

The apparent mass and glove transmissibility were acquired from a referred paper previously published [2]. The palm of the hand was placed on a wooden palm adapter containing an accelerometer measuring acceleration at the palm. An impedance head with a contact area of 25 mm was attached on the table of vibrator. The palm adapter was then placed on top of a contact area with diameter of 25 mm. The input acceleration and output force were measured by the impedance head. A 10-s random vibration with frequency ranging from 5 to 500 Hz was generated with a magnitude of 2 m/s^2. In the measurement of glove transmissibility, the glove material was placed between the palm adapter and the contact area.

This study only covered 5 subjects of 10 subjects from the referred paper. The subjects were required to place their hand on the palm adapter with a preload force of 20 N. They had to maintain the force during the vibration exposure.

The dynamic stiffness of the glove material on the other hand was acquired from a referred paper previously published [2]. The dynamic stiffness was measured using an indenter rig. An impedance head was used to measure the output force due to the input acceleration from vibration generated by a shaker placed opposite to the fixed force cell.

2.2. Mechanical Impedance Model

The mechanical impedance model was used to predict glove transmissibility to the palm [4] and given by:

$$T(f) = \frac{\omega^{-2}S(f)}{\omega^{-2}S(f) - M_{hw}(f)} \quad (1)$$

IOP Conf. Series: Journal of Physics: Conf. Series **1262** (2019) 012028 doi:10.1088/1742-6596/1262/1/012028

Where $S(f)$ is the dynamic stiffness of the material, $M_{hw}(f)$ is the apparent mass of the hand, $T(f)$ is the transmissibility of the glove material to the hand.

By rearranging the equation, the dynamic stiffness of the glove material can be predicted when the apparent mass of the hand and the glove transmissibility were acquired.

3. Results and Discussion

3.1. Measured apparent mass of the hand and the measured glove transmissibility

The apparent mass of the palm of the hand used in the prediction of dynamic stiffness is shown in Figure 1 for all 5 subjects from 20 to 200 Hz. The apparent mass decreased with increasing frequency of vibration. There were two resonances observed, one at frequency about 20 Hz and another at frequency about 50 Hz. Previous studies have shown that the resonances in the apparent mass of the hand are expected to be at around 15 Hz and 50 Hz. The apparent mass was much less than the apparent mass measured in some studies [12].

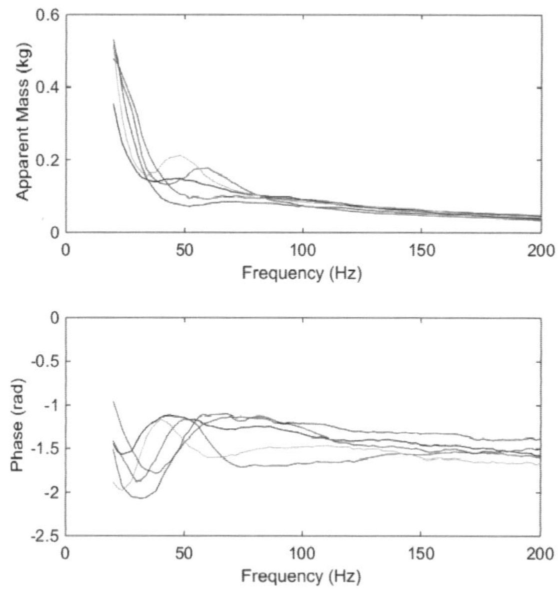

Figure 1. Apparent mass of the palm of the hand measured experimentally used for the prediction of material dynamic stiffness, from the referred published paper. Individual data.

Figure 2 shows the glove transmissibility to the palm of the hand used in the prediction of material dynamic stiffness for all 5 subjects from 20 to 200 Hz. The glove transmissibility increased before decreased at frequencies greater than the resonances. The resonance was found to occur at frequency about 50 Hz depending on subjects.

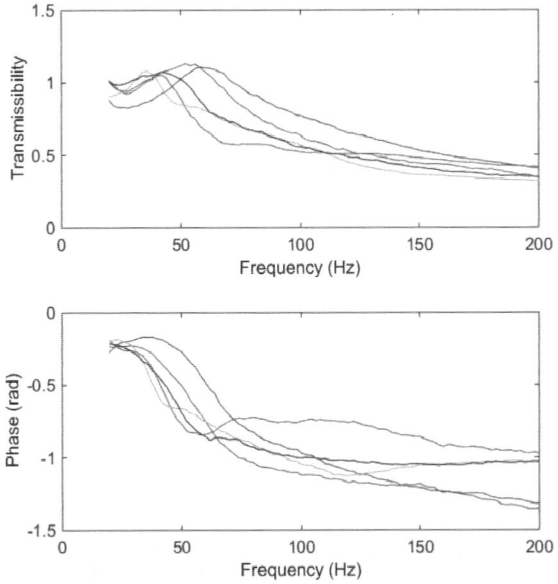

Figure 2. Glove transmissibility to the palm of the hand used for the prediction of material dynamic stiffness, from the referred published paper. Individual data.

3.2. Predicted and measured dynamic stiffness of the glove material

The dynamic stiffnesses of glove material measured experimentally and predicted using the mechanical impedance model are shown in Figure 3. The measured material dynamic stiffness was lower than the predicted material dynamic stiffness at frequencies greater than about 100 Hz. At frequencies less than 100 Hz, the measured material dynamic stiffness was similar to the predicted material dynamic stiffness with an exception to two subjects at frequencies less than about 40 Hz.

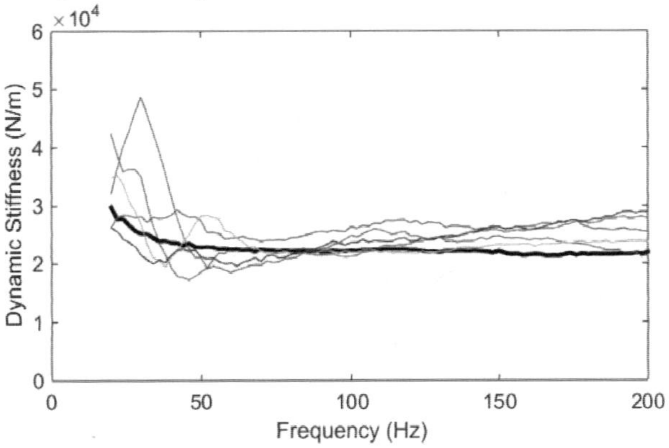

Figure 3. Measured and predicted material dynamic stiffness. Black line is measured dynamic stiffness. Colour lines are predicted dynamic stiffness of all 5 subjects.

3.3. General discussion

The material dynamic stiffness was predicted by using the mechanical impedance model and compares well with the measured dynamic stiffness. The limitation of the indenter rig is that it can only measure dynamic stiffness in flat condition whilst in real situation, anti-vibration gloves are used in various situation, with vibration exposure from different axis. It is expected that with this finding, researchers

IOP Conf. Series: Journal of Physics: Conf. Series **1262** (2019) 012028 doi:10.1088/1742-6596/1262/1/012028

can predict dynamic stiffness of glove material in a condition that is difficult to be represented by using an indenter rig.

4. Conclusions

The predicted dynamic stiffness of the glove material was similar to the dynamic stiffness experimentally measured at frequencies less than 100 Hz but was overestimated at frequencies greater than about 100 Hz. This shows that the mechanical impedance model is capable of predicting the material dynamic stiffness.

Acknowledgements
The authors would like to be obliged to Universiti Putra Malaysia for providing laboratory facilities and financial assistance under project no. GP-IPM 9499300.

References

[1] International Standard ISO10819 (2013) Mechanical vibration and shock - Hand-arm vibration - Method for the measurement and evaluation of the vibration transmissibility of gloves at the palm of the hand
[2] Rezali KAM and Griffin MJ (2017) Transmission of vibration through gloves: effects of contact area. Ergonomics, 60(1), 69-81
[3] Rezali KAM and Griffin MJ (2018) Transmission of vibration through gloves: effects of contact force. Ergonomics, 61(9), 1246-1258
[4] Rezali KAM and Griffin MJ (2016) Transmission of vibration through gloves: effects of material thickness. Ergonomics, 59 (8), 1026-1037
[5] Laszlo HE and Griffin MJ (2011). The transmission of vibration through gloves: effects of push force, vibration magnitude and inter-subject variability. Ergonomics 54(5): 488-496.
[6] Adewusi SA, Rakheja S, Marcotte P, (2012). Biomechanical models of the human hand-arm to simulate distributed biodynamic responses for different postures. International Journal of Industrial Ergonomics 42 249-260.
[7] Rezali KAM (2015). Measuring and predicting the transmission of vibration through gloves to the hand. Thesis for the degree of Doctor of Philosophy. University of Southampton.
[8] O'Boyle M and Griffin MJ (2004) Predicting the effects of push force on the transmission of vibration through glove materials to the palm of the hand. 10th International Conference on Hand-Arm Vibration, 07-11 June 2004 Las Vegas, Nevada, USA.
[9] International Standard ISO13753 (2008) Mechanical vibration and shock - Hand-arm vibration - Method for measuring the vibration transmissibility of resilient materials when loaded by the hand-arm system.
[10] Koton J, Kowalski P, Szopa J (1998) An attempt to contruct anti-vibration gloves on the basis of information of the vibration transmissibility of materials. 8th International Conference on Hand-Arm Vibration, 9-12 June 1998, Umea, Sweden.
[11] Fairley, T.E. and Griffin, M.J. (1886) A test method for the prediction of seat transmissibility. Society of Automotive Engineers International Congress and Exposition, 24th – 28th February 1986, Detroit, USA.
[12] Dong RG, Rakheja S, Mcdowell TW, Welcome DE, Wu JZ, Warren C, Barkley J, Washington B and Schopper AW (2005) A method for assessing the effectiveness of anti-vibration gloves using biodynamic responses of the hand–arm system. Journal of Sound and Vibration, 282, 1101-1118

Stability analysis of regenerative vibration in turning operation using I-kaz3D™ signal processing approach

R Samin[1, 2*], M Z Nuawi[1], S M Haris[1] and J A Ghani[1]
[1] Department of Mechanical and Material Engineering, Faculty of Engineering and Build Environment, Universiti Kebangsaan Malaysia, 43600 Bangi, Selangor, Malaysia
[2] Department of Mechanical and Manufacturing Engineering, Faculty of Engineering, Universiti Putra Malaysia, 43400 Serdang, Selangor, Malaysia.

[*]Corresponding author: zali@upm.edu.my

Abstract. This paper presents a stability analysis to predict regenerative vibration during turning process. Although the stability of machining process has been widely studied in metal cutting it still presents a challenge as surface roughness, tool wear and productivity have to be considered. If the tool is vibrating as it removes material, these vibrations are imprinted on the workpiece surface as a wavy profile. However, the relative phasing between the surface waviness from one pass to the next determines the level of force variation and whether the operation is stable or unstable. Then we investigate the unstable and stable phenomena during the machining process by discussing the number of time-domain simulation to determine the force and surface roughness values these stability thresholds. A signal processing was carried out using I-Kaz3D technique to identify the effect between the unstable and stable phenomena. The results show that the relationship between I-kaz3D coefficients and surface roughness values can be considered very highly correlated.

1. Introduction

Machining process is involved dynamic instability that influence from interaction between the structural dynamics of the system and the metal cutting process. Effect of machine tool and workpiece system provides the regenerative vibrations due to self-excited vibrations which occur during cutting processes. Although regenerative vibration has been widely studied in machining processes, it still presents a challenge as cutting parameters such as cutting speed, feed rate and depth of cut have to be considered for optimum cutting conditions. Siddhpura and Paurobally [1] and Shrivastava et al. [2] identified regenerative vibration as regenerative chatter which it contributes the most important causes of inconsistent in the cutting process stability and the most destructive among all other vibrations have to be predicted and detected before it occurs. It also certainly guarantees the machine productivity causes reduces tool life and poor surface finish. Researchers have investigated and correlated between regenerative vibration and cutting parameters in turning operations. Okonkwo et al. [3] studied that the comparative analysis from regenerative vibration of turning AISI 4340 alloy steel which feed rate was very significant for higher regenerative vibration (chatter) frequency 89.84% meanwhile, cutting speed 5.79% and depth of cut 3.92%. Taylor et al. [4] analysed at higher feed rates and low surface speeds, the cutting processes exhibited high amplitude vibrations due to the chip segmentation phenomenon. Patel and Gandhi [5] concluded that among the cutting parameters, feed rate is most significant parameter affecting the surface roughness followed by depth of cut and cutting speed. Sivaiah and Chakradhar [6] investigated the interaction effect of cutting variable on surface roughness. Lowest surface roughness can be obtained at the highest level of cutting velocity and lowest level of feed rate respectively. Meanwhile, lowest level of depth of cut and highest level of cutting velocity could produce low surface roughness respectively at the given condition. It is well known that the chatter stability depends on cutting conditions and tool geometry such as depth of cut and the spindle speed, it also depends on tool path/posture to dynamically cutting direction [7]. Several researchers have studied the correlation between chatter and surface roughness. Quintana et al. [8] observed that regenerative vibrations can be effected from the loud noise and the poor surface integrity due to chatter marks. Amin et al. [9] investigated that regenerative vibration has direct effect on surface roughness measured data (Ra) which it is much higher with chatter compared to without chatter condition (at stable cutting process).

The objective of this paper was to investigate the effect of surface roughness due to the regenerative vibration by using the Integrated Kurtosis-Based Algorithm for Z-Notch Filter (I-kaz™). In stability analysis, surface roughness has been used for determining the chatter condition such as chatter marks, chatter prediction and etc. as mentioned by previous researchers. Force and vibration signals is choose by the researchers for the reason that it provides thorough insight into the dynamics of the cutting processes and it is very useful in the monitoring of the condition of the machining processes. The I-Kaz™ statistical analysis technique as a signal processing was developed by Nuawi et al. in 2008 [10]. This technique was applied in cutting processes such as tool wear detection, prediction and monitoring using force, vibration and acoustic signals in a series of papers [11-14] and was not been used for stability analysis in surface roughness measurement. In practice the I-kaz coefficients would have high values with increasing in flank wear values and amplitudes of force and vibration signals in cutting processes. Therefore, the research indicates the statistical significance of the I-kaz3D coefficient effects is able to increase or decrease the stability of the regenerative vibration.

2. Methodology

2.1 Materials and Equipment

In this paper, a workpiece material, medium carbon steel S45C of 59 HRB hardness was used as a round bar with the dimension of the work specimen was 75 mm diameter and 250 mm length. The nominal material composition of the workpiece of C = 0.42% - 0.48%, Si = 0.15 % - 0.35%, Mg = 0.6% - 0.9%, P = 0.03% and S = 0.035%. Figure 1 shows the experiments were performed on the CNC turning with the workpiece was mounted with the tailstock of the CNC Mazak 200MY turning machine and Mazatrol 640T PC-Fusion-CNC control software as shown in Figures 1(a) and 2(b). The cutting parameters under investigation were cutting speed, feed rate and depth of cut with the ranges selected as per the cutting tool manufacturer's recommendation. The turning processes were conducted under dry cutting condition using CVD coated carbide insert, CNMG120404N-GU AC2000 with carbide tool holder, ECLNR-2020K12 designated by ISO was used to hold the insert.

The vibration measurement during the machining process was measured using LMS Test.Lab and data acquisition system, LMS Scadas Mobile with 4 channels using 2048 Hz bandwidth, 163384 spectral lines and 0.125 Hz resolution. A tri-axial accelerometer (model Dytran 3263M8 sensitivities X axis = 100.55 mV/g, Y axis = 100.07 mV/g and Z axis = 99.62 mV/g) was mounted on the tool holder. A single-axial accelerometer (model Dytran 3145A sensitivity = 101.3 mV/g) was used as a reference in cutting or tangential direction to the signals. Then, the surface roughness measurement of the machined workpiece was measured by using a portable PS2 MarSurf Psi gage.

2.2 Experimental Procedures

In this work, the experimental procedure was done as shown in Figure 1. Firstly, the operator enters the numerically controlled (NC) program to CNC machine as shown in Figure 1(a). Secondly, install the insert and tool holder on machine tool drive, then setup the tri-axial accelerometer on tool holder and single-axial accelerometer on the CNC machine as shown in Figure 1(b), Thirdly, the experimental design involves cutting parameters ranges of three factors such as cutting speed (50–250 m/min), feed rate was kept constant (0.1 mm/rev) and depth of cut (0.5–1.0 mm). An effort has been made in this work to determine whether the cutting parameters can influence the vibration signal responses in axial (*X*), tangential (*Y*) and radial (*Z*) directions and surface roughness arithmetic mean (R_a) values obtained from the experiments. Lastly, the vibration signals for thrust, cutting and feed forces were measured during the cutting processes as shown in Figure 1(c) and surface roughness was measured after the cutting processes as shown in Figure 1(d). The vibration signals were obtained in the time responses data from LMS DAQ, then the data was exported in text format. The data of R_a value was obtained by averaging surface roughness values at three location points around the circumference of the workpiece. All data were processed and analyzed with signal statistical analysis using MATLAB

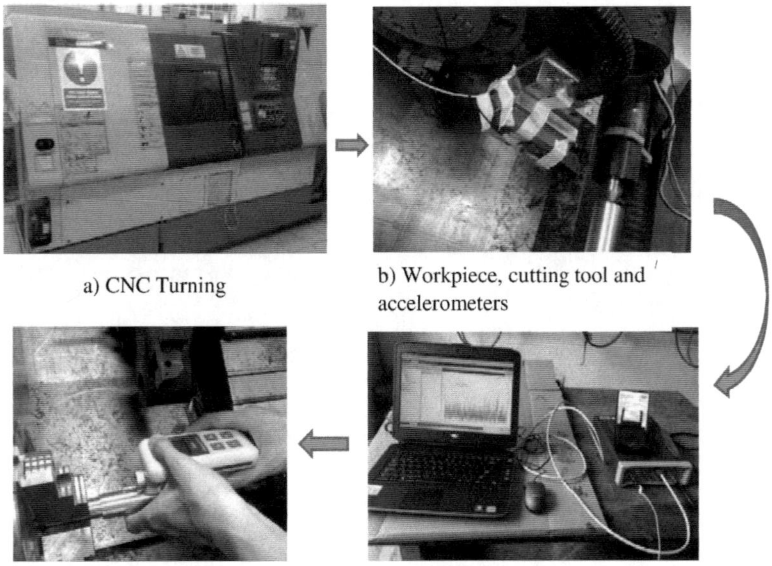

a) CNC Turning

b) Workpiece, cutting tool and accelerometers

d) Surface roughness measurement

c) LMS system for vibration measurement

Figure 1. Experimental setup.

Table 1. Experimental procedure

Test run	Cutting speed, V_c *(m/min)*	Feed rate, *f* (mm /rev)	Depth of cut, *d* (mm)
1	50	0.1	0.5
2	100	0.1	1.0
3	150	0.1	0.5
4	200	0.1	0.5
5	250	0.1	0.5

2.3 Signal Processing using I-kaz3D Statistical Analysis

A signal processing using I-kaz3D statistical analysis was developed based on I-kaz^{TM} (Integrated kurtosis-based algorithm for Z-filter) technique. Some researchers investigated whole body vibration for driving and transportation were used I-kaz3D. This technique based on the signal statistical analysis approach of mean value, the standard deviation value, the root-mean-square (*RMS*) value, the

Kurtosis (*K*) and the crest factor (*CF*). The standard deviation values measures the spread of the data about the mean value. The *RMS* value, which is the 2nd statistical moment is used to quantify the overall energy content of the signals. The kurtosis, *K* which is the fourth order statistical moment of signals, which is highly sensitive to the spikiness of the data scatteration about its mean. The crest factor which is commonly encountered in engineering applications is defined as ratio between the maximum value in the time domain and *RMS* value. On the other hand, the I-kaz3D technique provides a 3D graphical representation using the scattering of variance data distribution X-, Y- and Z-axes respectively. A tri-axial accelerometer for vibration signals measurement was also provided data of channels X-axis, Y-axis and Z-axis for radial force, cutting force and feed force respectively. The radial force that tends to push the tool away from the workpiece being machined, the cutting forces that forces acting on a cutting tool in turning and the thrust or feed force that in the direction of feed. For a discrete data set the *RMS*, Kurtosis and crest factor values are defined as:

$$RMS = \sqrt{\frac{1}{N}\sum_{j=1}^{N} x_j^2} \tag{1}$$

$$K = \frac{1}{N(RMS)^4}\sum_i^N (x_i - u)^4 \tag{2}$$

$$CF = \left|\frac{x_{,i\max}}{RMS}\right| \tag{3}$$

The variance, σ^2 obtains the average magnitude deviation, $(x_i - u)$ of the instantaneous points, x_i with respect to the mean value, u as follows:

$$\sigma_X^2 = \frac{\sum_i^N (x_{X,i} - u_X)^2}{N},\ \sigma_Y^2 = \frac{\sum_i^N (x_{Y,i} - u_Y)^2}{N}\ and\ \sigma_Z^2 = \frac{\sum_i^N (x_{Z,i} - u_Z)^2}{N} \tag{4}$$

where σ_X^2 , σ_Y^2 and σ_Z^2 are the variances, $x_{X,i}$, $x_{Y,i}$and $x_{Z,i}$ are data for *i*–sample of time, and μ_X, μ_Y and μ_Z are the means of each frequency band in the *X*-, *Y*- and *Z-axes* respectively, and *N* is the number of data points. From equation (1), the sum of the root-mean square variance can be expressed as the sum of the intensities, each of which influences the *RMS* variance is known as I-kaz3D coefficient, Z_{3D}^{∞} can be written as:

$$Z_{3D}^{\infty} = \sqrt{\sigma_X^4 + \sigma_Y^4 + \sigma_Z^4} \tag{5}$$

$$Z_{3D}^{\infty} = \sqrt{\frac{\sum_i^N (x_{X,i} - u_X)^4}{N^2} + \frac{\sum_i^N (x_{Y,i} - u_Y)^4}{N^2} + \frac{\sum_i^N (x_{Z,i} - u_Z)^4}{N^2}} \tag{6}$$

Equation (6) can be simplified as:

$$Z_{3D}^{\infty} = \frac{\sqrt{K_X s_X^4 + K_Y s_Y^4 + K_Z s_Z^4}}{N} \tag{7}$$

Equations (1), (2), (3) and (4) are considered in terms of the sum of *RMS* of the Kurtosis, *K* and the standard deviation, *s* for each number of the measured signal data, where, K_X, K_Y and K_Z are the kurtosis of the vibration signals and s_X^2, s_Y^2 and s_Z^2 are the standard deviation in *X*-, *Y*- and *Z*-axes respectively. Equation (5) is used for displaying the scattering of statistical signal analysis using I-kaz[TM] technique in three-dimensional graphical representation. Meanwhile, equation (6) is used to measure I-kaz3D coefficient, Z_{3D}^{∞} of the vibration signals. The data were processed using MATLAB.

3. Results and Discussion

As shown in Figures 2 to 6, the time domain from the five tests were analysed to access the fast fourier transform (FFT), power spectral density (PSD), kurtosis, crest factor and I-kaz3D. All force signals from the time responses having different amplitude consist of the thrust, cutting and feed forces. In time domain, the cutting force for run 2 as shown in Figure 3 was highest amplitude, 2.45 g compared than other experiment runs. The purpose of the FFT was used is to find the frequency components of a signal buried in a noisy time domain signal. The highest amplitude 0.2 g for FFT was presented in Run 1 as shown in Figure 2. Meanwhile, the purpose of the PSD is to show the strength of the variation energy as a function of frequency from the experiment.

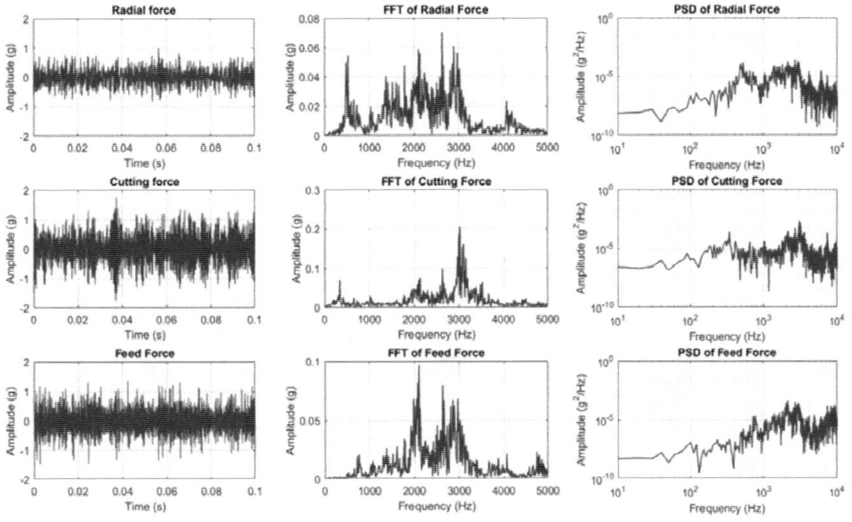

Figure 2. Force signals, FFT and PSD for radial, cutting and feed forces for Run 1

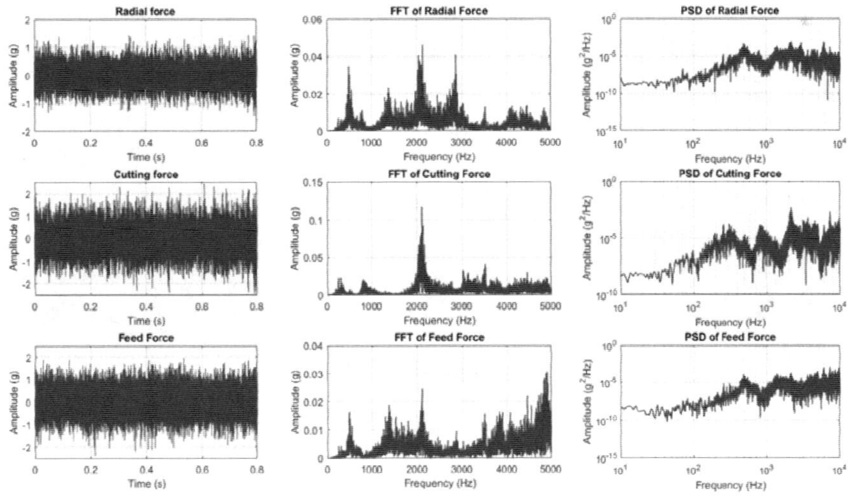

Figure 3. Force signals, FFT and PSD for radial, cutting and feed forces for Run 2

IOP Conf. Series: Journal of Physics: Conf. Series **1262** (2019) 012029 doi:10.1088/1742-6596/1262/1/012029

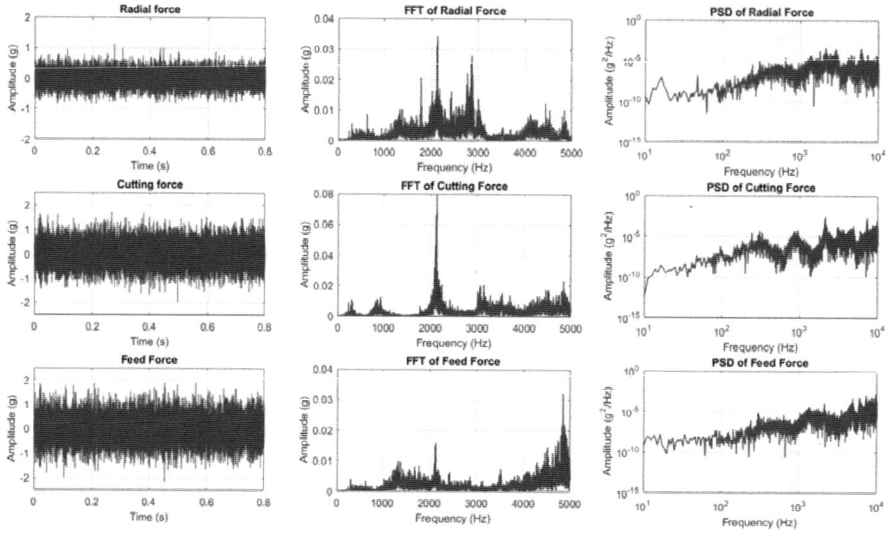

Figure 4. Force signals, FFT and PSD for radial, cutting and feed forces for Run 3

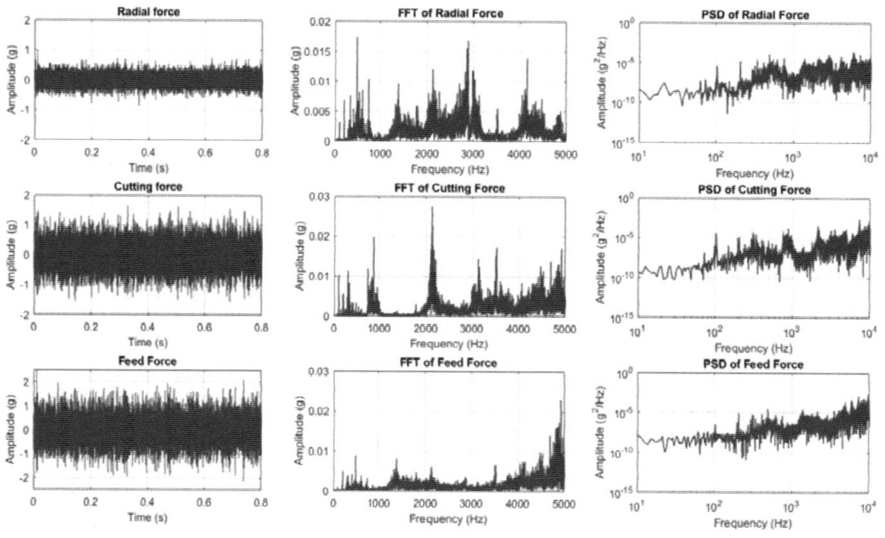

Figure 5. Force signals, FFT and PSD for radial, cutting and feed forces for Run 4

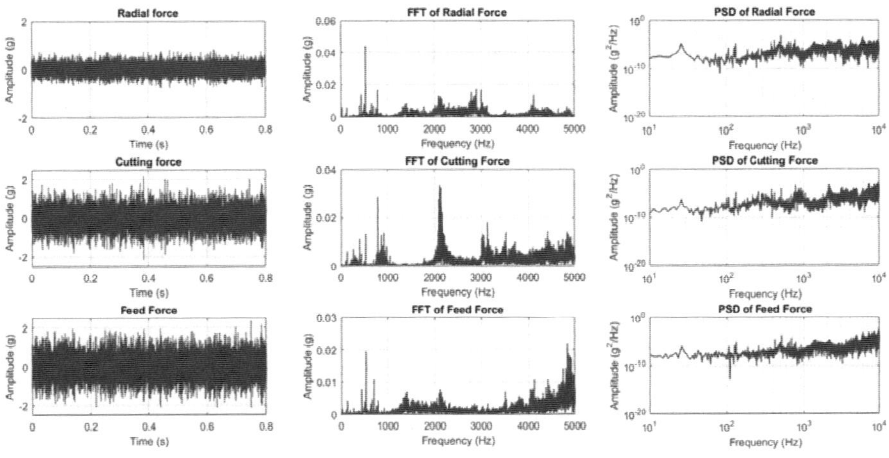

Figure 6. Force signals, FFT and PSD for radial, cutting and feed forces for Run 5

In PSD result, the vibration signals for run1 was produced the strong peaks as shown in Figure 2 compare to other experiment runs. The results for statistical analysis are shown in Table 2. Kurtosis and crest factor were measured for X-, Y- and Z-directions, then the average of the Kurtosis and crest factor were calculated. For kurtosis, the minimum and maximum values were 2.9173 and 3.018, and for crest factor the minimum and maximum values were 3.473 and 4.508. The different between kurtosis and crest factor are too small, therefore the relationship to vibration signals very difficult to predict and not suitable to be used for stability analysis based on the surface roughness measurement. I-kaz3D technique shows a significant difference between I-kaz3D coefficients and surface roughness, R_a as shown in Table 2. The minimum and maximum values of the I-kaz3D coefficients were 3.6748x10^{-5} and 27.620x10^{-5}, and surface roughness were 0.669 μm and 6.203 μm. Higher I-kaz3D coefficients will have higher surface roughness values. Figure 7 shows the three-dimensional graphical representation of the I-kaz3D technique for X-, Y- and Z-directions. It is noted that the I-kaz3D coefficient relation decreases with decrease of the surface roughness measurement. This can be summarized that the significance of the I-kaz3D effect which may increase or decrease regenerative vibration stability.

Table 2. Results for signal statistical analysis, I-kaz3D and surface roughness

Test run	Kurtosis				Crest factor				Z_{3D}^{∞} (10^{-5})	R_a (μm)
	K_x	K_y	K_z	K_{ave}	CF_x	CF_x	CF_x	CF_{ave}		
1	2.814	2.864	3.241	2.973	3.421	3.429	3.568	3.473	27.620	6.203
2	2.993	3.050	2.956	3.000	3.722	3.845	4.111	3.893	5.7811	1.832
3	3.000	2.996	2.964	2.987	4.589	4.236	4.095	4.307	5.6022	1.215
4	3.054	2.948	3.051	3.018	5.071	3.846	4.606	4.508	3.6748	0.669
5	2.977	3.009	2.986	2.991	3.789	4.273	3.951	4.004	4.6874	0.771

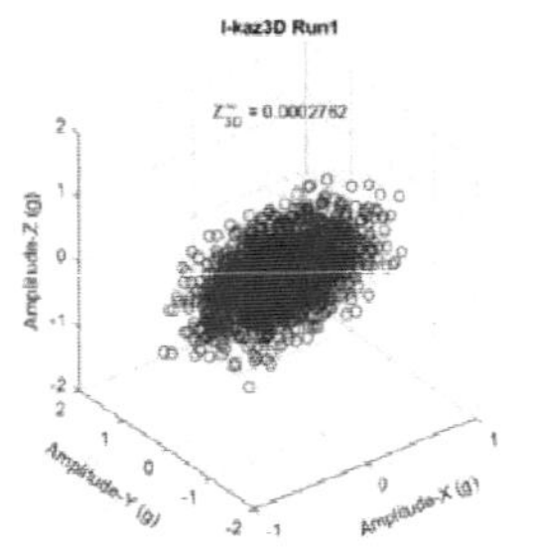

(a) I-kaz3D Run1

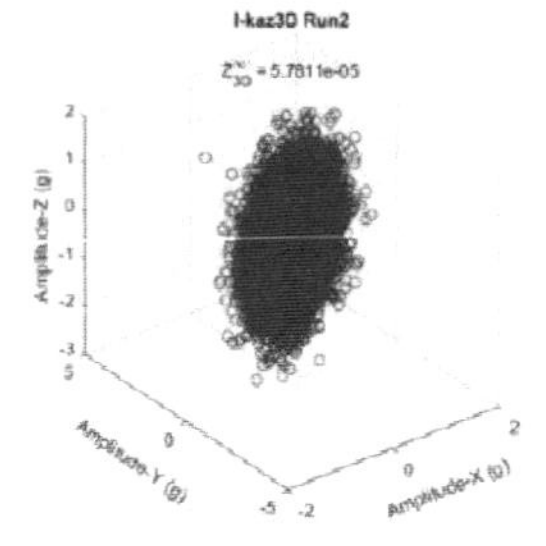

(b) I-kaz3D Run2

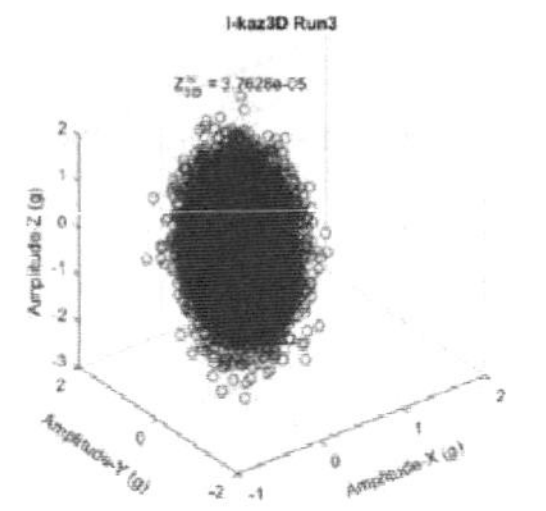

(c) I-kaz3D Run3

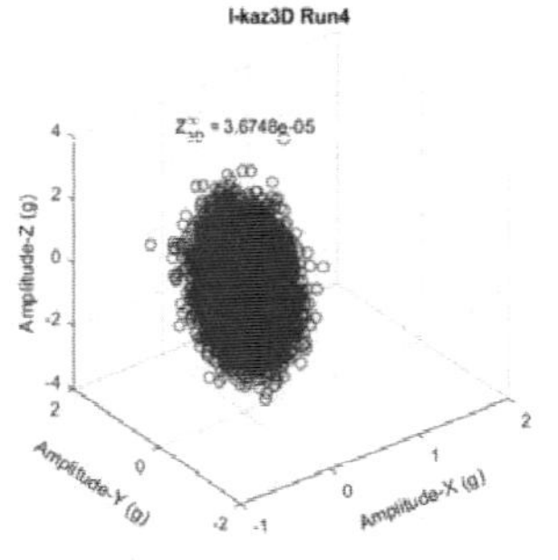

(d) I-kaz3D Run4

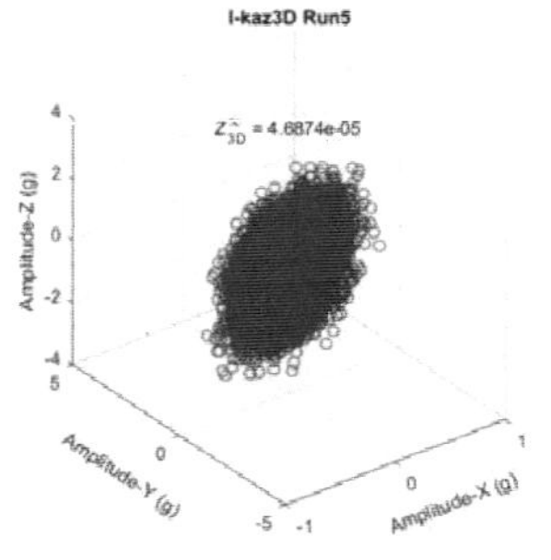

(f) I-kaz3D Run5

Figure 7. Three-dimensional graphical representations for I-kaz3D analysis

4. Conclusions

This paper presented regenerative vibration signals from turning process with cutting, feed and thrust forces. We find that the I-kaz3D technique is suitable for stability analysis in the turning process with surface roughness values monitoring. It is concluded that the relationship between I-kaz3D coefficients and surface roughness values is considered to be good correlation.

Acknowledgements

The authors gratefully to acknowledge the support of the Department of Mechanical and Manufacturing Engineering, Faculty of Engineering, Universiti Putra Malaysia and the support provided by Universiti Putra Malaysia grant GP/2017/9564400.

References

[1] M. Siddhpura and R. Paurobally, "A review of chatter vibration research in turning," *Int. J. Mach. Tools Manuf.*, vol. 61, pp. 27–47, 2012.

[2] Y. Shrivastava, B. Singh, and A. Sharma, "Identification of Chatter in Turning Operation using WD and EMD," *Mater. Today Proc.*, vol. 5, no. 11, pp. 23917–23926, 2018.

[3] U. C. Okonkwo, N. O. N, and I. P. Okokpujie, "Comparative Analysis of Chatter Vibration Frequency in CNC Turning of AISI 4340 Alloy Steel with Different Boundary Conditions," vol. 1, no. 1, pp. 13–30, 2018.

[4] C. M. Taylor, S. Turner, and N. D. Sims, "Chatter, process damping, and chip segmentation in turning: A signal processing approach," *J. Sound Vib.*, vol. 329, no. 23, pp. 4922–4935, 2010.

[5] N. Van Dijk, E. Doppenberg, H. Oosterling, and H. Nijmeijer, "Chatter control in the high-speed milling process using," vol. 20, no. 4, pp. 6121–6126, 2010.

[6] P. Sivaiah and D. Chakradhar, "Analysis and Modeling of Cryogenic Turning Operation Using Response Surface Methodology," *Silicon*, vol. 10, no. 6, pp. 2751–2768, 2018.

[7] E. Shamoto, S. Fujimaki, B. Sencer, N. Suzuki, T. Kato, and R. Hino, "A novel tool

path/posture optimization concept to avoid chatter vibration in machining - Proposed concept and its verification in turning," *CIRP Ann. - Manuf. Technol.*, vol. 61, no. 1, pp. 331–334, 2012.

[8] C. A. Rodrı, G. Quintana, and J. Ciurana, "Sound mapping for identification of stability lobe diagrams in milling processes," *Int. J. Mach. Tools Manuf.*, vol. 49, pp. 203–211, 2009.

[9] A K. M. N. Amin, A. U. Patwari, M. S. Sharulhazrin, and I. Hafizuddin, "Investigation of Effect of Chatter Amplitude on Surface Roughness during End Milling of Medium Carbon Steel," *Int. Conf. Ind. Eng. Oper. Manag.*, pp. 127–131, 2010.

[10] F. L. and C. K. E. N. M.Z. Nuawi, M.J.M. Nor, N. Jamaludin, S. Abdullah, "Development of integrated kurtosis based algorithm for Z-filter technique," *J. Appl. Sci.*, vol. 8, no. 8, pp. 1541–1547, 2008.

[11] J. A Ghani, M. Rizal, M. Z. Nuawi, C. Hassan, C. Haron, and R. Ramli, "Statistical Analysis for Detection Cutting Tool Wear Based on Regression Model," *Int. MultiConference Eng. Comput. Sci. Vol III*, vol. III, 2010.

[12] M. Rizal, J. A. Ghani, M. Z. Nuawi, and C. H. C. Haron, "The application of I-kaz-based method for tool wear monitoring using cutting force signal," in *Procedia Engineering*, 2013, vol. 68.

[13] M. A. F. Ahmad, M. Z. Nuawi, S. Abdullah, Z. Wahid, Z. Karim, and M. Dirhamsyah, "Development of tool wear machining monitoring using novel statistical analysis method, I-kaz," *Procedia Eng.*, vol. 101, no. C, pp. 355–362, 2015.

[14] M. Rizal, J. A. Ghani, M. Z. Nuawi, M. A. S. r Mohd Tahir, and C. H. Che Haron, "A Comparative Study of I-kaz Based Signal Analysis Techniques: Application to Detect Tool Wear during Turning Process," *J. Teknol.*, vol. 66, no. 3, pp. 99–105, 2014.

Simulation of Lamb wave interactions with defects in a thin plate

N Ismail[1*], Z M Hafizi[1], C K E Nizwan[1], S Ali[1].
[1]Advances Structural Integrity and Vibration Research (ASIVR), Faculty of Mechanical and Manufacturing Engineering, Universiti Malaysia Pahang (UMP), Pekan, 26600, Pahang, MALAYSIA.

[*]Corresponding author: nurazima@ump.edu.my

Abstract. The understanding of Lamb wave propagation and its interaction with defects are crucial part of SHM technology. Being of high relevance due to the costs of experimental equipment, the dynamic finite element analysis is performed in this study. FEA analysis was implemented to simulate propagation of Lamb waves in healthy and defective aluminium plate. The wavefield visualisation shows a high degree of difference between these two plates, which shows the interaction with defects. This study is useful for a good understanding of the Lamb wave interaction with defect before applying this method for a real application.

1. Introduction

Structural health monitoring (SHM) has gained more attention among researcher due to its ability to allow a continuous supervision of the structural integrity of any operational systems. Among the various developed techniques, Lamb wave has shown capability to be utilised as one of the SHM approaches. The principle advantage of this method is performing the inspection over long distances from a single probe position [1]. They are very sensitive to any changes of the structural properties even at a tiny scale or underneath structural surfaces and able to detect any types and sizes of defects [2, 3]. Moreover, they feature prominent merits like low attenuation, low energy consumption and convenience in actuation or acquisition. These characteristics make Lamb wave as one of the promising diagnostic tool. However, there are several challenges when utilising Lamb wave method in the SHM system. Multimodal and dispersive nature makes the Lamb wave interpretation difficult [4]. Multimodal property refers to the existence of at least two modes that coexist at any given frequency, which leads to multiple wave packets in the acquired signal [5]. On top of that, other wave characteristics such as wave dispersion, which is associated with the velocity of each mode which, varies with respect to its frequency have added complexity to the analysis of the signal [6]. This complexity may lead to the wrong diagnosis and false alarm in the inspection. Thus, the understanding of Lamb waves propagation plays an important role in the successful application of Lamb wave for damage detection. This paper presents the visualisation simulation of the Lamb wave responses for an intact and damaged isotropic plate. This visualisation will help the beginner in the Lamb wave based SHM filed to fully understand the propagation of the Lamb wave which contain various wave modes. The modes were clearly appeared and their interaction with the defects was successfully simulated.

Visualisation of the wave propagation generated from the simulation is used to study in details about the effect of defects to the Lamb wave propagation. The changes of the wave contour when interacting with the defects may be used as an indicator for the existence of damage. The extracted features can be used as the defects signature for the SHM diagnostic system.

2. Theory of Lamb wave

Lamb wave is ultrasonic guided wave, which is induced by free surfaces of plate, shell and tube, and propagates along the directions of length and width. They can be classified as symmetric and anti-symmetric modes, which exist simultaneously and propagate independently of each other [7]. The symmetric modes are symmetric with respect to the mid-plane of the plate and are noted S_i, where i is the order or number of the mode. The anti-symmetric modes are anti-symmetric with respect to the mid-plane of the plate and are noted A_i. Fundamental symmetrical, S_0 and anti-symmetrical, A_0 mode is the lowest order of the mode and followed by and A_1, A_2, A_3, ... A_∞, respectively. Symmetric mode having symmetric deformation about the mid-plane and the particles inside the plate primarily have in-plane motions. The displacements in x direction are larger than the displacements in y direction. Particles at the top and at the bottom experiencing the same direction of displacements in the x component and opposite direction in the y component as illustrated in Figure 1(a).

For the A mode, the deformation of the plate is anti-symmetric and the particles inside the plate have out-of-plane motions. The displacements in the y direction are larger than the x direction. The displacements for the y direction are in the same direction and the x component in the opposite directions for particles at the top and bottom [8] as shown in Figure 1 (b).

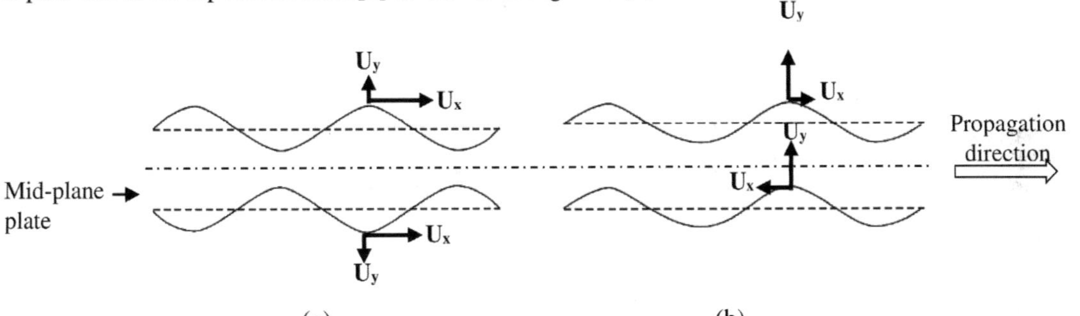

Figure 1. Propagation of the (a) symmetric mode and (b) anti-symmetric mode

The number of existing modes of a plate structure depends on the product of excitation frequency and thickness (*fd*). An excitation frequency must be decided properly before performing any analysis regarding Lamb wave propagation in order to minimize the number of generated modes. For signals simplification, it is a common practice to apply low excitation frequencies, so that only fundamental modes (S_0 and A_0) exist. The reference for the appropriate excitation frequency can be referred to theoretical dispersion curve and it was generated using equation below [9]

$$\frac{\tan qh}{\tan ph} + \frac{4k^2pq}{\left(q^2 - k^2\right)^2} = 0 \quad (1)$$

$$\frac{\tan qh}{\tan ph} + \frac{\left(q^2 - k^2\right)^2}{4k^2pq} = 0 \quad (2)$$

Where the parameters p and q are defined as follows:

$p^2 = \frac{\omega^2}{c_L^2} - k^2$, $q^2 = \frac{\omega^2}{c_T^2} - k^2$, $h = \frac{d}{2}$ and $k = \frac{\omega}{c_p}$, where ω is circular wave frequency, k is wavenumber, c_p is phase velocity, c_L and c_T are the propagation velocities of bulk waves. From equation 1 and 2, the phase velocity, c_p dispersion curves can be plotted. The other useful plot is group velocity dispersion curves which can be derived from the phase velocity curves using equation (3). The c_p is propagating velocity of the wave with a single frequency and c_g is the propagation velocity of the wave packet with adjacent multi frequencies.

$$c_g = \frac{d(kc_p)}{dk} = c_p + k\frac{dc_p}{dk} \tag{3}$$

Referring to Equation (1) and (2), different material with different plate thickness will produce different dispersion curves. Figure 2 depicts the dispersion curve for aluminium plate with thickness of 2 mm. This theoretical curve is generated using PACshare Dispersion Curves software. It can be seen that the number of modes increases with the excitation frequency. Thus, infinite number of symmetric and anti-symmetric modes can exist. Referring to Figure 2, excitation frequency more than 800 kHz can generate more than two modes leading to the complexity of the recorded signals. For this study, the excitation frequency of 200 kHz is chosen for the Lamb wave actuation, which is actuated only two modes that is S_0 and A_0. The characteristics of these two modes has been identified for a better understanding of their interactions with defects. The existing feature signatures influenced by the defects were extracted for further studies.

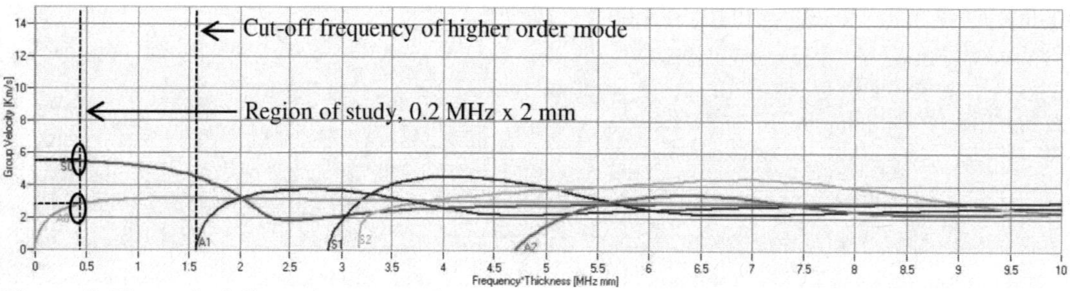

Figure 2. Theoretical dispersion curve of group velocity against frequency for aluminium plate with thickness of 2 mm

3. Finite element modelling

A 90 cm x 90 cm x 0.2 cm aluminium plate with stress-free boundary condition was considered in this study. The properties of the aluminium are: density = 2580 kg/m^3, Young's Modulus = 72.7 GPa and Poisson's ratio = 0.33. A commercial ABAQUS finite element software package was used to perform the dynamic simulation. A three dimensional FE method using eight-node brick solid elements was used to model the geometry and a square element type was applied for meshing generation. Figure 3 shows the healthy and damaged plate model used in this study. The damage slot was introduced with the size of 0.2 cm x 2 cm as presented in Figure 3 (c).

(a) (b)

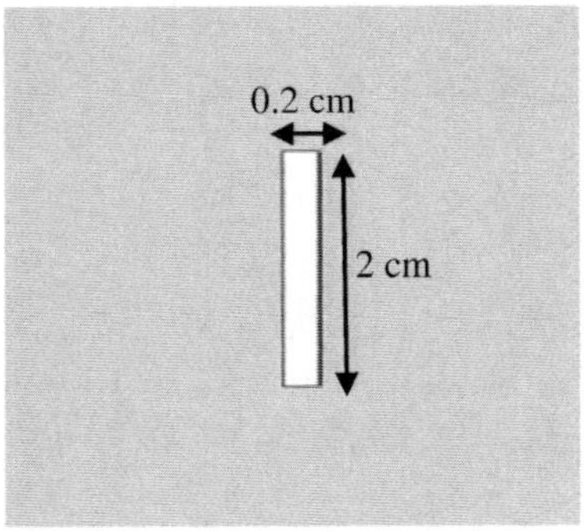

(c)

Figure 3. Plate model for (a) healthy plate, (b) damaged plate and (c) close-up for damage slot

3.1. Defect plate simulation

Dynamic simulation of the healthy plate was performed in order to obtain the baseline signals that will be used as the reference for the comparison analysis for damaged plate. The interaction of the Lamb wave with the defects was extracted by using this baseline signals. Referring to the wavefield visualisation in Figure 4(a), two main wave packets were appeared for the healthy plate. Theoretical dispersion curve in Figure 2 was used to identify the types of mode of that wave packets. At the given excitation frequency (200 kHz), there is only two modes exist, S_0 and A_0. Based on that projection, it was assumed the first wave packet is the S_0 and the second wave packet is the A_0. This assumption is due to *So* mode is moving faster than *Ao* mode based on dispersion curve in Figure 2. Healthy plate shows a smooth contour of the waves propagation as presented in Figure 4 (a).

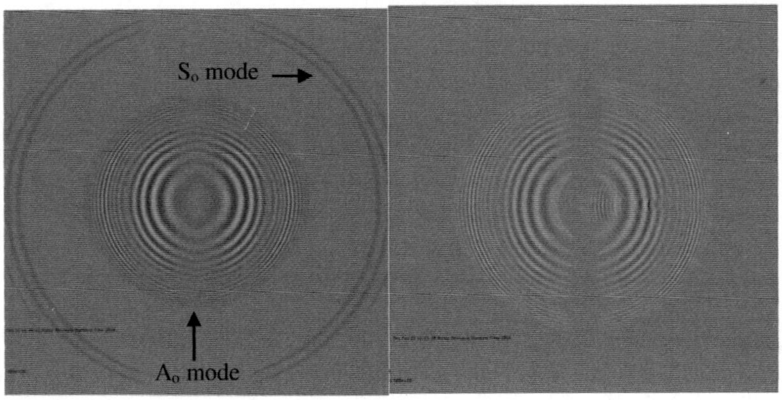

(a) (b)

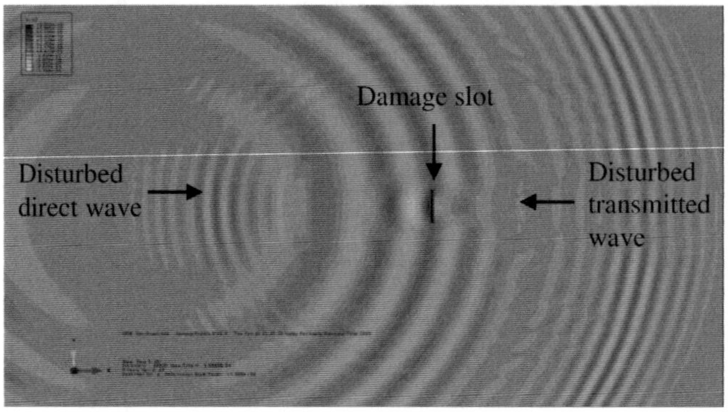

(c)

Figure 4. Visualisation of the wave propagation for (a) intact plate, (b) damaged plate and (c) close-up of the disturbed propagated wave for damaged plate

A through damaged slot was introduced at distance of 15 cm from the wave exciter, as presented in Figure 3(b) and (c). The same simulation set up was applied for damaged plate simulation. The obtained wavefield visualisation shows a high degree of difference compared with the healthy plate. The S_0 mode seems to attenuate when interacting with defect, leaving A_0 mode to dominantly propagate in the plate. Thus, only one mode appeared in the visualisation as presented in Figure 4(b). Close-up figure shows that A_0 mode contour experiencing disturbances when interacting with defect. Distorted contour was obviously noticeable which is happened before and after the defects. They are labelled as disturbed direct wave and disturbed transmitted wave in Figure 4(c). This sensitivity characteristic of Lamb wave with the changes of structural properties is useful in the SHM application.

4. Conclusions

Lamb wave was attracted by many researchers due to their advantages. However due to multimode and dispersive nature, well understanding of their behaviour is crucial for acquiring accurate results for SHM system. Simulation work was performed in this present work for a better understanding of its interaction with the defects. A smooth propagation of A_0 and S_0 modes was observed in the healthy plate. On the other hand, only A_0 mode was appeared in the damaged plate but in distorted condition compared with healthy plate. S_0 mode is assumed to attenuate when interacting with defects. Existence of defects in the structure alters the behaviours of the wave. It can be concluded that Lamb wave is one of the promising tool for the diagnostic analysis due to their sensitivity to changes of the structural properties.

Acknowledgment

The authors would like to express their gratitude to Universiti Malaysia Pahang through the fund of PGRS180376 for supporting these research activities.

References

[1] Han S, Palazotto A N and Leakeas C L 2009 Finite-Element Analysis of Lamb Wave Propagation in a Thin Aluminum Plate *Journal of Aerospace Engineering* **22** 185-97
[2] Guo N and Cawley P 1993 The interaction of Lamb waves with delaminations in composite laminates *The Journal of the Acoustical Society of America* **94** 2240-6
[3] Rose J L 2002 A Baseline and Vision of Ultrasonic Guided Wave Inspection Potential *Journal of Pressure Vessel Technology* **124** 273-82

[4] Olson S E, DeSimio M P and Derriso M M 2006 Analytical Modeling of Lamb Waves for Structural Health Monitoring In: *3rd European Workshop on Structural Health Monitoring.*, (Granada, Spain

[5] SeJin Han B 2007 Finite Element Analysis of Lamb Waves Acting Within a Thin Aluminum Plate. In: *Department of Aeronautical and Astronautical Engineering*: Air Force Institute of Technology, Air University)

[6] Crespo B H, Courtney C R P and Engineer. B 2018 Calculation of guided wave dispersion characteristics using a three-transducer measurement system *Applied Sciences* **8**

[7] De Pauw B, Goossens S, Geernaert T, Habas D, Thienpont H and Berghmans F 2017 Fibre Bragg Gratings in Embedded Microstructured Optical Fibres Allow Distinguishing between Symmetric and Anti-Symmetric Lamb Waves in Carbon Fibre Reinforced Composites *Sensors (Basel, Switzerland)* **17**

[8] Feng B, Ribeiro A L and Ramos H G 2018 Interaction of Lamb waves with the edges of a delamination in CFRP composites and a reference-free localization method for delamination *Measurement* **122** 424-31

[9] Rose J L 2014 *Ultrasonic Guided Waves in Solid Media*: Cambridge University Press)

Training Feedforward Neural Networks for Structural Health Monitoring of an Aircraft Structure

Tuan A Z Rahman[1,*], K A M Rezali[1], N A Abdul Jalil[1], A As'arry[1] and Raja Kamil[2]

[1] Sound and Vibration Research Group, Department of Mechanical and Manufacturing Engineering, Faculty of Engineering, Universiti Putra Malaysia 43400 UPM Serdang, Selangor, Malaysia

[2]Department of Electrical and Electronic Engineering, Faculty of Engineering, Universiti Putra Malaysia 43400 UPM Serdang, Selangor, Malaysia

[*]Corresponding author: tuanzahidi@upm.edu.my

Abstract. In the past decades, structural health monitoring (SHM) has become an emerging research area for globally monitoring expensive aircraft and bridge structures. This paper presents the application of stochastic fractal search (SFS) algorithm and its chaotic-enhanced variants to train feedforward neural networks (FNNs) for monitoring an aircraft structure based on vibration data. An experimental spectral testing was carried out to obtain the normal and damaged condition data of a laboratory stiffened panel structure which imitated wingbox of an aircraft. Added mass as pseudo-fault was employed to simulate damage condition with three different damage levels at three different locations. Vibration signature features were generated based on measured frequency response functions (FRFs) and principle component analysis (PCA). Then, metaheuristic-based FNNs approach were applied to localize and predict the severity of damage on the structure. The results reveal that the proposed approach produces high classification and localization accuracy as parameters of the FNNs were optimized systematically using metaheuristic algorithms. In conclusion, the Sine chaos-enhanced SFS algorithm highlights better convergence performance and results accuracy compared to other contested metaheuristic approaches.

1. Introduction

Structural Health Monitoring (SHM) can be referring as the process of implementing a damage detection strategy for engineering infrastructures such as aerospace, civil and mechanical structures [1]. This process involves essentially 3 main stages which are:

- Surveillance of dynamical response measurement of structure periodically over time
- Feature extraction and selection of damage sensitive traits
- Determination of present condition of structure health using statistical analysis based on selected feature/s

Implementation of SHM technologies can bring benefits in terms of increasing economic outputs by minimizing failures incidence, reducing maintenance expenses, increasing the efficiency of inspection and most importantly preventing catastrophic failure that can lead to loss of life. SHM strategies possess two main approaches which are model-based and data-driven approaches. The latter approach has advantages over the former with regard to accuracy and less complexity.

As vibration-based SHM can be treated as a pattern recognition problem, several machine learning approaches have been employed i.e. artificial neural networks (ANNs), support vector machines (SVMs), k-nearest neighbours (k-NN) and many more. The feedforward multilayer perceptron (MLP) neural networks is a popular example of ANNs categories to solve pattern recognition problems [2]. However, training neural networks in order to find best set of weight and bias parameters to the problem is a very challenging and hard process as entangled with many local optimal [3]. Thus, many metaheuristic optimization approaches were employed to train ANNs for clustering and regression problems.

In this paper, training a feedforward neural networks (FNNs) classifier model using stochastic fractal search (SFS) optimization algorithm and its chaotic variants is proposed. The rest of the paper organized as follows; **Section 2** shows the experimental setup and vibration data collection of an aircraft structure for SHM. The metaheuristic algorithms to train FNNs model is explained in details in **Section 3**. **Section 4** discussed and summarized the results before conclusions are drawn in **Section 5**.

2. Experimental Setup

The wingbox structure was constructed by using an aluminium sheet (750 x 500 x 3 mm) as upper surface, two ribs composed of C-channel riveted to the short edges and two stiffening L-shape stringers that secured with bolts along the length of the sheet. The wingbox structure is shown in Figure 1. Free-free boundary conditions are approximated experimentally by suspending the wingbox from a substantial frame using springs and nylon line attached at the corners of the top sheet.

The acquisition system used during the test was a DIFA SCADAS III controlled by LMS software running on a Dell desktop PC. The SCADAS III is an 8-channel, portable and high-speed data acquisition system. All measurements were recorded within a frequency range of 0-1024 Hz with a resolution of 0.25 Hz. Then, the wingbox was excited with a periodic chirp signal through a shaker at location A (80, 250) and the responses were measured using three unidirectional PCB resonant piezoelectric accelerometers attached vertically at three different locations shown in Figure 2. The damage states were introduced by using added mass bolted to the front stringer [4]. The $Hv(w)$ estimator is used throughout all the testing.

Figure 1. The experimental setup

IOP Conf. Series: Journal of Physics: Conf. Series **1262** (2019) 012031 doi:10.1088/1742-6596/1262/1/012031

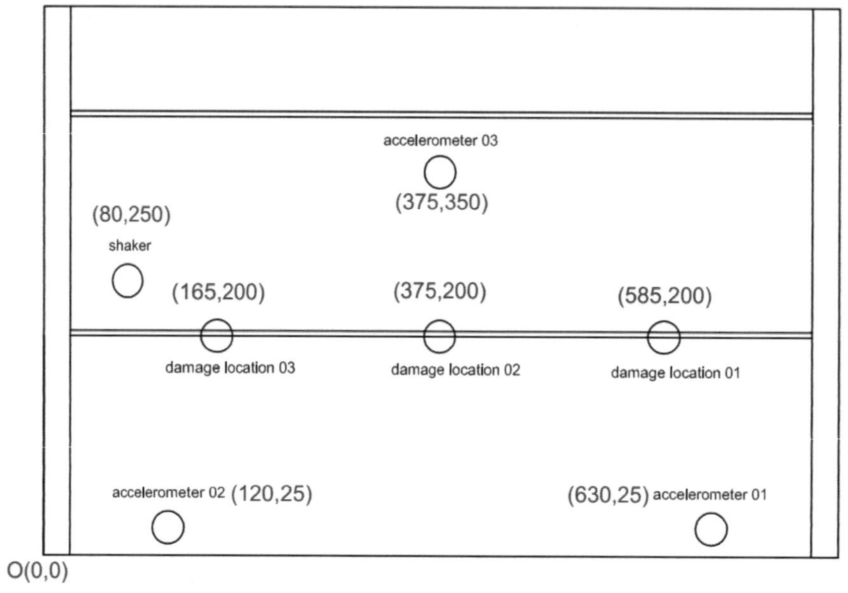

Figure 2. Location of the excitation and sensors (top view)

The test order for the programme and configuration was as follows:

- Normal Condition
- 10g Mass Added at Location 01
- 30g Mass Added at Location 01
- 50g Mass Added at Location 01
- Normal Condition
- 10g Mass Added at Location 02
- 30g Mass Added at Location 02
- 50g Mass Added at Location 02
- Normal Condition
- 10g Mass Added at Location 03
- 30g Mass Added at Location 03
- 50g Mass Added at Location 03
- Normal Condition

Each configuration consisted of a clean signal acquired using 40 averages, to be used as a reference for the features selection as shown in Figure 3. Spectral lines between 187 to 194 Hz were selected from each accelerometer as input feature to the FNNs classifier.

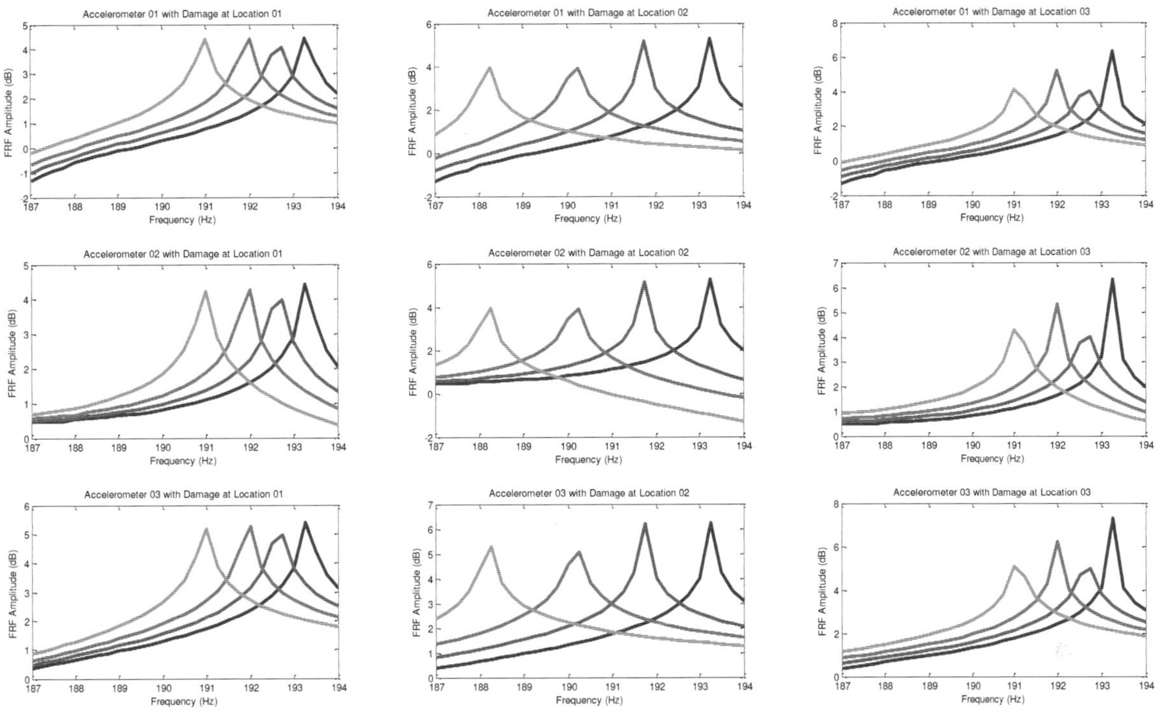

Figure 3. Selected input features

For each configuration, the data were copied and then corrupted with Gaussian noise with variance of 0.0025 to produce another 100 observations data. Figure 4 shows the 3-D visualization of all data using Principal Component Analysis (PCA). All undamaged condition is clearly separable with the damage condition introduced. However, the observation of damage data was mixed especially for damage at location 01 (black) and location 03 (green).

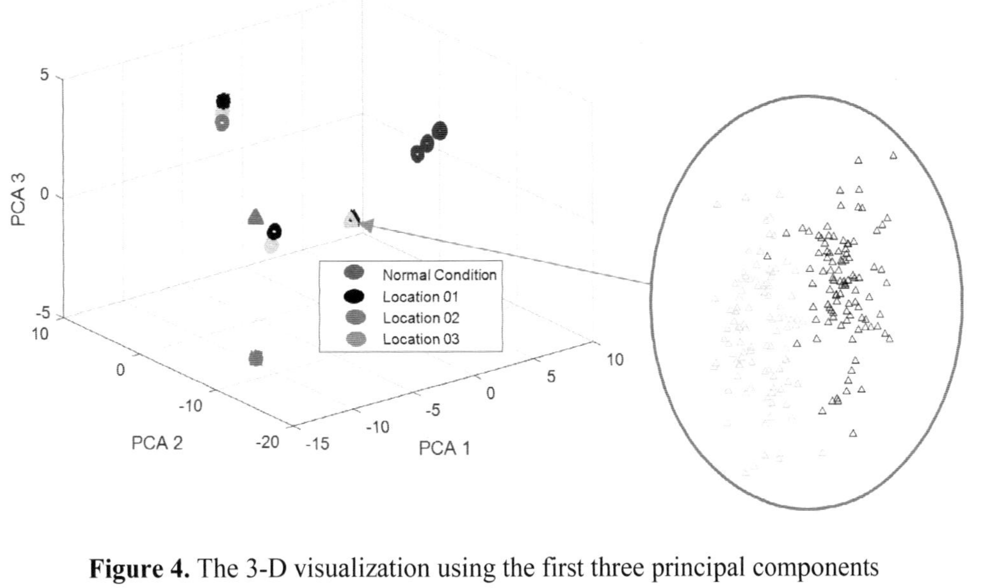

Figure 4. The 3-D visualization using the first three principal components

3. SFS-based FNNs Classifier for SHM

The process of training FNNs model mainly consists of determining the weight and bias parameters between the neurons by minimizing the error between actual and predicted outputs. Figure 5 shows three layered FNNs with one hidden layer employed in this study. The first layer is called input layer, while the last layer is called the output layer. Other layers between the input and output layers are called hidden layer. Each layer contains neurons. The neurons in input layer are connected to neurons in hidden layer. And, the hidden layer with output layer. In FNNs, the information moves in only one-way and one-directional connections between their neurons. Information flow from input layer to output layer is achieved by hidden layers using weight and bias parameters. Each weight determines the influence of an input on the neuron and bias controls the amplitude of the output of the neuron.

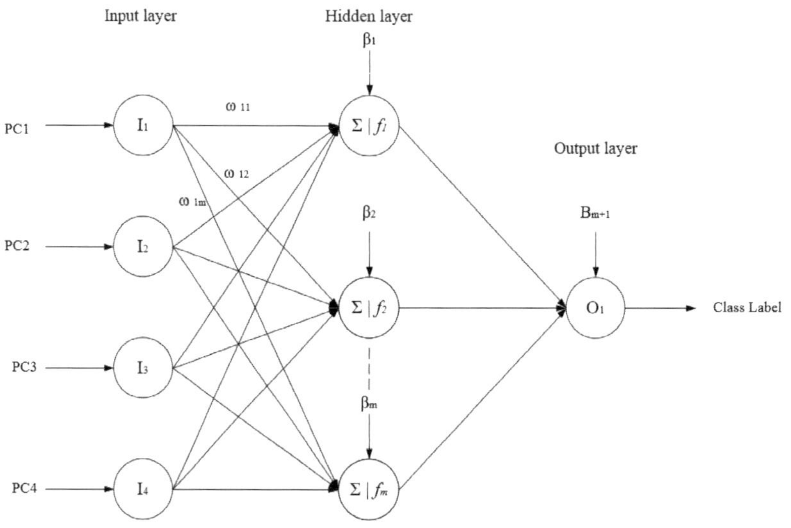

Figure 5. Feedforward neural networks with one hidden layer [5]

The mathematical expression of summation function of an FNNs is as shown in Eq. (1).

$$S_j = \sum_{i=1}^{n} w_{ij} I_i + \beta_j \tag{1}$$

where w_{ij} is the connection weight the input i and neuron j, β_j is the bias term of the neuron, n is number of inputs. A sigmoid function usually employed as activation function as in Eq. (2).

$$f_j(x) = \frac{1}{1+e^{-S_j}} \tag{2}$$

Eq. (3) expressed the final output of the FNNs where m is the number of hidden neurons and W_i is the connection weight between neuron i and the output neuron. β_{m+1} is the bias of the output neuron.

$$\hat{y} = \sum_{i=1}^{m} W_i f_i + \beta_{m+1} \tag{3}$$

Finally, the learning error, *MSE* between actual and predicted output is calculated as fitness function for metaheuristic algorithms as in Eq. (4) where p is the number of outputs, o_i the actual output and $\hat{o}_i$ is the desired output.

$$MSE = \sum_{i=1}^{p} (o_i - \hat{o}_i)^2 \tag{4}$$

IOP Conf. Series: Journal of Physics: Conf. Series **1262** (2019) 012031 doi:10.1088/1742-6596/1262/1/012031

4. Implementation and Results

An FNNs model with 4-15-10 structure was employed in the study. The first four principle components were selected as input to the FNNs model and ten classes of wingbox condition as its output. The computing platform of a Dell desktop PC with Intel Xeon 3.4 GHz CPU, 8 GB RAM, a window 7 operating system with the Visual Studio 2010 and MATLAB 2014a development environment was used in all calculation. The weight and bias parameters of the FNNs model were optimized using stochastic fractal search (SFS) algorithm and its two chaotic variants which are enhanced using Gauss/mouse and Sine maps. The more details regarding the SFS algorithm and its improved chaotic version can be referred in [6, 7]. The FNNs model was trained for ten independent runs using these three metaheuristic algorithms. The data was divided into two groups which are training (60%) and testing (40%).

Figure 6 shows the convergence curve of metaheuristic algorithms in training the FNNs model over 1000 iterations. The Sine-SFS algorithm outperformed standard SFS and Gauss/mouse-SFS in terms of average and best achieved MSE. The performance comparison of SFS and its chaotic variants in clustering the damage location and severity of wingbox is highlighted in Figure 7. The best results are indicated in bold type. Both chaotic variants of SFS algorithm achieved 92.31% of best classification accuracy in comparison to standard SFS (84.62%). In regarding the average accuracy for 10 independent runs, the Sine-SFS algorithm yield better results compared to other, as tabulated in Table 1. On the other hand, the original SFS algorithm has lowest standard deviation.

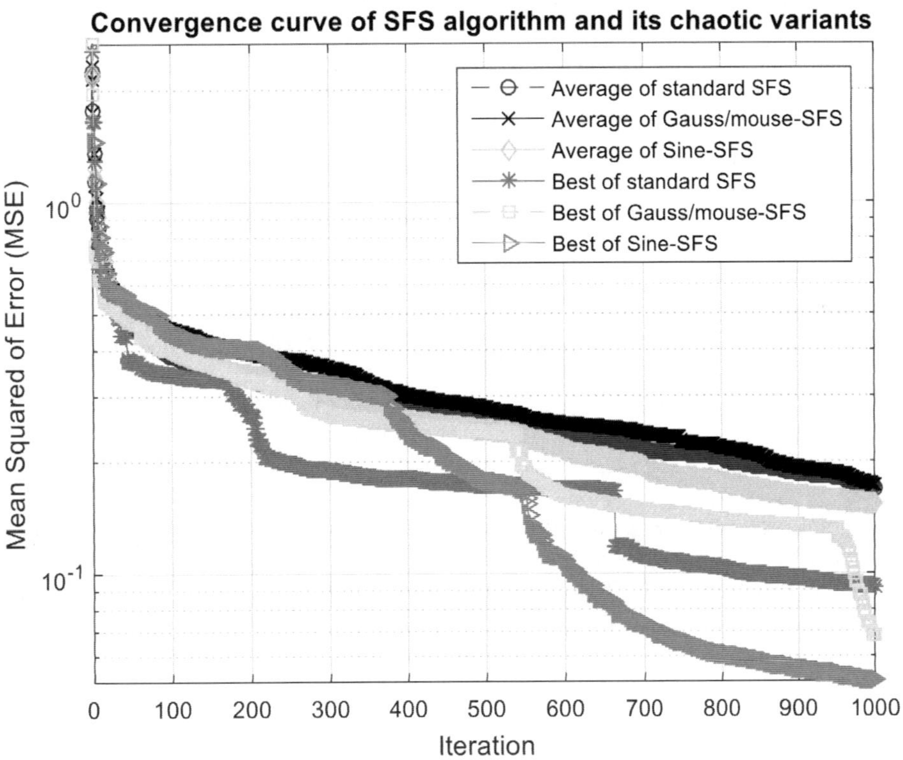

Figure 6. Convergence curve of SFS algorithm and its chaotic variants

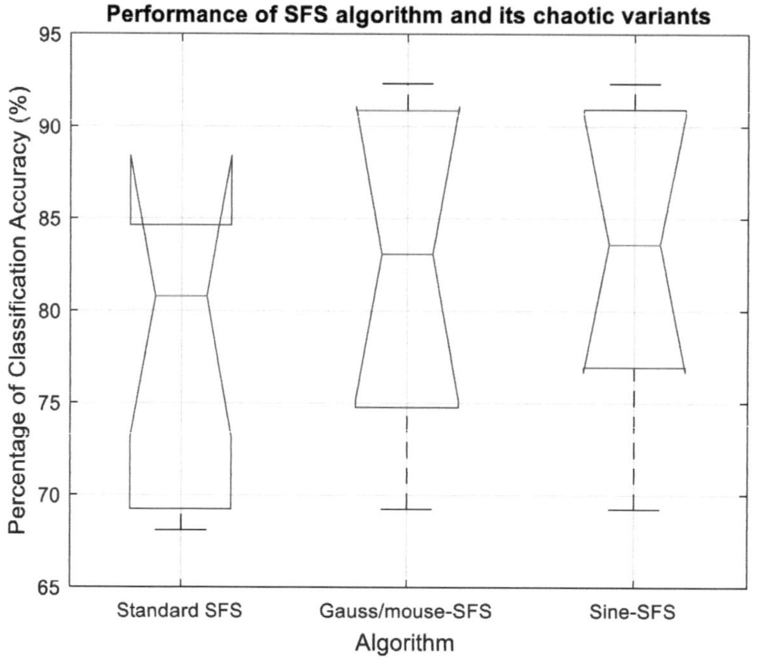

Figure 7. Performance of SFS algorithm and its chaotic variants

Table 1. Percentage of classification accuracy for each algorithm.

Algorithm	Best	Average	Worst	Std. Dev.
Standard SFS	84.62	78.24	68.08	7.27
Gauss/mouse-SFS	92.31	79.32	46.15	13.98
Sine-SFS	92.31	81.97	69.23	8.72

Based on the obtained results, the Sine-SFS algorithm possesses better searching capability in comparison to standard SFS and Gauss/mouse-SFS algorithms in training the FNNs model.

5. Conclusion

The weight and bias parameters selection of an FNNs has great influence to its performance. This study presents newly implemented stochastic fractal search (SFS) algorithm and its two chaotic variants to optimize the parameters of FNNs for SHM of an aircraft structure with their performance evaluation in terms of convergence speed and classification accuracy. The Sine-SFS algorithm highlights better convergence performance and results accuracy compared to other approaches. The proposed method will be employed to classify damage condition of real aircraft structure in future study.

Acknowledgment

The first author would like to express his gratitude to the Universiti Putra Malaysia (UPM) and Malaysian Ministry of Education (MOE) for PhD scholarship funding. This research was performed with financial assistance under Putra Grant IPM using Vot No. 9499300.

References

[1] Farrar C R and Worden K 2012 *Structural Health Monitoring: A Machine Learning Perspective* (Chichester: Wiley & Sons)

[2] Fine T L 1999 *Feedforward Neural Network Methodology* (New York: Springer)

[3] Blum A L and Rivest R L 1992 Training a 3-node neural network is NP-complete *Neural Netw* **5** 117-127

[4] Papatheou E, Manson G, Barthorpe R J and Worden K 2010 The use of pseudo-faults for novelty detection in SHM *J Sound Vib* **329** 2349-2366

[5] Faris H, Mirjalili S and Aljarah I 2019 Automatic selection of hidden neurons and weights in neural networks using grey wolf optimizer based on a hybrid encoding scheme *Int. J. Mach. Learn. & Cyber.*

[6] Salimi H 2015 Stochatic fractal search: a powerful metaheuristic algorithm *Knowl.-Based Syst.* **75** 1-18

[7] Rahman T A Z and Tokhi M O 2016 Enhanced stochastic fractal search algorithm with chaos *2016 7th IEEE Control and System Graduate Research Colloquium (ICSGRC)* (Shah Alam: IEEE) pp 22-27

IOP Conf. Series: Journal of Physics: Conf. Series **1262** (2019) 012032 doi:10.1088/1742-6596/1262/1/012032

Biodynamic Modelling of Hand for Glove Transmissibility Prediction using Artificial Neural Networks

Tuan A Z Rahman[1*], K A M Rezali[1], A As'arry[1] and N A A Jalil[1]
[1]Sound and Vibration Research Group (SVRG), Department of Mechanical and Manufacturing Engineering, Universiti Putra Malaysia (UPM) 43400 UPM Serdang Selangor, Malaysia

[*]Corresponding author: tuanzahidi@upm.edu.my

Abstract. Prediction of glove transmissibility to the hand plays an important role in order to improve its vibration reduction capability. This study presents a non-parametric biodynamic modelling of the hand based on artificial neural networks (ANNs) model for predicting the apparent mass in order to estimate the transmissibility of a glove to the hand. An experimental investigation was carried out to obtain the input-output vibration data using five subjects. Then, an ANNs model was used to map between the input and the output with its weight and bias parameters optimized using chaos-enhanced stochastic fractal search (CFS) algorithm. The results indicate that the developed ANNs hand model capable to predict the apparent mass of the human hand with an average accuracy of 97.67%.

1. Introduction

Hand-transmitted vibration (HTV) deal with mechanical vibration entering the body at the fingers or the palm of the hands which produced by powered tools or processes. Prolonged exposure to HTV can lead to the vascular, neurological and osteoarticular disorder of the upper limbs [1]. Thus, the development of anti-vibration gloves plays important roles in order to attenuate vibration transmitted to the hand. If the transmissibility of the glove to the hand can be predicted, improvement of a glove capability to reduce vibration transmitted to the hand can be performed [2]. Previously, many attempts by researchers around the world to model the hand system and to study its biodynamic response to HTV using multiple approaches i.e. lumped parameter, distributed-parameter, mechanical impedance and etc. However, these modelling approaches have limitation due to entangled with highly non-linear effects and many parameters to be considered [3, 4].

Artificial Neural Networks (ANNs) is an established supervised modelling approach in machine leaning research area. Many ANNs-based biodynamic models have been previously developed. Gohari *et al.* [5] develop ANNs-based biodynamic model to simulate transmitted vibration to head for seated human body. An indoor vertical vibration experiments using five males was conducted. The developed ANNs-based biodynamic model possess high accuracy in comparison to other lumped models. An ANNs-based model was developed by Kumar *et al.* [3] to measure biodynamic response seat to head transmissibility (STHT) for seated human subjects exposed to random whole-body vibration (WBV). The constructed model capable to predict STHT with high accuracy when compared with experimental findings. Taghavifar and Rakheja [4] employed ANNs to predict seated body apparent mass under different levels of vibration excitations with two different sitting postures were considered. The model has the ability to predict the seated body response within the range of trained data. In this paper, the development of biodynamic hand model using an ANNs-based non-parametric system identification approach for predicting apparent mass is proposed.

2. Experimental Setup

An experimental investigation was conducted to collect vibration response data of human hand. Figure 1 shows the direction of the dynamic force and acceleration for the calculation of the apparent mass of the hand to vibration based on ISO 5349-1; 2001. The participated human subject's height and weight was tabulated in Table 1. Five healthy male subjects were selected for the experimental test. The participants placed the palm of their hand on the circular wooden adapter without any support on the arm as in Figure 2. They pushed the circular wooden support plate downward while maintaining a force of 10 N. A 10-s period of random vertical vibration was generated using MATLAB with the *HVLab* toolbox (version 1.0). More details regarding the experimental setup and procedures can be found in [6].

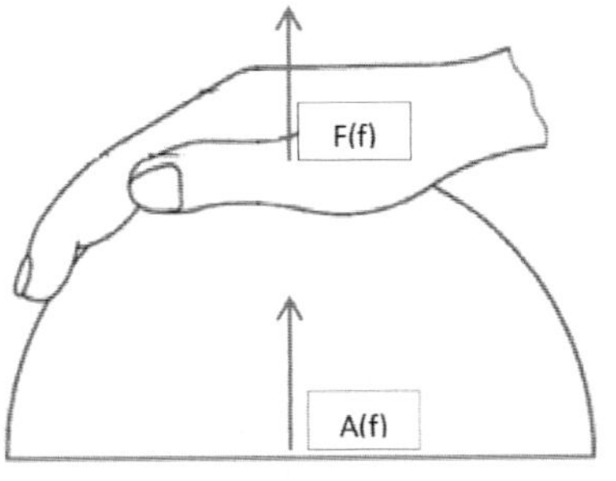

Figure 1. The direction of the input output data based on ISO 5349-1;2001.

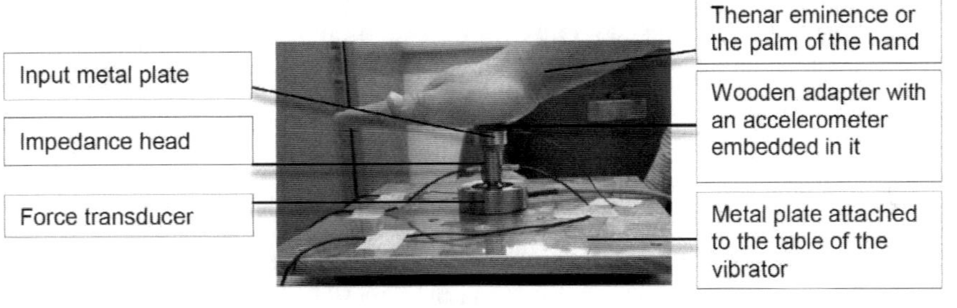

Figure 2. Experimental setup.

Table 1. Height and weight value of human subject.

No. of subject	1	2	3	4	5
Height (cm)	172	161	168	181	163
Weight (kg)	68	56	72	77	65
BMI	22.99	21.6	25.51	23.50	24.46

Figure 3 illustrated the input output data for the modelling of biodynamic response of the hand for Subject 5.

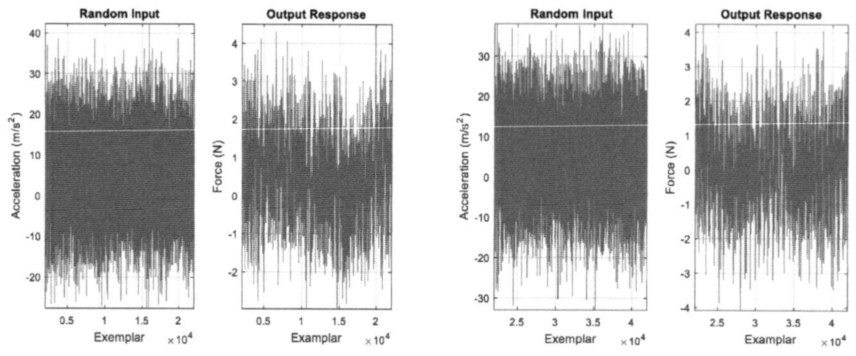

a) Training Data b) Testing Data

Figure 3. Sample of recorded data for Subject 5 for training and validation phases.

3. ANNs based Modelling and Validation

System identification is a well-established area in automatic control for constructing models to be used for simulation, prediction and control design for dynamical systems [7]. In this study, an ANNs-based non-parametric model is developed using experimental input-output data.

3.1. Modelling Phase

In system identification, there are two main processes involved; model structure selection and estimation of model parameters. A nonlinear auto-regressive model with exogenous input (NARX) structure was employed in this study due to its lowest complexity and without noise term incorporated. However, this model structure possesses capability to adapt with the nonlinearity [8]. The NARX model can be defined as in Eq. 1.

$$y(t) = f(y(t-1), y(t-2), \ldots, y(t-n_y), u(t-1), u(t-2), \ldots, u(t-n_u)) \tag{1}$$

where, the next value of the output signal, $y(t)$ is regressed on previous values of the output $y(t\text{-}n_y)$ and input $u(t\text{-}n_u)$ signals. The function $f()$ can be approximated using feedforward neural networks (FNNs). Figure 4 shows the implementation of FNNs-based NARX model to estimate output force response of the hand. A 4-10-1 of FNNs architecture was employed with model order, $n_y = n_u = 2$ and number of hidden neurons of 10. The details regarding the FNNs matrix encoding strategy used in the study can be referred in [9].

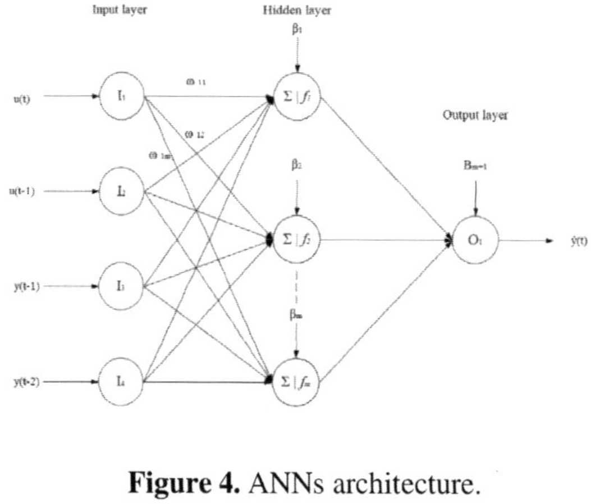

Figure 4. ANNs architecture.

A total of 61 weight and bias parameters of the FNNs were optimized using a metaheuristic algorithm named chaos-enhanced stochastic fractal search (SFS) [10] within the searching range of [-5, 5]. The Sine chaotic map was embedded in Diffusion and First Updating Processes in order to improve the searching capability of SFS algorithm [11]. Sigmoid activation function was employed in both hidden and output layers. The input-output dataset was divided into two sets as shown in Figure 3. The first 2/3 of dataset was used to train the FNNs model and the balance unseen 1/3 of dataset was used to validate the model. The mean squared error (MSE) between the actual and estimated output was used as objective function. The population size and maximum iteration were set to 30 and 150, respectively. The maximum diffusion number (MDN) of 3 and first Gaussian walk was selected.

3.2. Validation Phase

In this study, several validation tests were used to ensure the adequacy of developed FNNs model which are one-step-ahead (OSA) prediction, mean square error and correlation tests. Model validation was done in both time and frequency domains by mapping and comparing between the actual and predicted outputs. Correlation tests are a statistical test that indicates the degree of relationship between two variables [12]. An adequate model can be indicated by the unbiased prediction error with all linear and nonlinear combination of past inputs and outputs and lay within 95% confidence limits. Finally, apparent mass (AM) prediction accuracy is computed for each human subject. Apparent mass can be calculated using Eq. 2 as follows;

$$AM(f) = \frac{F_{io}(f)}{A_{ii}(f)} \qquad (2)$$

where, F_{io} is the output dynamic force and A_{ii} is the power spectral density of the input acceleration.

4. Results and Discussion

The computational calculation was performed using MATLAB 2014a environment on a PC with Intel Xeon 3.4 GHz, 8Gb RAM, a window 7 operating system. Figure 5 shows the convergence curve of the chaos-enhanced SFS (CFS) algorithm in training FNNs model of the biodynamic hand. A value of MSE = 0.0075 has been achieved in modelling phase. Table 2 tabulates the MSE between experimental and predicted force output by the FNNs model in modelling and validation phases while Figure 6 highlights the developed model performance. The FNNs model able to predict the force output with small MSE values for each subject. The average of MSE equal to 0.0089 was found in validation phase.

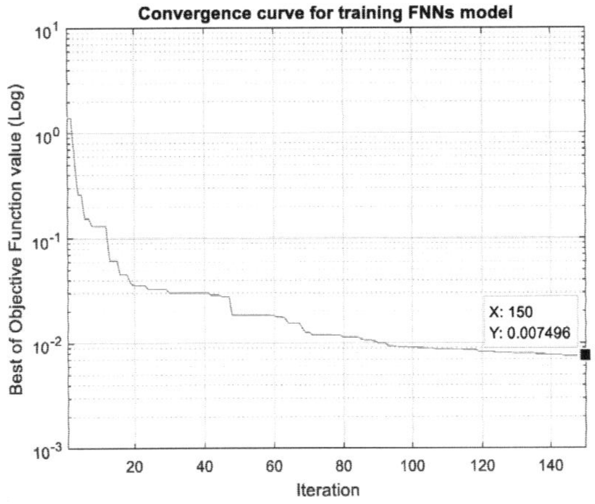

Figure 5. Convergence curve of training FNNs model using CFS algorithm.

Table 2. Mean squared of error (MSE) of between actual and predicted force outputs.

Subject No.	1	2	3	4	5	Average
Modelling	0.0055	0.0090	0.0136	0.0056	0.0034	0.0075
Validation	0.0059	0.0113	0.0170	0.0048	0.0055	0.0089

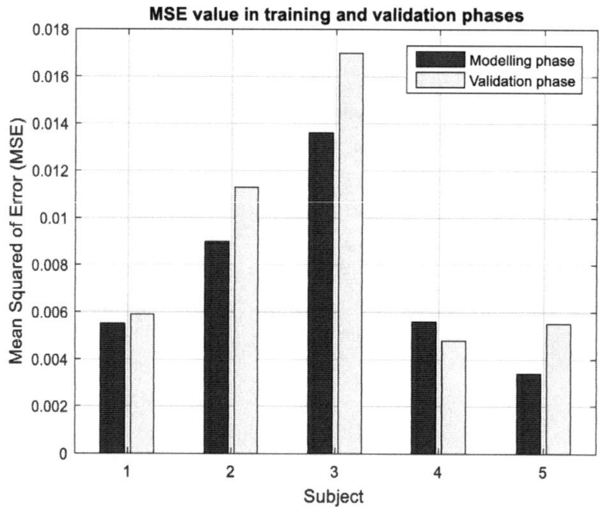

Figure 6. Performance of CFS-FNNs model to predict the force output response.

The mapping between experimental and CFS-FNNs predicted force output for each subject were plotted in Figure 7. It is clearly can be seen that the generalized CFS-FNNs model possesses excellent capability to predict the force output response of the hand for every participant.

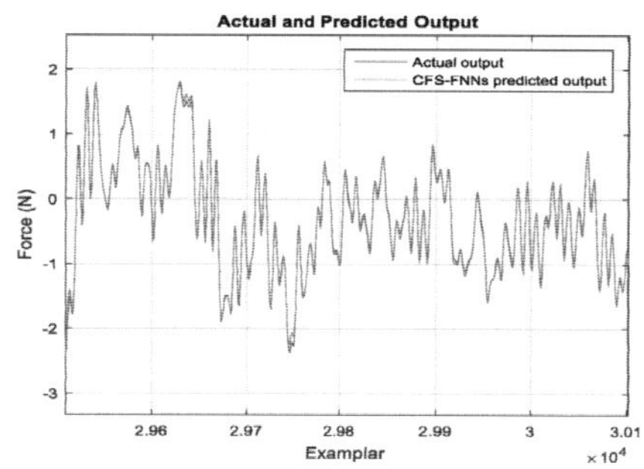

a) Subject 1

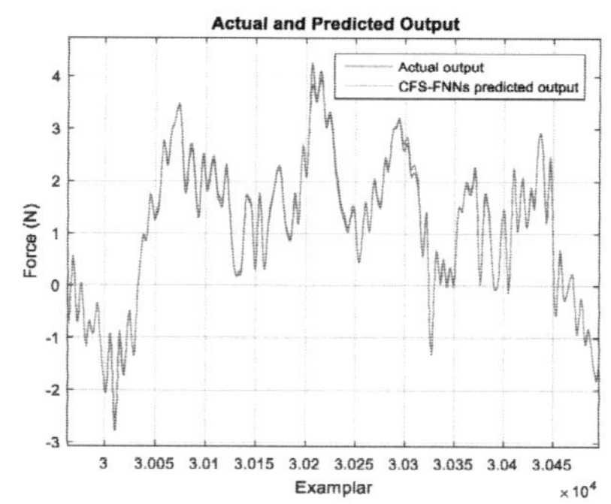

b) Subject 2

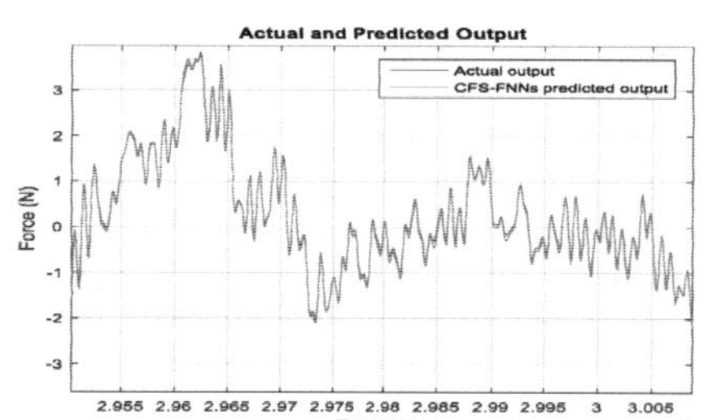

c) Subject 3

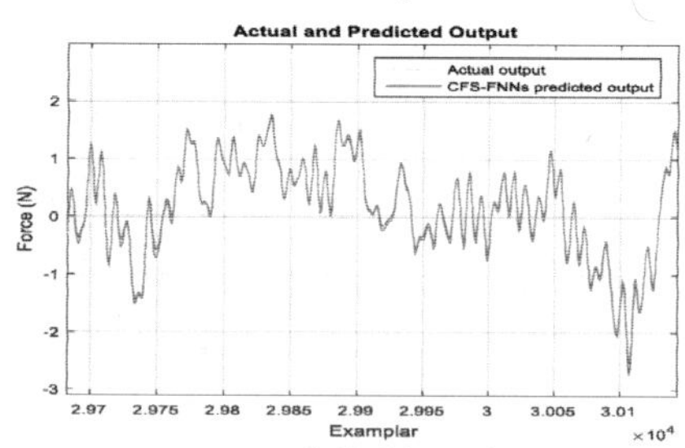

d) Subject 4

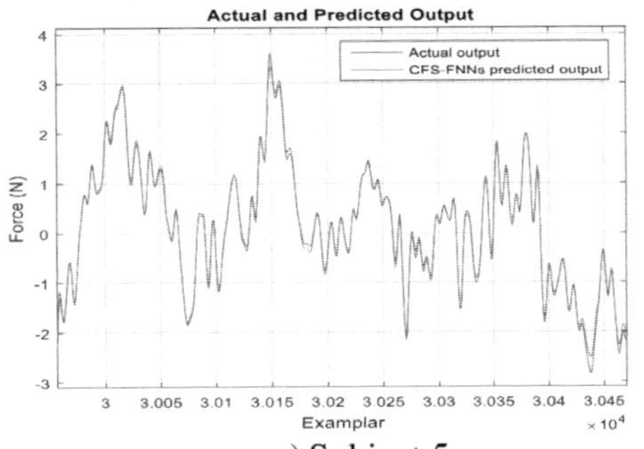

e) Subject 5

Figure 7. The comparison between actual and CFS-FNNs model predicted force outputs.

Next, the CFS-FNNs model was validated using correlation tests in order to evaluate its adequacy. Based on Figure 8, three out of five subjects successfully passed the auto-correlation and cross correlation tests.

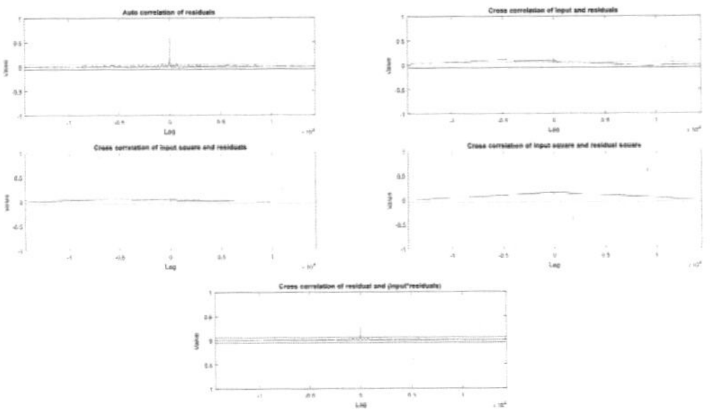

a) Subject 1

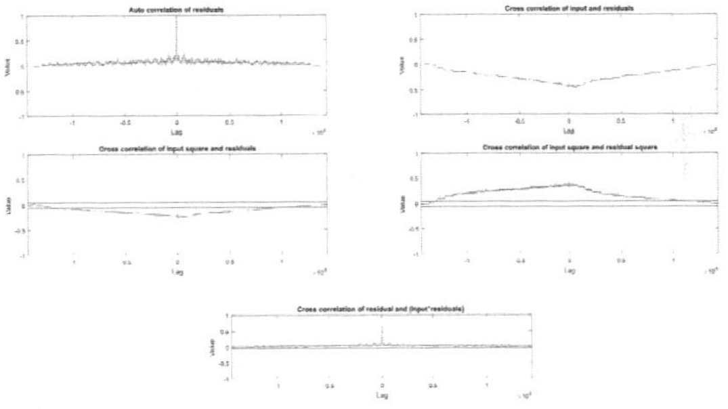

b) Subject 2

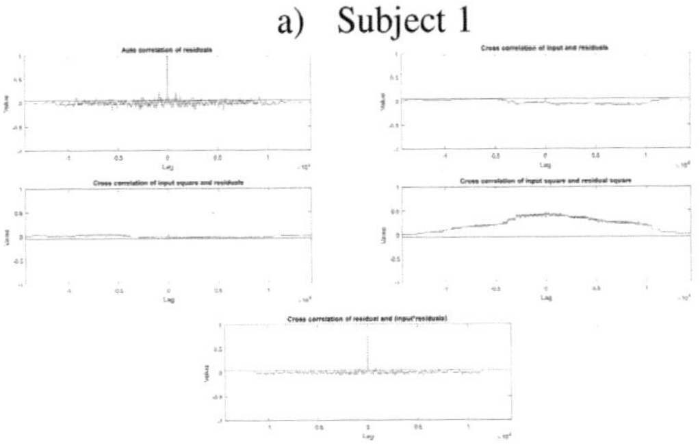

c) Subject 3

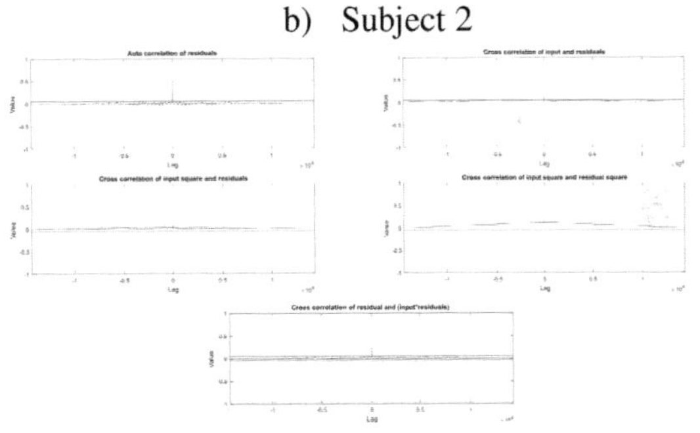

d) Subject 4

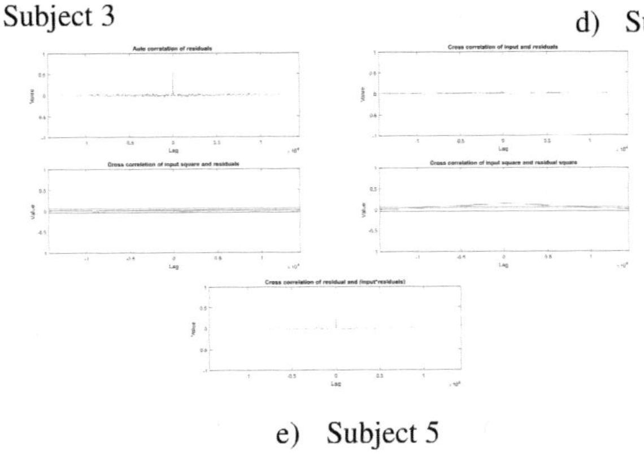

e) Subject 5

Figure 8. Correlation tests for CFS-FNNs model for each human subject.

However, the CFS-FNNs model incapable to generate unbiased prediction error for Subjects 2 and 3 as correlation values lay outside the 95% confidence limits for cross correlation tests. Then, the estimated apparent mass values were computed using MATLAB *HVLab* toolbox based on force output response from CFS-FNNs model prediction and compared with experimental results. Figure 9 shows the performance of CFS-FNNs model to predict apparent mass of the hand for each subject within the interested frequency range of [5, 200] Hz. The CFS-FNNs model shows superiority performance to predict the biodynamic hand response with high accuracy as highlighted in Table 3. The average of prediction accuracy equal to 97.67% was achieved by the constructed CFS-FNNs model to predict the apparent mass of the hand for five human subjects.

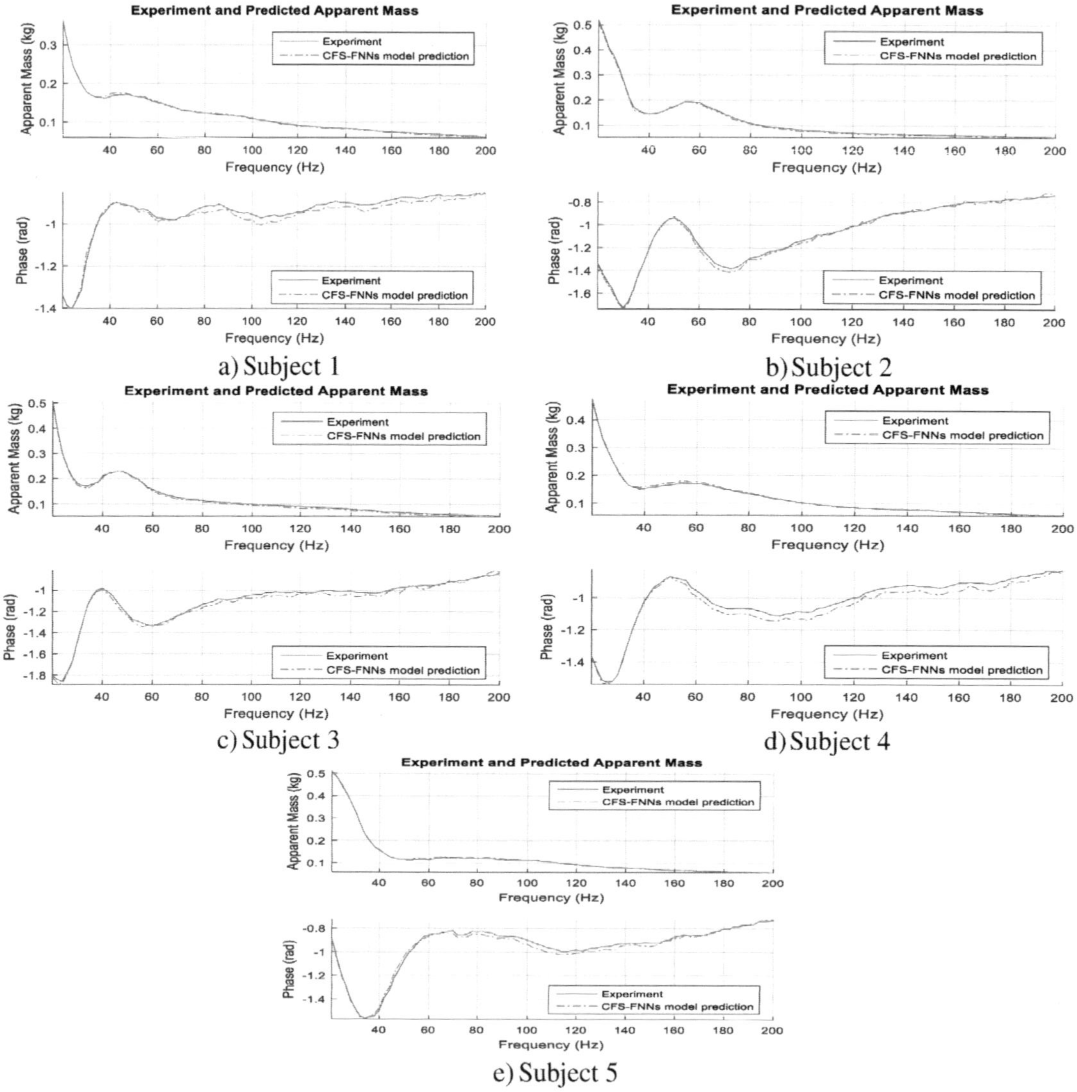

Figure 9. Apparent mass prediction by CFS-FNNs model.

Table 3. Overall performance of CFS-FNNs model to predict apparent mass.

Subject	1	2	3	4	5	Average
Prediction Accuracy (%)	98.08	98.22	97.42	96.81	97.82	**97.67**

5. Conclusion

An ANNs-based nonparametric model identification approach is proposed to model the biodynamic response of the hand expose to HTV. The generated CFS-FNNs model were evaluated with various validation tests. It is found that the model has performed well in approximating the hand response for five human subjects in this study. Future works intended to develop generalized model of the human hand using the proposed approach for glove transmissibility prediction. Moreover, the developed hand model can also be used in designing active controller to attenuate unwanted HTV from the human hand system.

Acknowledgments

The first author would like to express his gratitude to the Universiti Putra Malaysia (UPM) and Malaysian Ministry of Education (MOE) for PhD scholarship funding. This research was performed with financial assistance under Putra Grant IPM using Vot No. 9499300.

References

[1] Bovenzi M 2005 Health effects of mechanical vibration *G. Ital. Med. Lav. Ergon.* **27** 58–64
[2] Rezali K A M, As'arry A, Zulkefli Z A, Samin R and Jalil N A A 2016 Predicting the transmissibility of a glove material to the palm using a simple lumped parameter model of the hand and the glove *J. Teknol.* **78** 17-22
[3] Kumar A Saran V H and Harsha S P 2017 Biodynamic modelling of seated human body under whole body vibration exposure using ANN *Int. J. Vehicle Noise and Vibration* **13** 187-99
[4] Taghavifar H and Rakheja S 2018 Supervised ANN-assisted modelling of seated body apparent mass under vertical whole body vibration *Measurement* **127** 78-88
[5] Gohari M, Rahman R A, Raja R I and Tahmasebi M 2012 A novel artificial neural network biodynamic model for prediction seated human body head acceleration in vertical direction *J. Low Freq. Noise V. A.* **31** 205-16
[6] Rezali K A M 2015 Measuring and predicting the transmission of vibration through gloves to the hand (Southampton: University of Southampton)
[7] Ljung L 1987 *System Identification: Theory for the User* (New Jersey: Prentice Hall)
[8] Darus I Z M and Al-Khafaji A A M 2012 Non-parametric modelling of a rectangular flexible plate structure *Eng. Appl. Artif. Intell.* **25** 94-106
[9] Mirjalili S, Hashim S Z M and Sardroudi H M 2012 Training feedforward neural networks using hybrid particle swarm optimization and gravitational search algorithm *Appl. Math. Comput.* **22** 11125-37
[10] Salimi H 2015 Stochastic fractal search: a powerful metaheuristic algorithm *Knowl. Based Syst.* **75** 1-18
[11] Rahman T A Z and Tokhi M O 2016 Enhanced stochastic fractal search algorithm with chaos *2016 7th IEEE Control and System Graduate Research Colloquium (ICSGRC)* (Shah Alam: IEEE) pp 22-27
[12] Billings S A and Voon W S F 1986 Correlation based model validity tests for non-linear models *Int. J. Control* **44** 235-44

Review on the Vibration Perception Threshold to Early Detect Diabetes Mellitus

M. Taufiq Zaliran[1], Khairil Anas Md Rezali[1*], Ng Yee Guan[2], Azizan As'arry[1] and Nawal Aswan Abdul Jalil[1]

[1] Sound And Vibration Group Research, Department Of Mechanical Engineering, Faculty of Engineering, Universiti Putra Malaysia (UPM), 43400 Serdang, Selangor, Malaysia.

[2] Department of Environmental and Occupational Health, Faculty of Medicine and Health Sciences, Universiti Putra Malaysia (UPM), 43400 Serdang, Selangor, Malaysia.

*Corresponding author: khairilanas@upm.edu.my

Abstract. The total population of diabetic patients increases from 4.7 % to 8.5 % in the adult population worldwide in only three decades. In Malaysia, more than 3 million of total population have been inflicted with diabetes. Vibration Perception Threshold is a non-invasive technique that can measure loss of sensation due to diabetic peripheral neuropathy. The threshold mainly depends on the frequency of vibration, the area of contact, the applied force, and the location of measurement. This paper will review on the factors influencing the vibration perception threshold and recent studies on the use of VPT to detect diabetes.

Keywords: diabetes mellitus; vibration perception threshold (VPT); diabetes neuropathy

1. Introduction

In three decades from 1980 to 2014, the world population diagnosed with diabetes increased from 108 million to 422 million people, rising nearly half percent from 4.7 % to 8.5 % [1]. One of the main factors for the increase was due to the change in daily diet, which led to overweight, apart from other equivalence factors such as the lack of diabetes knowledge [2]. The prevalence of diabetes has risen faster in low- and middle-income countries than in high-income countries [3]. In 2012 alone, about 1.5 million people died due to diabetes, more than other diseases such as cardiovascular disease [4].

Diabetes is a disease caused when the level of glucose in blood is too high. When glucose stays in a body for too long, it will lead to other diseases. Thus, patient with diabetes will have to monitor their diet so that the glucose level can be controlled [1]. The most common types of diabetes are Type 1, Type 2, and Gestational diabetes [1]. Over three million cases had been reported in Malaysia or nearly 17 % of population in Malaysia has been diagnosed by diabetes [5].

Generally, method for diagnosing diabetes can be divided into two distinct approaches which are invasive and non-invasive. Insistent screening for diabetes is crucial for early detection and consequently increases benefits which can be gained from quick intervention [6]. A study suggests the necessity for early intervention and treatment of diabetes through routine screening to prevent severe complication in later stages [7].

Diabetes is conventionally screened by assessing the level of glucose and haemoglobin in blood, coupled with urine analysis and symptoms scores [8]. Three notable blood glucose tests namely random plasma glucose (RPG), fasting plasma glucose (FPG) and oral glucose tolerance (OGT). RPG is the most convenient since it does not require fasting in advance compared to the other twos. The blood glucose tests will require blood sample to be taken from the patient and analysed for its glucose content before being rated accordingly. Although the process can be quick, the method is invasive.

Several researchers have reviewed the implementation of haemoglobin AIC test as a screening test for Type 2 Diabetes [9]. Several drawbacks associated with the former glucose tests were eliminated through the introduction of hba1c test [10]. Once blood sample is taken, it will be analysed to measure the amount of sugar attached to the red blood cell. Fasting is not required and long-term hyperglycaemia is better reflected [11]. However, this test is still an invasive test.

Recent advancement in device technology has enabled researchers to propose another method for diagnosing diabetes. Quantitative Sensory Testing (QST) is a non-invasive technique developed to detect early and precisely, changes in sensory nerve fiber that often linked to the diabetic peripheral neuropathy (DPN). Diabetic peripheral neuropathy is a progressive manifestation of diabetes with length-dependent and significant damage of nerve fibers. Diabetic neuropathy is now thought to be the most common form of peripheral neuropathy. Loss of sensation due to DPN can be assessed through various testing methods including current peripheral threshold (CPT), thermal peripheral threshold (TPT), and vibration perception threshold (VPT). QST has been criticized by many as being highly subjective [12].

Over years of having diabetes, diabetic patients will have loss of sensation of vibration, leading to a vibration perception threshold (VPT) that is greater than healthy patient. Symptoms of diabetes are commonly unnoticed at the early stages, thus conventional diabetes tests are deferred [13]. At the time of diagnosis, patient commonly exhibits the presence of diabetics neuropathy, significantly that the clinically inception of the diseases ensued at an earlier period, presumably several years preceding diagnosis [citation needed]. Hence, the method of VPT has potential to be further studied to early detect diabetes. This paper will review on the factors influencing the vibration perception threshold to diagnose diabetes mellitus.

2. Vibration Perception Threshold

A study proposes that the decrease of vibratory sensation could be an indicator of DPN, which is one of the most common complication associated with the diabetes [14]. It has been showed that VPT can be used as a sensitive measure of DPN in Type 1 Diabetes, and came up with a comparable conclusion for Type 2 diabetes subjects [15]. Several researchers agreed that the diagnosis peripheral sensory neuropathy can be confirm with the VPT [16]. Developed a novel method with high accuracy and reliability in monitoring [17]. Higher VPT value of patient suffering from foot ulceration compared to those without were recorded by using a Biothesiometer [18].

The measuring of vibration perception threshold starts with the subject or patient is exposed to a vibration. The vibration magnitude increases gradually and when the subject perceives the vibration, he or she will give feedback by pressing the feedback button. Once the feedback button is being pressed, vibration magnitude will decrease gradually then when subject does not perceive the vibration, he or she will release the feedback button. The procedure repeats until a threshold is acquired based on an algorithm. The complete procedures of measuring VPT is published in the International Standard ISO13091.

A vibrotactile perception threshold device contains a vibrator, a force cell, an accelerometer, a feedback button, a probe, and a set of data acquisition device (Figure 1). The vibrator generates vibration either random or sinusoidal at certain frequency of vibration. Force cell is used to indicate whether appropriate force is applied onto the probe based on the international standard. The probe is designed so that it will provide optimized vibration exposure depending on the contact area and contact force. The accelerometer measures vibration exposed to the subject.

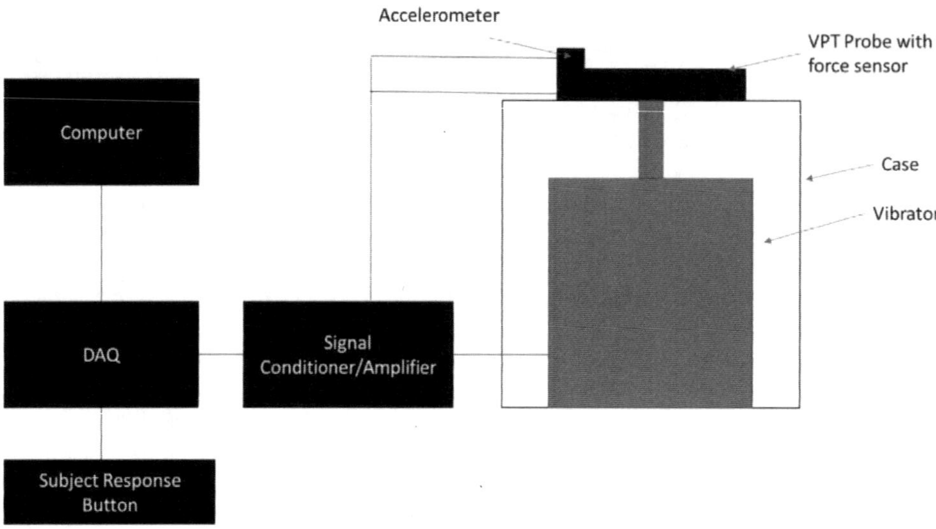

Figure 1 Schematic diagram of vibration perception device

3. Factors Influencing the Vibration Perception Threshold

The International Standard ISO13091 suggests three frequencies of vibration exposure as stimulus for the assessment of VPT which are 4, 31.5, and 125 Hz. Studies have shown that the prominent factors influencing the dynamic response of the human hand or finger are the contact area and contact force [19, 20], apart from the location of measurement, vibration magnitude and posture of the hand. The International Standard ISO13091 also proposes guidance for the applied force and contact area. Vibrotactile is lowest of the fingertip when compared to volar forearm, large toe and heel [21]. For this review, the effects of frequency of vibration, applied force and the area of contact on VPT measured at the fingertip (due to the lowest vibrotactile threshold found by previous study) are further discussed especially since both are the prominent factors that can influence the VPT.

Mechanoreceptors in the hand are responsible for the perception of vibration. The identification of mechanoreceptors that reflect to the vibration perception at certain frequencies has been the purpose of many fundamental studies conducted previously. Fast adapting (FA) units include those sensitive at frequencies between 5 Hz to 50 Hz (FAI) and frequencies above 50 Hz (FAII). A study conducted early 1990 ranging from 16 Hz to 800 Hz, has found that the vibration perception threshold increases as the frequency of vibration is increased [22]. A curve inflection was found at 63 Hz -80 Hz which though to be due to the FAI units for frequencies 16 Hz and 31.5 Hz, and FAII units at 125 Hz, 250 Hz and 500 Hz [22].

The sensitivity threshold depends on the mechanoreceptor and sensory nerve fibres. Previous study has shown that the dynamic response of the hand at frequencies greater than the resonance increases with increasing contact force [23]. The increase in the dynamic response of the hand can lead to a lower threshold since the motion could be vibration perception threshold. Hence, a study conducted previously has found similar observation when applied force is increased, although the reduction of threshold is only found to be significant at frequencies greater than 125 Hz [24].

The area in contact with the hand influences the dynamic response of the hand [23]. Increasing the contact area will increase the sensation for which the amount of vibration transmitted to the finger will be larger than smaller contact area. However, it is not as direct as that. Low frequency threshold is expected to be lower with smaller contact area whilst high frequency threshold is expected to be lower with larger contactor [21]. With surround condition, the vibrotactile threshold has been found to be lower than without surround condition at frequencies less than 63 Hz [22].

4. Diagnosing diabetes using VPT

The ability of the hand to feel vibration is determined by the mechanoreceptors responsible to detect the sensation exposed. Recent study has shown that the VPT can be a good method to detect Type 2 diabetes patients [25-29]. In a study that used a case-control comparison between diabetic patients and non-diabetic patients has shown that the prevalence of peripheral sensory neuropathy (PSN) screened by VPT is 24.5% in diabetic patients and 8.5% in non-diabetic patients with significant determinants of PSN by abnormal VPT were diabetes status, age, heart rate, second-hand smoking and body height in all the participants [26]. VPT used for early detection of neuropathy is highly sensitive and accurate when compared to gold standard NCV. However, the VPT can depends on the contact area and contact location [21] and should not be compared [30]. Increased VPT has been linked with diminished sensation in diabetes patients [25- 28]. Diabetic patients with less than or equal to 5 years of diagnosis have significantly higher perception threshold compared to the healthy group subjects [31]. It was found that there is a positive correlation between the number of years suffering from diabetes and the degree of threshold shift [31]. VPT is a non-invasive method that has the potential to detect diabetes at early stage. A study conducted in 2004 has found that vibration perception threshold of Malaysians is consistent with reference data of the ISO13091-2 [32].

5. Conclusions

Vibration perception threshold technique has been used widely to early detect peripheral neuropathies. However, the assessment depends on factors such as the contact area, contact force, frequency of vibration and location of measurement. Early study on VPT to detect diabetes has shown a good result to characterise diabetic and non-diabetic patient. The recommendation by testing the different prominent parameter by build the VPT experimentally can test the best condition to detect diabetes mellitus.

Acknowledgments

The authors would like to be obliged to Universiti Putra Malaysia for providing laboratory facilities and financial assistance under project no. GP-IPM 9499300.

References

[1] Abdelhafiz, A. H., & Sinclair, A. J. (2018). Diabetes in the elderly Key points. Medicine, 1–4.
[2] Manivannan, M., Periyasamy, R., & Narayanamurthy, V. B. (2008). Vibration perception threshold and the law of mobility in diabetic mellitus patients, 3, 17–21.
[3] Rubio-Cabezas O, Hattersley AT, Njølstad PR, et al. ISPAD Clinical Practice Consensus Guidelines 2014. The diagnosis and management of monogenic diabetes in children and adolescents. Pediatric Diabetes.2014;15(suppl 20):47–64.
[4] Carmody D, Støy J, Greeley SAW, Bell GI, Philipson, LH. A clinical guide to monogenic diabetes. In: Weiss RE, Refetoff S, eds. Genetic Diagnosis of Endocrine Disorders. 2nd ed. Philadelphia, PA: Elsevier; 2016:21–30.
[5] Chew, B., Lee, P., Cheong, A., & Ismail, M. (2016). Messages from the Malaysian Diabetes Registries on Diabetes Care in Malaysian public healthcare. Primary Care Diabetes, 10(5), 383–386.
[6] Kolber MR, Scrimshaw C. Family history of cardiovascular disease. Canadian Family Physician.2014;60(11):1016.
[7] American Diabetes Association. Classification and diagnosis of diabetes. Diabetes Care. 2018;41(suppl 1):S22–S24. [5] Shigley J E and Mischke C R 2006 *Mechanical Engineering Design* (New York, USA: McGraw-Hill)
[8] Huo X, Gao L, Guo L, et al. Risk of non-fatal cardiovascular diseases in early-onset versus late-onset type 2 diabetes in China: a cross-sectional study. The Lancet Diabetes & Endocrinology.

2016;4(2):115–124.
[9] Carmody D, Støy J, Greeley SAW, Bell GI, Philipson, LH. A clinical guide to monogenic diabetes. In: Weiss RE, Refetoff S, eds. Genetic Diagnosis of Endocrine Disorders. 2nd ed. Philadelphia, PA: Elsevier; 2016:21–30.
[10] Pihoker C, Gilliam LK, Ellard S, et al. Prevalence, characteristics and clinical diagnosis of maturity onset diabetes of the young due to mutations in HNF1A, HNF4A, and glucokinase: results from the SEARCH for Diabetes in Youth. The Journal of Clinical Endocrinology and Metabolism. 2013;98(10):4055–4062.
[11] Martin, L. B., Hopkins, W. A., Mydlarz, L. D., & Rohr, J. R. (2010). The effects of anthropogenic global changes on immune functions and disease resistance, 1195, 129–148.
[12] Letourneau LR, Carmody D, Wroblewski K, et al. Diabetes presentation in infancy: high risk of diabetic ketoacidosis [published online August 4, 2017]. Diabetes Care. 2017:e1–e2.
[13] National Centers for Disease Control and Prevention. National diabetes statistics report, 2014. https://www.cdc.gov/diabetes/data/statistics/statistics-report.html . Accessed December 22, 2016.
[14] Leavitt, P. R., Fritz, S. C., Anderson, N. J., Baker, P. A., Blenckner, T., Bunting, L., … Werne, J. (2009). Paleolimnological evidence of the effects on lakes of energy and mass transfer from climate and humans, 54, 2330–2348.
[15] R Daud, S Maeda, Nnm Kameel, My Ripin, N Bakrun , Rm Zein, M Kido, And K Higuchi. 2004. A Pilot Study of Reference Vibrotactile Perception Thresholds on the Fingertip Obtained with Malaysian Healthy People Using ISO 13091-1 Equipment. Industrial Health 42: 189-195.
[16] Bowling, B., & Phillips, C. (2007). Disproportionate and Discriminatory : Reviewing the Evidence on Police Stop and Search, 70, 936–961.
[17] Tucker, S., Bowen, W. D., Iverson, S. J., Blanchard, W., & Stenson, G. B. (2015). Sources of variation in diets of harp and hooded seals estimated from quantitative fatty acid signature analysis (QFASA), (July).
18] Jude, E. B., Eleftheriadou, I., & Tentolouris, N. (2010). Review Article Peripheral arterial disease in diabetes — a review, 4–14.
19] Rezali KAM and Griffin MJ (2017) Transmission of vibration through gloves: effects of contact area. Ergonomics, 60(1), 69-81
[20] Rezali KAM and Griffin MJ (2018) Transmission of vibration through gloves: effects of contact force. Ergonomics, 61(9), 1246-1258
[21] Morioka M and Griffin MJ (2008) Vibrotactile threshold at the fingertip, volar forearm, large toe, and heel. Somatosensory and Motor Research. 25 (2). 101-112.
[22] Harada N and Griffin MJ (1991) Factors influencing vibration sense threshold used to assess occupational exposures to hand transmitted vibration. British Journal of Industrial Medicine. 48, 185-192
[23] Bowling, B., & Phillips, C. (2007). Disproportionate and Discriminatory : Reviewing the Evidence on Police Stop and Search, 70, 936–961.
[24] Merlino, G., Valente, M., Serafini, A., & Gigli, G. L. (2007). Restless legs syndrome : diagnosis , epidemiology , classification and consequences.
[25] Shah A, and Shah D. 2 018. A Study to compare detection of peripheral neuropathy in type 2 diabetes using nerve conduction velocity and biothesoiometry. Paripex - indian journal of research
[26] Yeboah K, Puplampu P, Boima V, Antwi D, Gyan B, and Amoah A. 201 6. Peripheral sensory neuropathy in type 2 diabetes patients: A case control study in Accra, Ghana. Journal of Clinical and Translational Endocrinology. 5, 26-31
[27] Coppini D, Young P, Weng C, Macleod A, and Sönksen P. 1998. Outcome on diabetic foot complication s in relation to clinical examination and quantitative sensory testing: A case-control study. Diabetic Medicine. 15(9) 765-771
[28] Manivannan M, Periyasamy R, Narayanamurthy V. 2009. Vibration perception threshold and the law of mobility in diabetic mellitus patients. Primary Care Diabetes. 3(1), 17-21
[29] Lee CC, Perkins BA, Kayaniyil S, Harris SB, Retnakaran R, Gerstein HC, Zinman BH, Anthony

J. 201 5. Peripheral neuropathy and nerve dysfunction in individuals at high risk for type 2 diabetes: The PROMISE cohort. Diabetes Care 2015.

[30] A Sulistyo, SA Adam, KAM Rezali, and NAA Jalil. 2015. Vibration Perception Threshold for Early Diabetic Syndrome Detection. RCMME.

[31] Gu C, Griffin MJ. 2013. Spatial sum mation of vibrotactile sensations at the foot. Med Eng Phys 2013.

[32] E.H. Weber, De pulsu, resorptione, auditu et tactu, in: H.E. Ross, D.J. Murray (Eds.), E.H. Weber On the Tactile Senses, Taylor & Francis, Hove (UK), 1834.

Application of the Structural Dynamic Modification Method to Reduce the Vibration of the Vehicle HVAC System

MHA Satar[1], AZA Mazlan[1*], MH Hamdan[1], MS Md Isa[1], S Man[2], MAR Paiman[2], MZ Abd Ghapar[2] and MSA Sulaiman[2]
[1]TheVibrationLab, School of Mechanical Engineering, Universiti Sains Malaysia, 14300, Nibong Tebal, Pulau Pinang, Malaysia
[2]Vehicle Comfort, Testing & Development, Vehicle Development & Engineering, Proton Holdings Berhad

[*]Corresponding author: zhafran@usm.my

Abstract. Heating, ventilating and air conditioning (HVAC) system is one of the major sources for the vehicle noise and vibration, which subsequently contribute to the bad acoustical environment. The components of the HVAC system can produce a significant level of vibration during the operation and contributed to an unwanted noise. As the vibration of HVAC components get worse, it will be transferred to other components and excites the natural frequencies of the HVAC system. Natural frequencies of the components are depending on the mass and stiffness of the HVAC components and the value can been modify using these two parameters. This study is focusing on the specific type of HVAC noise and vibration problem and the counter measure has been perform by implementing the structural dynamic modification (SDM) method to the air conditioning (AC) pipe. The lab-scale of the vehicle HVAC system is set up to represent the actual HVAC system with the real vehicle operation. Noise and vibration of the HVAC system in the system level are measured and compared. From this data, the Dynamic Vibration Absorber (DVA) is designed and applied at the AC pipe of the HVAC system. The result shows that, the natural frequency of the AC pipe can be shifted which resulting a wide effective frequency range of 100-500 Hz. The effect of DVA on the HVAC system is observed during operation whereby a significant vibration attenuation has been achieved for operating frequency range of 100-300 Hz.

1. Introduction

Heating, ventilation and air conditioning (HVAC) system is a crucial section in the vehicle system which can affected the passenger comfort. Interestingly, it has influences the psychological and physical perception of the passenger and vehicle quality [1]. However, it has drawback whereby it produced significant contribution to the interior noise level [2], [3]. The vibration is the source of an uncomfortable issue and might come from the components of HVAC system [4]. Therefore, many studies aim to reduce the vibration of the HVAC system to the lowest possible in ensuring the comfortable of the passenger.

Structural dynamic modification (SDM) is one of the techniques that can be used to reduce the vibration of the structure or system. Generally, the SDM is used to change the dynamic behaviour (i.e. natural frequencies, mode shapes and frequency responses) of the structure [5]. This technique can be done by predicting the modified behaviour of the structure such as lumped masses, rigid links, and dampers or by adjusting the configuration parameters [6]. In term of application, this technique has been applied to an active suspended handle to overcome the displacement limitation of a piezoelectric actuator by increase the stiffness of the lower beam of the handle. The finding from this study shows that the SDM able to alter the stiffness of the handle and shifted the modes of the structure beyond operating frequency range [7].

Dynamic vibration absorber (DVA) is one of the scheme under SDM that consist of a mass attached to the structure to eliminate the vibration of a harmonically excited system [8]. This scheme produced a significant reduction in force transmissibility on flexible plate structure with vibration reduction up to 92 % [9]. In an automotive application, the DVA has been applied to the suspension structure to reduce the vibration from the road disturbances. The finding shows the RMS of the body acceleration and suspension can be decrease by 4 % and 16 %, respectively [10]. This study investigates the application of DVA to the HVAC system to reduce the vibration and noise produced as it will increases the comfortability of the passenger in vehicle.

2. Methodology

2.1. Design and Implementation of the HVAC system

The HVAC components are assembled and mounted in position (similarly as a real vehicle assembly) as shown in Figure 1 and the real operational condition of the HVAC is implemented to provide the similar system condition for the reliability of the test results. Generally, this system is operated with two main power supplies, which is direct power supply to the motor and battery for the HVAC system. The system has an activation button to control the engagement of compressor, which also connected with the battery as power supply. The control panel has a blower speed controller to control the blower speed that flow the air throughout the air ducting.

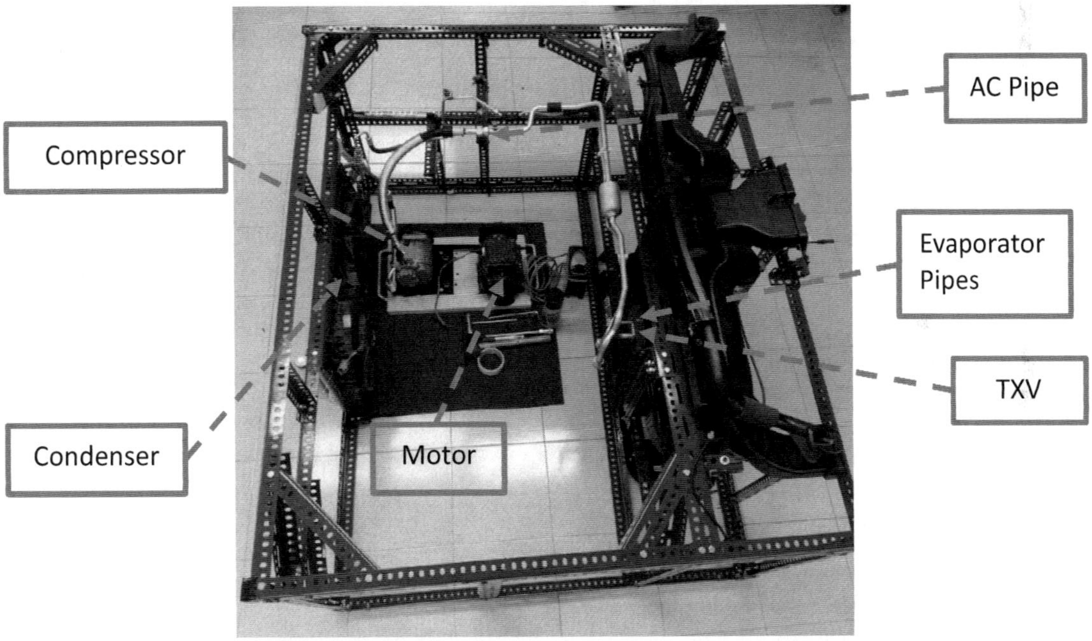

Figure 1. Overall view of the HVAC system

2.2. Noise and vibration measurement

Figure 2 shows the flowchart of the noise and vibration measurement conducted on the rig of the HVAC system. Two types of sensors are used which is accelerometer for vibration measurement and tachometer for rotational acceleration measurement. The accelerometer is attached on the six components as highlighted in Figure 1 to record the vibration responses. The tachometer is mounted on the rotating motor belt to measure the motor (engine) speed.

All the accelerometers are connected through LMS SCADAS Mobile and the results were analysed with LMS Test.Lab. The software used will record the data in time and frequency domains. In this study, there are two condition of measurements, which is idle (850 rpm) and tracking (850-1800 rpm) conditions. The idle condition is used as a reference condition whereas the tracking condition to replicate the actual running vehicle.

IOP Conf. Series: Journal of Physics: Conf. Series **1262** (2019) 012034 doi:10.1088/1742-6596/1262/1/012034

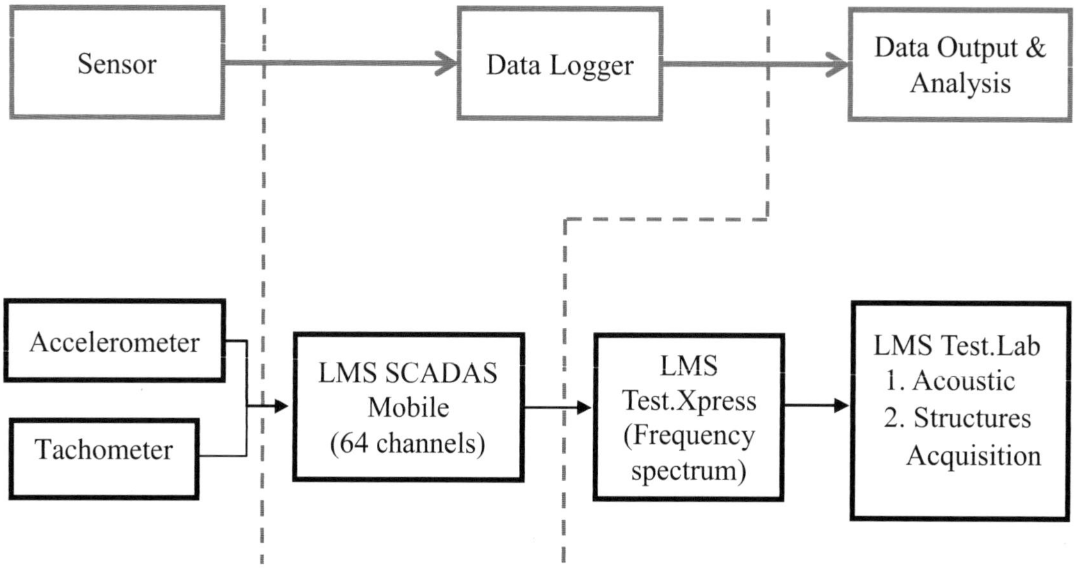

Figure 2. Experimental flow on vibration measurement

Next, an Experimental Modal Analysis (EMA) is carried out to obtain the natural frequencies of the HVAC components. There are two patterns of test which is free-free boundary condition and in-complete system structures. For the free-free boundary condition, the all the HVAC components are hanged individually using the rubber band. An accelerometer is attached on each component and impact hammer is used to apply external force on the component. Natural frequencies and mode shapes for each component are determined using the LMS software. EMA is then performed with same equipment in-complete system conditions. The HVAC components for EMA is summarized in Table 1 including the detail component arrangement and geometry modelling.

Table 1. Test set up and geometry for the assembly and individual components of EMA.

No.	Component	Test setup	Geometry
1	Overall rig structure		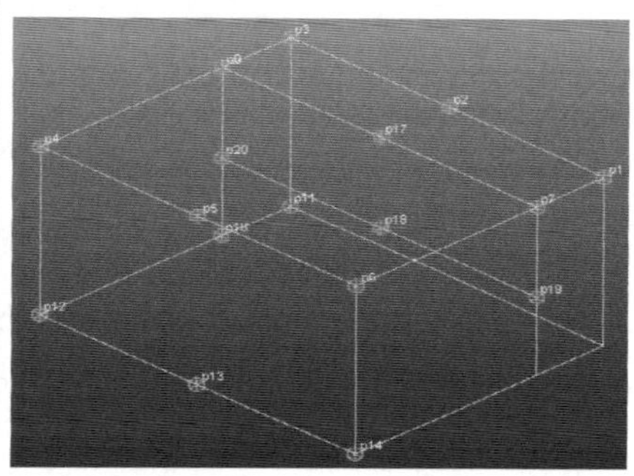

2	Compressor	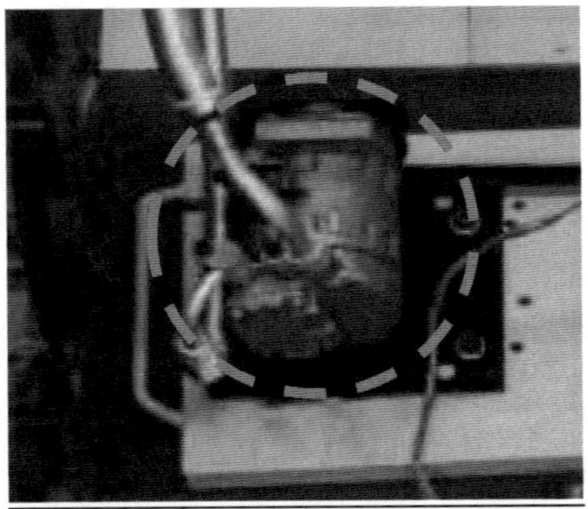	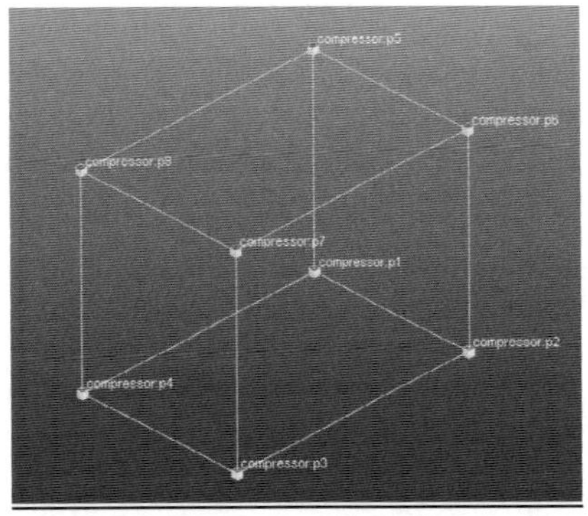
3	Evaporator pipe		
4	TXV		Single point measurement without geometry

2.3. Modelling and Implementation of DVA.

The air conditioning (AC) pipe is selected among the four components due to the highest contribution of humming type noise that obtained in Table 2. Again, the EMA is conducted to obtain the natural frequency before and after implementation of SDM. The DVA is modelled as a second spring-mass system which added to the primary mass as illustrated in Figure 3. In this testing, the SDM is applied at point 2, which is selected as a point for SDM attachment as shown in Figure 4. The total mass prediction of the absorber with 0.132 kg is included in the modification.

IOP Conf. Series: Journal of Physics: Conf. Series **1262** (2019) 012034 doi:10.1088/1742-6596/1262/1/012034

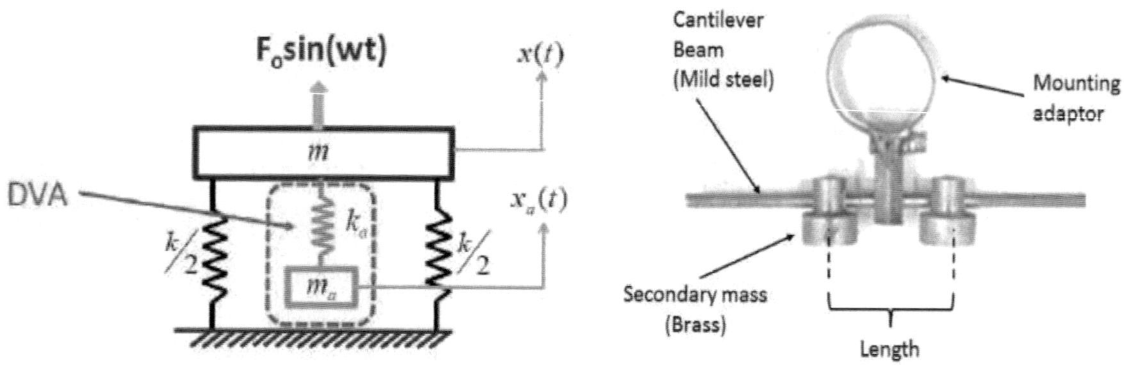

(a) (b)

Figure 3. (a) Model and (b) designed of the DVA in the system.

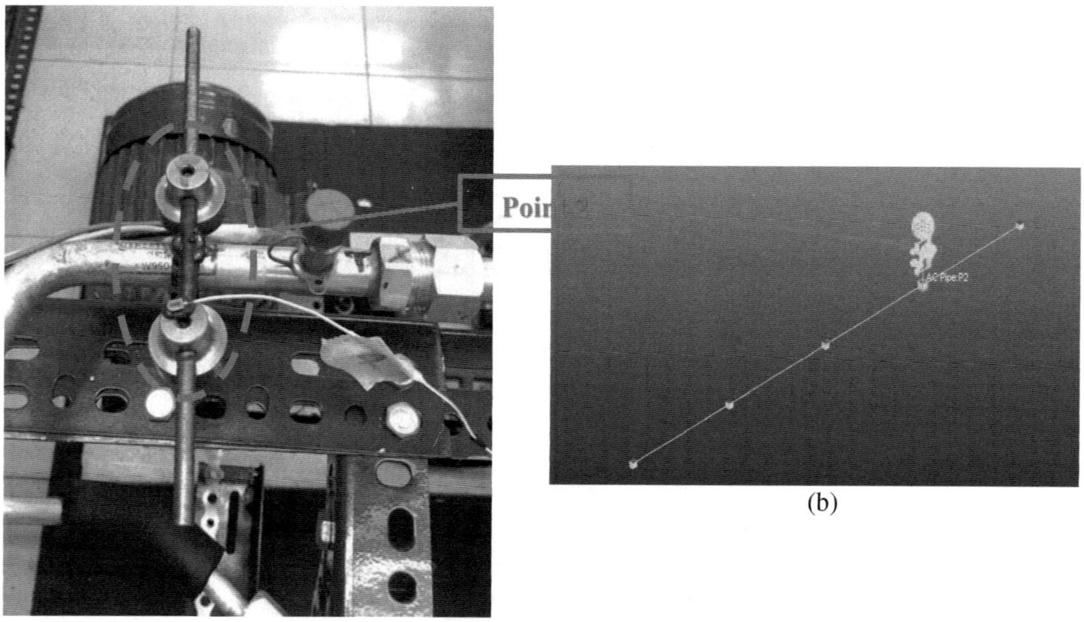

(b)

(a)

Figure 4. Location of point 2 in (a) actual AC pipe and (b) geometry drawn in impact testing.

Next step is the tuning of the DVA. The tuning is based on the theoretical calculation to determine the accurate distance between the two masses from the centre rod as shown in Figure 3(b). This distance becomes the first reference before the DVA is tuned into the targeted frequency. The theoretical calculation is shown as follows:

$$\text{Force, } F = k\partial \tag{1}$$

The deflection of cantilever beam,

$$\partial = \frac{FL^3}{3EI} \tag{2}$$

$$L^3 = \frac{3EI}{k} \tag{3}$$

The natural frequency,

$$\omega^2 = \frac{k}{m_a} \tag{4}$$

$$k = \frac{(2\pi f)^2}{m_a} \tag{5}$$

Substitute Eq. (4) into Eq. (3), the distance required is obtained as follows:

$$L^3 = \sqrt[3]{\frac{3EI}{(2\pi f)^2 \times m_a}} \tag{5}$$

Second moment of inertia of rod is given by:

$$\mathrm{I} = \frac{1}{4\pi r^4} \tag{6}$$

Where, m_a is the mass of each secondary mass (kg), r is the radius of the cantilever beam (m), L is the length of the cantilever beam (m), f is the natural frequency (Hz) and E is the Young's modulus of elasticity of the cantilever beam ($\mathrm{N/m^2}$). By considering Eq. (1) to (5), the estimated length, L for the given secondary mass values can be calculated.

3. Results and Discussion

3.1. Noise and vibration measurement

Table 2 summarizes the noise and vibration measurement results in the system level. Both test patterns for idle and tracking with four type of noises are obtained. Hissing noise is an audible noise that involves a high frequency range of 4000 to 6000 Hz at idle and tracking conditions. From the measurement, the highest contribution is from the evaporator pipe inlet. The noise may induce by the action of refrigerant expansion at orifice tube, as the R134a gas is forces to flow through a fine restriction orifice.

Humming noise is produced at three different operating frequencies, which is at 300 to 400 Hz for idle condition, 400 to 500 Hz for tracking condition (AC off) and 100 to 300 Hz (AC on). Differently for the humming noise, two components show the highest contribution, which is compressor and AC pipe. This result gives a clear insight that the operating vibration of the system is transferred through the compressor to the AC pipe in sequential transmission. However, when AC is turned off, the vibration amplitude is detected within frequency range of 400-500 Hz, which transferred from the compressor up to the evaporator location.

The clicking noise is the effect of the compressor clutch impact. When the clutch is engaged, the compressor shaft will rotate, and refrigerant circulate the internal components which containing the field coil. Field coil is energized with the pressure plate towards it and causing the compressor internals to turn. Thus, the compressor is the highest contribution of this noise at frequency range of 200 to 300 Hz. The fourth noise found is air rush-type. The noise between both locations of outlets is induced due to the turbulence effect of airflow at the HVAC blower. As the velocity of fluid flow increases, the chance of turbulences is high. The noise is detected at frequency range of 1400 to 1700 Hz for both locations.

IOP Conf. Series: Journal of Physics: Conf. Series **1262** (2019) 012034 doi:10.1088/1742-6596/1262/1/012034

Table 2. HVAC noise indentation results in system level.

No	Noise	Test method	Engine	Air condition (AC)	Blower speed	Accelerometer setup location	Noise Frequency (Hz)	Component with highest contribution
1	Hissing	Idle	On	On	1	1. Motor 2. Compressor 3. AC pipe 4. Evaporator pipe inlet 5. TXV	4000-6000	Evaporator pipe inlet
		Rpm tracking	On	On	1			
2	Humming	Idle	On	On	1	1. Motor 2. Compressor 3. AC pipe 4. Evaporator pipe inlet 5. TXV	300-400	Compressor and AC pipe
		Rpm tracking	On	Off	-		400-500	
			On	On	1		100-300	
3	Clicking (Compressor Clutch Engagement)	Idle	On	On	1	1. Motor 2. Compressor 3. AC pipe 4. Evaporator pipe inlet 5. TXV	200-300	Compressor
4	Air rush	With Vents	Off	Off	1		1400-1700	HVAC Blower
		Without Vents	Off	Off	1			

IOP Conf. Series: Journal of Physics: Conf. Series **1262** (2019) 012034 doi:10.1088/1742-6596/1262/1/012034

3.2. Experimental Modal Analysis

Table 3 shows four components of HVAC system with different natural frequencies. The natural frequency of AC pipe is 148.9 Hz which is the highest natural frequency compared to other components. This is due to the structure is not mounted rigidly and can be easily deflected when the external force applied. Second highest is TVX with 128 Hz. The structure is a hollow aluminium pipe and linked with other components which significantly differ from AC pipe. Compressor and overall rig structure natural frequencies are at 99.78 Hz and 54.64 Hz, respectively. The compressor and rig are lower due to the structures have high mass.

Table 3. EMA on assembly structure and individual components of HVAC

No	Component	Geometry point	Frequency range, Hz	Mode number	First natural frequency, Hz
1	Overall rig structure	19	0 - 1000	11	54.64
2	Compressor	4		5	99.78
3	AC pipe	3		5	148.9
4	TXV	2		7	128

3.3. Vibration measurement with application of DVA

Again, the EMA and vibration measurement is conducted for the second time after the implementation of DVA. Figure 5(a) and (b) shows the vibration spectrum and frequency response function (FRF) of the AC pipe before and after implementation of the DVA. Before implementation of DVA, the natural frequency of the AC pipe is obtained at 148.9 Hz while after the DVA implementation, the first natural frequency is shifted to frequency ranges of below 100 Hz and above 500 Hz as shown in Figure 5(b). In term of vibration spectrum in Figure 5(a), the vibration acceleration is changed and greatly reduced after implementation DVA. The figure clearly shows the vibration reduction and the effective frequency range for DVA is found from 100 to 500 Hz.

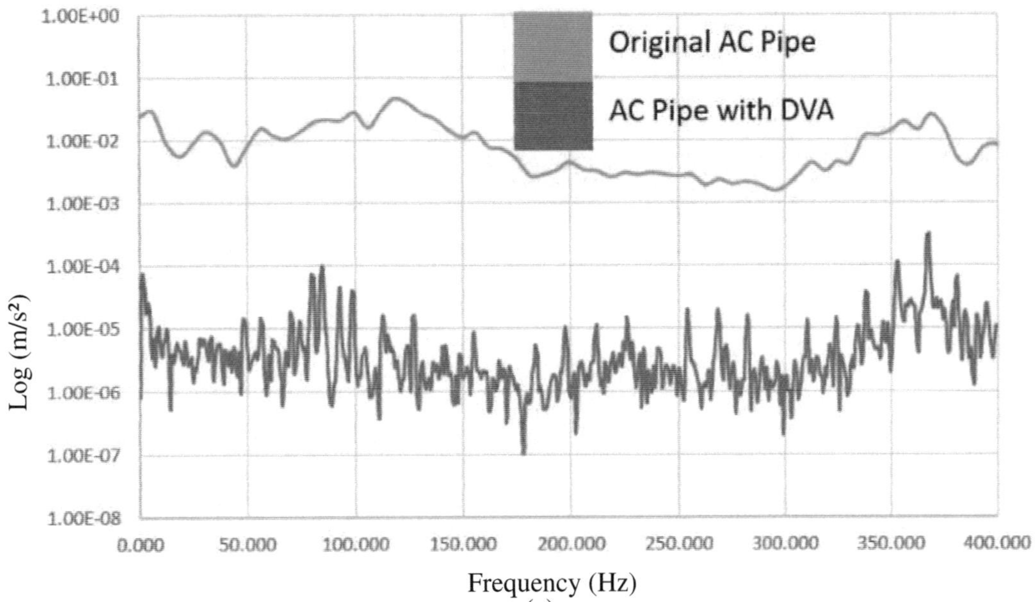

(a)

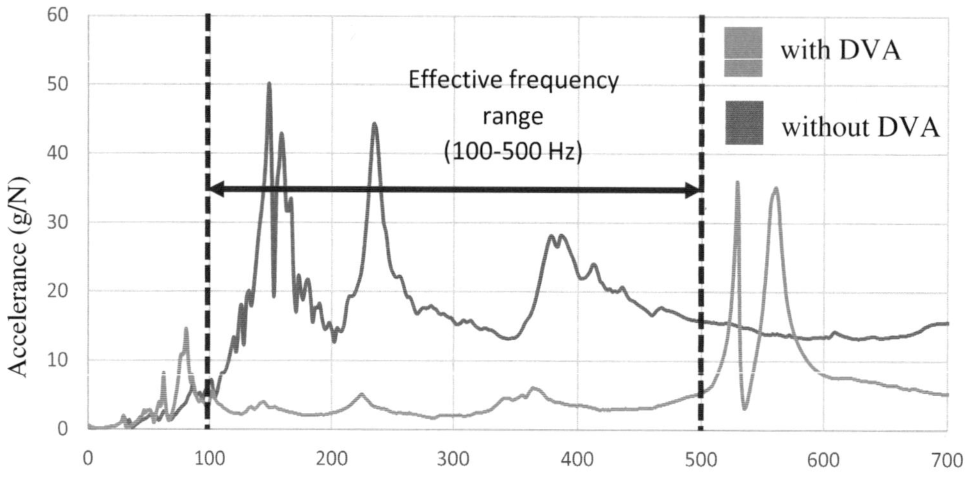

(b)

Figure 4. Natural frequencies of AC pipe (a) with DVA and (b) without DVA.

4. Conclusions

From this paper, the following conclusions can be made:

1. There are four type of vibration and noise found on the HVAC system which is hissing, humming, clicking and air-rush vibration and noises. Each noise has found at different operating frequency range and different highest component contribution.
2. The main component that has significant contribution for humming type noise and vibration in HVAC system is the compressor and AC pipe between which is between 100 to 300 Hz of operating frequency. The first natural frequency of the AC pipe is 148.9 Hz, which is within this noise range.
3. The DVA scheme of SDM method has been successfully applied to the AC pipe. With the implementation of DVA scheme, the resonance phenomenon can be avoided when the HVAC is running and subsequently reduced the vibration of the overall system. The effective frequency range of the AC pipe is found at 100 to 500 Hz after implementation of DVA.

Acknowledgement

The authors would like to acknowledge Universiti Sains Malaysia and Kementerian Pengajian Tinggi for providing the financial assistance under Short-term and FRGS grants (304/PMEKANIK/60313052 and 203/PMEKANIK/6071370) and Proton Holdings Berhad for their laboratory facilities and technical support for this project.

References

[1] H. B. Huang, X. R. Huang, M. L. Yang, T. C. Lim, and W. P. Ding, "Identification of vehicle interior noise sources based on wavelet transform and partial coherence analysis," *Mech. Syst. Signal Process.*, vol. 109, pp. 247–267, 2018.

[2] S. St. Hill, S. Watkins, S. P. Mavuri, X. Wang, and D. Weymouth, "An investigation of vehicle HVAC cabin noise," in *Proceedings of SAE World Congress*, 2008, pp. 1–12.

[3] E. Canepa, A. Cattanei, and F. M. Zecchin, "Aerodynamic noise from cooling and HVAC systems," in *International Conference on Connected Vehicles and Expo*, 2013, pp. 164–170.

[4] J. H. Yoon, I. H. Yang, J. E. Jeong, S. G. Park, and J. E. Oh, "Reliability improvement of a sound quality index for a vehicle HVAC system using a regression and neural network model," *Appl. Acoust.*, vol. 73, no. 11, pp. 1099–1103, 2012.

[5] A. Sestieri, "Structural dynamic modification," *Sadhana - Acad. Proc. Eng. Sci.*, vol. 25, no. 3, pp. 247–259, 2000.

[6] T. K. Kundra, "Structural dynamic modifications via models," *Sadhana - Acad. Proc. Eng. Sci.*, vol. 25, no. 3, pp. 261–276, 2000.

[7] A. Z. A. Ahmad Mazlan and Z. M. Ripin, "Structural dynamic modification of an active suspended handle with a parallel coupled piezo stack actuator," *Proc. Inst. Mech. Eng. Part I J. Syst. Control Eng.*, vol. 230, no. 2, pp. 130–144, 2016.

[8] H. L. Sun, H. B. Chen, P. Q. Zhang, X. L. Gong, and K. Zhang, "Application of dynamic vibration absorbers in structural vibration control under multi-frequency harmonic excitations," *Appl. Acoust.*, vol. 69, no. 12, pp. 1361–1367, 2007.

[9] Y. Du, R. A. Burdisso, and E. Nikolaidis, "Control of internal resonances in vibration isolators using passive and hybrid dynamic vibration absorbers," *J. Sound Vib.*, vol. 286, no. 4–5, pp. 697–727, 2005.

[10] Y. Shen, L. Chen, X. Yang, D. Shi, and J. Yang, "Improved design of dynamic vibration absorber by using the inerter and its application in vehicle suspension," *J. Sound Vib.*, vol. 361, pp. 148–158, 2016.

IOP Conf. Series: Journal of Physics: Conf. Series **1262** (2019) 012035 doi:10.1088/1742-6596/1262/1/012035

Finite Element Model Updating of Dissimilar Plate with Rivet Joint

H. A. Rahman[1,3], N. D. A. Khairi[1] and M. S. M. Sani[1,2]
[1]Advanced Structural Integrity and Vibration (ASIVR), Faculty of Mechanical Engineering, Universiti Malaysia Pahang, 26600 Pekan, Pahang, Malaysia.
[2]Automotive Engineering Centre, Universiti Malaysia Pahang, 26600 Pekan, Pahang, Malaysia.
[3]Faculty of Mechanical Engineering, Universiti Teknologi MARA (UiTM) Terengganu

[*]Corresponding author: mshahrir@ump.edu.my

Abstract. This study focusses on experimental modal analysis (EMA), finite element analysis (FEA) and model updating. The result of FEA and EMA will be validated and analysed. Then, the study extended to the early stage of model updating by assuming that the riveted dissimilar plates is a rigid body instead of two dissimilar plates. These modal parameters are crucial in providing various periods at which it will naturally resonate to avoid the structure continue to resonate and experience structural damage. Modal parameters of the dissimilar plate with rivet joint which include natural frequency and mode shapes were determined through EMA and FEA. Comparison of experimental data and the simulation data reveals a good correlation between the values of natural frequency. Minor discrepancies of the percentage of error between EMA and FEA are acceptable with the percentage of error is within the range of 0.4% to 13%. Furthermore, the conducted model updating able to reduce to an error range of 0.03% to 0.1%. Through this study, the best way of conducting finite element analysis and model updating for dissimilar plate material with rivet joint is also determined.

1. Introduction

Over the past years, the use of joining few dissimilar materials has increased particularly in the transportation industry, and other high-performance engineering application. Effective 2008, New European anti-pollution and energy saving laws imposed that requirement concerning the reduction of fuel consumption. One approach to partly tackle this issue is to reduce the total weight of the vehicle by using dissimilar materials as part of the structure. A vehicle's main composition, such as steel made chassis module can be joined with secondary structural elements of aluminum alloy materials [1]. Apart from reducing weight, it is also lower the center of gravity of the vehicle. Katayama concluded that development in joining aluminum alloy with other metals shown impressive high-performance structures. In the aircraft industry, weight reduction can also boost payload and increase range [2].

There are many joining technologies used in industry and mechanical fastening is one of the most commonly used joining technologies. This method allows dissimilar materials to be joined together, using only interference or interlocking at a microscopic or macroscopic level. Dursun & Soutis highlighted on the new development in welding, bonding, and extrusion emerged as a new trend in manufacturing and construction industry slowly replacing rivet joint [3]. Nevertheless, rivet joint is still considered as the most practical method to join dissimilar materials. With that, a reliable joining must be developed and evaluated. Over the years, many new developments in joining two dissimilar materials have been developed to fulfill the demand in the industry. Many riveted joints have been introduced in automotive assemblies and regard as the most effective solution. But even many riveted joints have been widely used in joining dissimilar material; there is no update on the performance data reported in the open literature as a guideline to provide better analysis of a structure.

Rivet joint has been used for years mainly in the automotive industry which imparts a better solution to join dissimilar material such as steel, aluminum, and magnesium [4]. Numerous researches have been done regarding the mechanical properties of the riveted joint. There are three types of structural rivet steel which are ASTM A502 grade 1, carbon rivet steel, ASTM A502 grade 2, high-strength structural steel rivets and ASTM A502 grade 3, similar to grade 2 but with enhanced atmospheric corrosion resistance. On the research discussed by Sun, Stephens, & Khaleel [5] on the effect of Self-Pierce Rivet (SPR) and Spot Resistant Welding (SRW) and concluded that SPR possessed 100% higher fatigue life compared to SRW which this conform to another previous research done by [6] regarding the comparison of fatigue life between SPR and SRW. Further research has been done in comparing two methods of the joint which is rivet joint and welding joint which proved that it rivets joint is more favorable for fixing a steel door impact beam which proved that rivet joint offers a strong joint compared to spot welding joint and more preferably in joining two dissimilar materials especially aluminum and steel as it does not produce brittle material [7].

Finite Element Analysis (FEA) is a method of a numerical technique for finding solutions to structural or performance issues. Basically, it is a branch of solid mechanics that studies the behaviour of solid materials. This method is very handy and convenient for complicated structures with unusual geometric shapes such as trusses, frames, and machine parts. Types of analysis include linear statics which is linear analysis with applied loads and constraints that are static and normal modes which are natural frequencies of vibration. Many types of research have been conducted by using FEA in terms of normal mode analysis[8-13].

A research on experimental investigation and numerical modeling of steel adhesive joint and reinforced by rivets [14] used finite element software to analyze the three types of joint behavior. When the maximum shear strength of the rivet is reached at any point of the rivet, the shearing process starts and finite elements are removed from the model. The numerical results were compared with the experimental data and the result suggested that the strengthening of double lap adhesive joints by rivets introduced by riveting improve both the static strength and the stiffness of joint.

Sani et al. [15] performed an experimental analysis of car chassis to determine the dynamics modal parameter for car chassis. Wira car chassis is drawn in MEscope software then simulated using ME' Scope VES software. The frequency of 40 Hz to 136 Hz is selected for impact hammer test and four mode shapes are obtained from the simulation. Mode shapes obtained shows that the bending occurs on the chassis as in mode 1 when dwelling at the first natural frequency. After that, the twisting occurs on the chassis as in mode 2 when dwelling at the second natural frequency. While in mode 3, dwelling in the third natural frequency produced second bending on the chassis. Lastly, dwelling on the fourth natural frequency displayed second twisting on the chassis on mode 4. On another researcher [16-18] which investigate the finite element analysis and modeling of the structure with bolted joints used shaker or impact hammer vibration in modal testing. From lists of natural frequencies obtained, the author concluded that the bolt models proposed in this study can be employed in a dynamic analysis as well as a static analysis.

This project used finite element analysis (FEA) and experimental modal analysis (EMA) by using impact hammer. These two methods define the dynamic properties of the element and to provide a vibrational analysis data for analysis of dissimilar plate materials with rivet joint. The purpose of this study is to compare and validate the result of modal parameters in dissimilar material using FEA and EMA. Furthermore, this study is extended to model updating so that a feasible model of test structure could be establish. However, this paper only establishes the dissimilar material structure with rivet joints as one rigid body due to the established model updating is at the early stage of this research.

2. Methodology

Figure 1 shows the flowchart of this study. Basically, this research can be summarized into four main phase which is test structure preparation, finite element analysis, model testing and finally, model updating.

2.1. Test Structure Preparation

The test structure was crucially designed at the riveted point of the structure to simulate the joining of two dissimilar material before the final design was selected. Figure 2 illustrates the completed design of riveted structure using SolidWorks 2017 software. The test structure was prepared by undergoing a few processes such as bending, drilling and riveting process. The two dissimilar material properties of the test structure can be shown in Table 1.

Table 1. Material Properties of Test Structure

Material	Young Modulus [GPa]	Density [kg/m^3]	Poisson Ratio
Galvanised Iron	210	7900	0.4
Electro Galvanised	74	3000	0.4

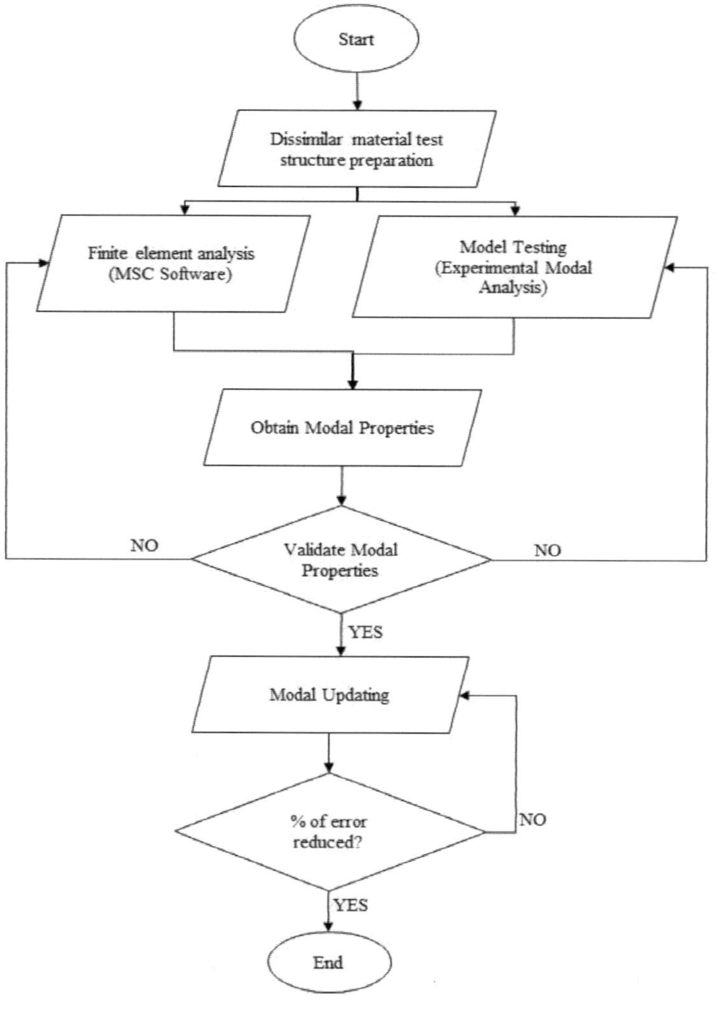

Figure 1. Flowchart of study

Figure 2. Designed Test Structure

2.2. Finite Element Analysis (FEA) and Experimental Modal Analysis (EMA)

In this project, MSC Nastran Patran software was utilized to perform FEA. The designed test structure was imported to be analyzed and the data or material properties from Table 1 was assigned for the FE model. In this study, a free-free boundary condition was applied and the normal mode analysis SOL103 was performed to obtain modal parameters of the test structure. Initially, FEA was performed earlier than EMA so that the range of natural frequencies and the mode shapes of each mode could be predicted and then compared accordingly to the results from EMA.

In order to obtain the modal parameters experimentally, EMA was performed by using impact hammer and the accelerometer was roved to designated points. The test structure was hanged to simulate free -free boundary condition. Figure 3 shows the schematic diagram of the modal testing and equipment used to perform the measurement.

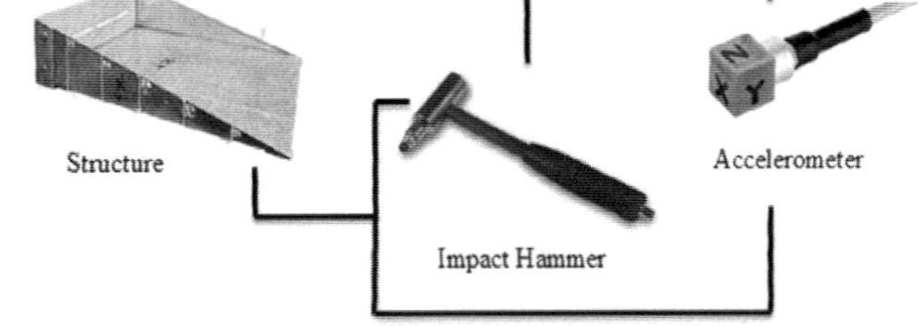

Figure 3. Schematic diagram of modal testing

Initially, the test structure was marked with 55 points and these points was set as measuring points, sketched in a wire-frame model as shown in Figure 4 using modal testing software, ME's Scope VES. The hammer was impacted on one fixed excitation point which is at point 23, while the accelerometer was roved to all 55 measuring points.

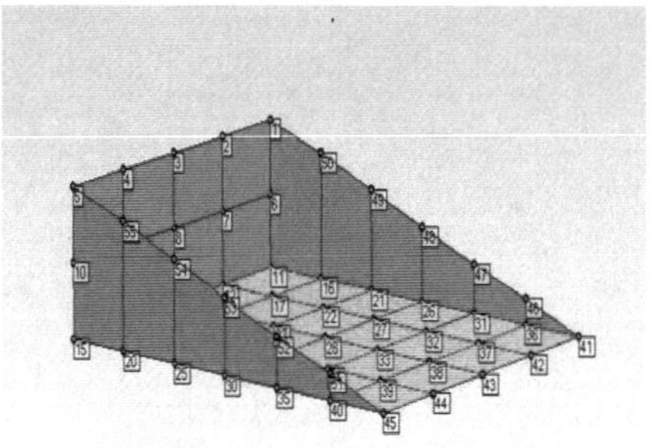

Figure 4. 55 measuring points of test structure

In order to obtained a conclusive experimental data, the signal processing need to be considered in the measuring process. The excitation force from the tip of the impact hammer was set at the magnitude sensitivity of 2.25 mV/g and axis of impact direction was also set accordingly. Piezoelectric tri-axial accelerometer was used to measure the output response of the testing. Then, both excitation force and output response were transferred to the data acquisition system (DAQ). NI 9234 Acoustic and Vibration Data Logger was used as data acquisition system to obtain Frequency Response Function (FRF). FRF was then extracted to obtain the modal parameters which are natural frequency and mode shape using curve fitting method. As a result, 5 modes were successfully extracted from the frequency range of 1 Hz to 1000 Hz. The selected modes were based on predicted results from FEA.

3. Results and Discussion

3.1 Correlation Between FEA and EMA

To validate both results from FEA as well as EMA, initially, mode shapes of each selected mode are compared and then the natural frequencies of each mode were also compared. The summarized correlation between FEA and EMA of the test structure can be seen in Table 2 and Table 3. Table 2 shows the correlation of mode shapes between FEA and EMA. Meanwhile, Table 3 shows the correlation of natural frequencies between FEA and EMA.

Table 2. Natural Frequencies Between FEA and EMA

Mode	Natural Frequencies [Hz]		Percentage Error [%]
	EMA	FEA	
1	133	152.74	12.9
2	212	199.26	6.0
3	408	416.49	2.0
4	468	466.28	0.4
5	488	516.75	5.6

IOP Conf. Series: Journal of Physics: Conf. Series **1262** (2019) 012035 doi:10.1088/1742-6596/1262/1/012035

Table 3. Mode Shapes Between FEA and EMA

Mode	Mode Shape	
	FEA	EMA
1	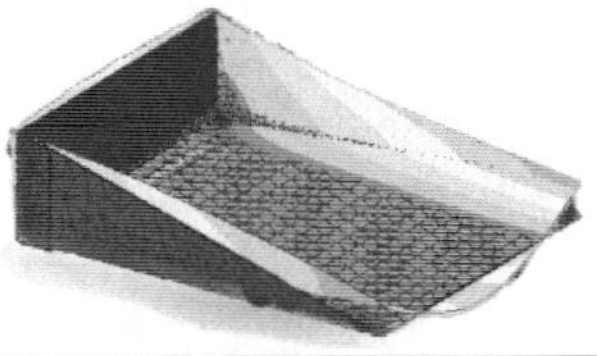	
2	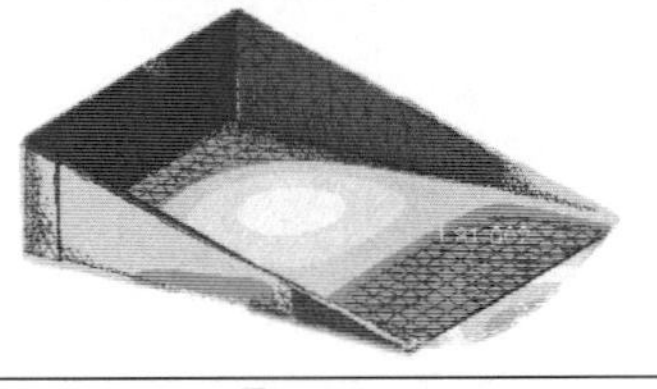	
3	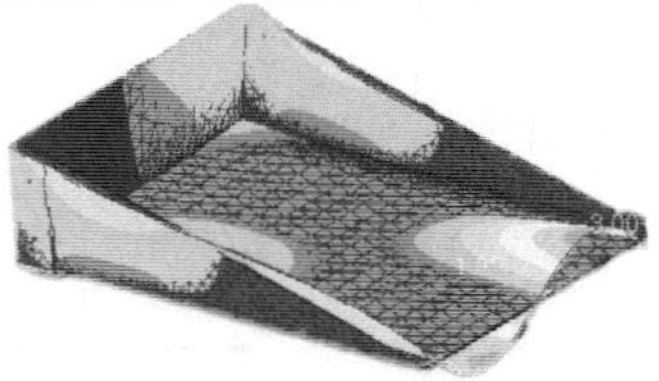	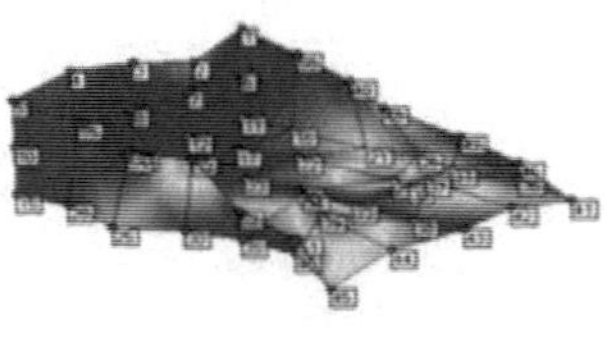
4	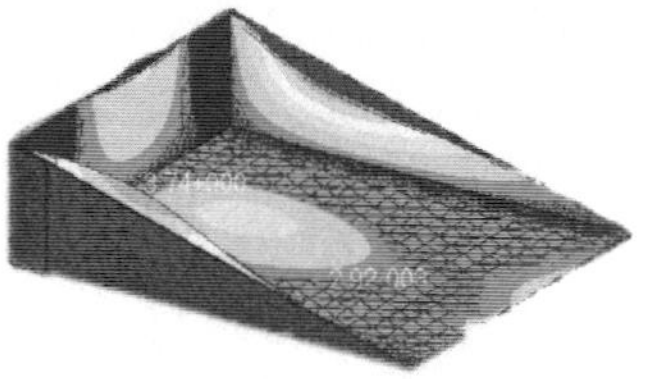	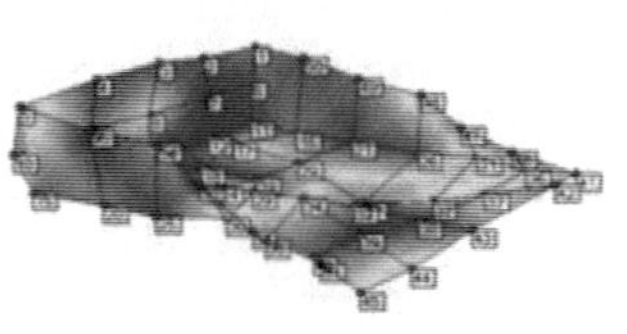
5	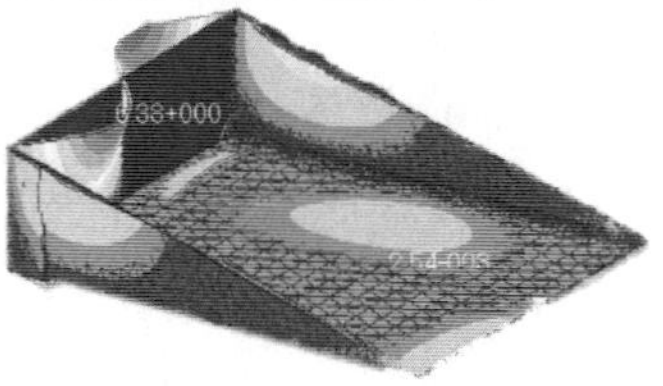	

Based on Table 2, percentage of error calculated depicted there is a discrepancy between the result of EMA and FEA. The highest percent of error is Mode 1 which is 12.9 and the lowest is Mode 4 which is 0.4. The other modes show a range of values which less than 10%. The high percent of error can be due to the error which occurred during analysis. While performing the analysis, there are numerous sources of error which can contribute to the discrepancy between FEA and EMA. Source of error in FEA can be categorized into two which are idealization error and discretization error. Idealization error mainly related to the mathematical structure of the model and those that can be improved through model updating [19]. Discretization error can be defined as an error related to element errors and global errors. Apart from the two main source of error, the joining process can also contribute to the error. Rivet joint is seen as the most reliable joining however, there is no established standard practiced in the industry for the riveting process with respect to riveting sequence, the distance between rivets, and the gap between sheets. Joining of the two dissimilar materials might not be perfectly joint due to the misalignment of the hole. The gap between the rivet and the material also contribute to the imperfect joining.

IOP Conf. Series: Journal of Physics: Conf. Series **1262** (2019) 012035 doi:10.1088/1742-6596/1262/1/012035

3.2 FE Model Updating

There are discrepancies between the FE or numerical models and modal testing dynamics result from inaccuracy in modelling, approximation of boundary condition, incorrect initial assumptions of geometry and material properties, and limitations of modelling structural connections [19]. Therefore, main objective of FE model updating is to improve a FE model by designing parameters of the model in the light of experimental data to an acceptable of accuracy [20]. Model updating parameter selection is a crucial step in order to avoid ill conditioning problem in optimization. Thus, the sensitivity analysis is required to identify which parameters are sensitive in model updating. The sensitivity analysis performed in this study and represented in Table 4. There are four parameters; Modulus's Young ($E1$) and Density ($\rho 1$) for galvanised iron, Modulus's Young ($E2$) and Density ($\rho 2$) for electro galvanised iron in sensitivity matrix form.

Table 4. Sensitivity matrix for four parameters

Mode	Young Modulus (E1)	Young Modulus (E2)	Density (ρ1)	Density (ρ2)
1	1.79	69.84	-1.95	-65.18
2	2.03	91.60	-7.41	-80.77
3	2.25	193.10	-4.86	-178.52
4	23.03	196.90	-6.06	-197.82
5	215.59	50.13	-199.10	-38.41

Table 5 shows the updated natural frequency of FE model obtained from output file then compared with the initial value of natural frequency. The discrepancy successfully reduced from initial value 5.38 % of total average error to 5.05 % of total average error.

Table 5. Comparison between initial FEA and updated FEA with measured data from EMA.

Mode	EMA [Hz]	Initial [Hz]	Percentage Error [%]	Updated [Hz]	Percentage Error [%]
1	133	152.74	12.90	147.89	11.20
2	212	199.26	6.00	193.41	8.77
3	408	416.49	2.00	404.08	0.96
4	468	466.28	0.40	449.12	4.03
5	488	516.75	5.60	489.35	0.28
		Average Error	5.38	Average Error	5.05

4. Conclusion

The dynamic characteristic correlation of dissimilar plate with rivet joint are investigated through numerical prediction method via finite element analysis (FEA) software. Experimental modal test was performed to determine the structural response in order to verify the predicted result computed in FEA. In order to reduce the discrepancy between predicted and its measured properties, FE model updating method was applied in this study with sensitive updating parameters. The discrepancy between EMA and FEA successfully minimized from 5.38 % of total average error to be 5.05 % of total average error after model updating process. As conclusion, FE model updating method is a crucial and feasible step in order to produce a reliable FE model in numerical analysis for further engineering analysis.

Acknowledgement
The authors of this paper would like to acknowledge a great support and encouragement by focus group of Advanced Structural Integrity of Vibration Research (ASIVR), Universiti Malaysia Pahang (UMP) for providing all the equipment used for this work Fundamental Research Grant Scheme (FRGS/1/2017/TK03/UMP/02-19) – RDU 170123.

References
[1] Mathieu A, Shabadi R, Deschamps A, Suery M, Matteï S, Grevey D & Cicala E 2007 Dissimilar material joining using laser (aluminum to steel using zinc-based filler wire) *Optics and Laser Technology* 39(3) 652–661.
[2] Katayama S 2004 Laser welding of aluminium alloys and dissimilar metals *Welding International* 18(8) 618–625.
[3] Dursun T & Soutis C 2014 Recent developments in advanced aircraft aluminium alloys *Materials and Design* 56 862–871.
[4] He X, Pearson I & Young K 2008 Self-pierce riveting for sheet materials: State of the art *Journal of Materials Processing Technology* 199(1) 27–36.
[5] Sun X, Stephens E V & Khaleel M A 2007 Fatigue behaviors of self-piercing rivets joining similar and dissimilar sheet metals *International Journal of Fatigue* 29(2) 370–386.
[6] Booth G S, Oliver C A, Westgate S A, Liebracth F and Braunling S B 2000 Self-Piercing Riveted Joints and Resistance Spot Welded Joints in Steel and Aluminium *SAE Technical Paper* 01-2681.
[7] Sakayama T, Murayma G, Naito Y, Saita K, Miyazakki Y, Oikawa H & Nose T 2013 Dissimilar metal joining technologies for steel sheet and aluminum alloy sheet in auto body *NIPPON Steel Technical Report* 393(103) 91–98.
[8] Fouzi M S M, Jelani K M, Nazri N A and Sani M S M 2018 Finite Element Modelling and Updating of Welded Thin-Walled Beam *International Journal of Automotive and Mechanical Engineering (IJAME)* 15(4) 5874-89
[9] Abdullah N A Z, Sani M S M, Hussain N A, Rahman M M and Zaman I 2017 Dynamics properties of a Go-kart chassis structure and its prediction improvement using model updating approach *International Journal of Automotive and Mechanical Engineering (IJAME)* 14 3887-97
[10] Jelani K M, Razip M H M, Nazri N A, Sani M S M and Yasar M 2018 Dynamics investigation on motorcycle chassis based on Finite Element (FE) modelling and updating *IOP Conf. Ser. Mater. Sci. Eng.* 469 012103
[11] Mohd Zin M S, Rani M N A, Yunus M A, Wan Iskandar Mirza W I I, Mat Isa A A and Mohamed Z 2017 Modal and FRF Based Updating Methods for the Investigation of the Dynamic Behaviour of a Plate *Journal of Mechanical Engineering* 4 175–89
[12] Zahari S N, Sani M S M, Husain N A, Ishak M and Zaman I 2016 Dynamic analysis of friction stir welding joints in dissimilar material plate structure *J. Teknol.* 78
[13] Sani M S M, Abdullah N A Z, Zahari S N, Siregar J P and Rahman M M 2016 Finite element model updating of natural fibre reinforced composite structure in structural dynamics *MATEC Web of Conferences* 83 (03007)
[14] Sadowski T, Kneć M & Golewski P 2010 Experimental investigations and numerical modelling of steel adhesive joints reinforced by rivets *International Journal of Adhesion and Adhesives* 30(5) 338–346.
[15] Sani M S M, Rahman M M, Noor M M, Kadirgama K & Izham M H N 2011 Identification of dynamics modal parameter for car chassis *IOP Conference Series: Materials Science and Engineering* 17(1) 012038.
[16] Kim J, Yoon J-C & Kang B-S 2007 Finite element analysis and modeling of structure with bolted joints *Applied Mathematical Modelling* 31(5) 895–911.
[17] Izham M H M, Abdullah N A Z, Zahari S N and Sani M S M 2017 Structural dynamic investigation of frame structure with bolted joints *MATEC Web of Conferences* 90 (01043).

[18] Izham M H M, Sani M S M, Rani M N A and Yunus M A 2019 Damage assessment on frame structure with bolted joints based on experimental modal analysis IOP Conference Series: Materials Science and Engineering 469 (012102)
[19] Mottershead J E, Link M and Friswell M I 2011 The sensitivity method in finite element model updating: A tutorial *Mech. Syst. Signal Process.* 25 2275–96
[20] Fouzi M S M, Sani M S M, and Muchlis Y 2019 "Finite Element Modelling and updating of welded joint for dynamic study of exhaust structure" *IOP Conference Series: Materials Science and Engineering* vol. 469 pp. 1-10